Environmental Chemistry

Environmental Chemistry

Edited by **Clayton Murphy**

Willford Press
New York

Published by Willford Press,
118-35 Queens Blvd., Suite 400,
Forest Hills, NY 11375, USA
www.willfordpress.com

Environmental Chemistry
Edited by Clayton Murphy

International Standard Book Number: 978-1-68285-129-6 (Hardback)

Printed in the United States of America.

Contents

Preface

Environmental chemistry has emerged as a distinct field of study in the recent times. It is focused on understanding the chemical & biochemical processes and interactions that occur in the environment. The topics included in this text, such as relationship between chemistry, biomaterials and biological matters, speciation, bioavailability, etc., are of utmost significance and bound to provide incredible insights to readers. In this book, using case studies and examples, constant effort has been made to make the understanding of the difficult concepts of environmental chemistry as easy and informative as possible. It is highly recommended for all associated with the study of environmental chemistry and allied disciplines.

This book has been the outcome of endless efforts put in by authors and researchers on various issues and topics within the field. The book is a comprehensive collection of significant researches that are addressed in a variety of chapters. It will surely enhance the knowledge of the field among readers across the globe.

It gives us an immense pleasure to thank our researchers and authors for their efforts to submit their piece of writing before the deadlines. Finally in the end, I would like to thank my family and colleagues who have been a great source of inspiration and support.

Editor

Nitrate anion levels in water from selected wells and points along Kimondi River, Nandi

Magut Hillary and Terer Erick Kipngetich*

Department of Chemistry, University of Eastern Africa, Baraton, P. O. Box 2500 Eldoret, Kenya.

This paper focuses on the comparative determination of nitrate anion concentration from selected wells and points along the Kimondi River using UV-Visible spectrophotometric method. The areas monitored were Tulon, Sitatunga swamp, Sironoi, Kimondi Bridge, kipchabo tea factory and Samoo. This research was to determine whether nitrate anions in water are beyond the threshold limit which is harmful to both plant and animal life. Relatively high concentrations of NO_3 usually have their origin in processes of organic pollution and excessive use of inorganic fertilizers. In the case of agricultural areas for example, Nandi County where our analysis was based, these activities may generate great quantities of nitrates. The water was sampled from both the river point and one selected borehole in the respective areas. Three samples were obtained from each of the sampling points and were analyzed for nitrate using the UV-Visible spectrophotometer set at 420 nm. The average of the three-absorbance values was computed and converted to concentration in mg/L. The research showed that levels of nitrate ion in both river and well waters were below the threshold limits.

Key words: Nitrate anion, threshold limit, UV-Visible spectrophotometric method.

INTRODUCTION

The determination of nitrate (NO_3^-) is a difficult task because of the relatively complex procedures involved, the high probability that interfering constituents will be present and the limited concentration ranges of the various techniques. An analytical technique that measures the absorbance is suitable for screening uncontaminated water (low in organic matter). This method is applicable to the analysis of drinking river water, borehole water and surface water.

According to Villa et al. (2010), the modern civilization, industrialization, urbanization and increase in population have led to fast degradation of our groundwater quality. As water is the most important component of eco-system, any imbalance created either in terms of amount, which is presence of impurities added to it can harm the whole eco-system (Hem, 1961).

According to World Health Organization (WHO), the permissible limit of nitrate value is in the range of 40 to 50 mg/L. The Indian Council of Medical Research has recommended desirable limit of 20 mg/L of nitrate for drinking water Nitrate is a problem as a contaminant in drinking water (primarily from groundwater and wells) due to its harmful biological effects (Hallberg and Keeney, 1993).

Research shows that 97% of the world water is saline and is thus, non-drinkable, while 2% is locked in glaciers and polar ice caps. This leaves 1% to meet humanity needs (Elliot et al., 2008).

Nitrates exist largely due to the presence of animal manure in the water bodies. The excess of nitrates has contributed to the high levels of eutrophication along River Kimondi and the entire water bodies in Nandi County. There is great evidence from the fact that, there is an intensive growth of papyrus and other plant species which if not monitored, may in future clog the whole Kimondi River.

Increasing population size, climate change and pollution will exacerbate the nitrate pollution situation (Jagessar, 2011). The results showed that the concentrations of nitrates were not as high and are below

*Corresponding author. E-mail: e_terer@yahoo.com.

the internationally accepted threshold values. The applicable range of concentrations using the stated method is in the range of 0.1 to 2 mg/L NO_3^-. A maximum level of 45 mg/L is established as worldwide guidance for nitrate concentration in water.

In Europe, the maximum permitted levels of nitrate in potable water is 50.0 mg/L, while in the US-Environmental Potential Agency (EPA) has established a guideline for the maximum level of nitrate-nitrogen of 10 mg/L.

MATERIALS AND METHODS

Apparatus, materials and reagents

The apparatus, materials and reagents used in this study are UV spectrophotometer, hot plate, volumetric pipettes (2.5 and 10 ml), calibrated pipette, fume hood, analytical balance, volumetric flask (25 to 1000 ml), weighing dish, funnel, 100 ml beaker, distilled water, ammonium molybdate, sodium sulphide, sodium hydrogen phosphate, measuring cylinder, sulfuric acid, phenol, hot water bath, centrifuge, conical flask, silver sulphate, phenoldisulphonic acid, potassium hydroxide, ammonia, anhydrous potassium nitrate and filter paper.

Procedure for water sample collection and determination of nitrates

Sampling plan was specific for each sampling site (APHA, 1999). Water samples were collected in brown, resistant borosilicate winchester bottles, which had previously been washed thoroughly with detergent, rinsed with HCl, followed with distilled water for a prolonged period. They were filled to the brim and then sealed with Teflon lined caps. The choice of the brown bottles was to prevent decomposition through light. Sampling bottles were kept closed and much caution was taken not to contaminate the inner surfaces of stoppers, caps and necks of bottles. They were transferred to the laboratory in big plastic cooler. The samples were then acidified using 1 M HCl acid in order to prevent interference from hydroxide or carbonate concentrations up to 1000 mg $CaCO_3$/L. 1 M sodium arsenite and 1 M hydrogen peroxide were added to all the samples to stop potential interference with oxidizing and reducing agents, respectively. The samples were stored in the refrigerator at 4°C. They were then analyzed immediately for nitrates, using UV-Vis method by a DR 5000 Spectrophotometer following the method developed by Eaton et al. (1995). The same sampling procedure was used in sampling water from the boreholes corresponding to the points of river points.

The preparation of reagents

The nitrate standard (1000 ppm) was prepared using potassium nitrate that had been dried in oven for 1 h at 105°C and stored in desiccators until cool. 3.6107 g of potassium nitrate (KNO_3 was weighed and washed over into a 500 ml flask with distilled water, 1000 ppm std). 10 ml of distilled chloroform was added to the prepared standard solution to stop the effect of interfering agents. It was stopped, shaken and then labeled. 50 ml was pipette into a 500 ml volumetric flask. The sample was then diluted to 500 ml with distilled water (100 ppm stock std.) and again stopped, shaken and labeled. 10 ml was then pipette from the 100 ppm stock solution into 100 ml volumetric flask and was made up to mark (10 ppm std), stopped, shaken and then labeled.

Table 1. Nitrate standards concentration (ppm) and absorbance reading at 420 nm.

Nitrate concentration (ppm)	Absorbance
0.1	0.031
0.2	0.047
0.3	0.076
0.4	0.105
0.5	0.143

Preparation of samples

5 ml of distilled water was pipette into a 150 ml beaker (blank). 5 ml of sample was then filtered and added into the 150 ml beaker (sample volume), and placed on a hot plate and taken just to dryness. 2 ml of phenoldisulphonic acid was added and the sides were washed down lightly, warmed on hot plate, removed and allowed to cool. 10 ml of concentrated ammonium hydroxide (NH_4OH) was then added carefully. The reaction was violent. The absorbance of the samples was measured with the aid of a UV/Visible spectrophotometer set at 420 nm (Clessicens et al., 1995). The same procedure was followed for standards. The following are the quantities of standards that were placed in a beaker: 1 ml of 10 ppm was pipette into 150 ml beaker = 10 ppm, 2 ml of 10 ppm was pipette into a 150 ml beaker = 20 ppm, 3 ml of 10 ppm was pipette into 150 ml beaker = 30 ppm and 4 ml of 10 ppm was pipette into 150 ml beaker = 40 ppm. A standard calibration graph was then prepared for nitrate (Table 1 and Figure 1).

RESULTS AND DISCUSION

Table 2 and Figure 2 show the average concentration of nitrate in water from the randomly selected points in the river Kimondi and the wells adjacent to the river sampling point. The wells were almost of same depth averaging 50 feet. From the results, generally the river water contained less amounts of nitrates than the well. The flow of well water is slow and hence contaminants are not diluted and washed away as they are in a swiftly moving river (Girard, 2005). The eight river points had a mean concentration of 0.40, 0.66, 0.34, 0.18, 0.35, 0.33, 0.30 and 0.41 mg/L, respectively, while the eight selected wells had an average of 0.34, 0.66, 0.78, 0.27, 0.45, 0.17, 0.64 and 0.68 mg/L, respectively. Well water flow is slow. Hazardous chemicals from dump sites and other sources seep through the ground; some pollutants are filtered by soil and travel only short distances. Nitrates are soluble ions which percolate downward into groundwater from septic tanks, fertilized farms and feedlots.

From the data analysis of Kipchabo Tea factory area, the well sample had the highest concentrations of nitrates. According to Hill et al. (1991), Nolan (1996) and Speiran et al. (1996), raw waste effluents from factories and sewage treatment usually have high levels of ammonia and nitrogenous wastes. Additionally, prevailing temperatures within factories and their effluent encourage

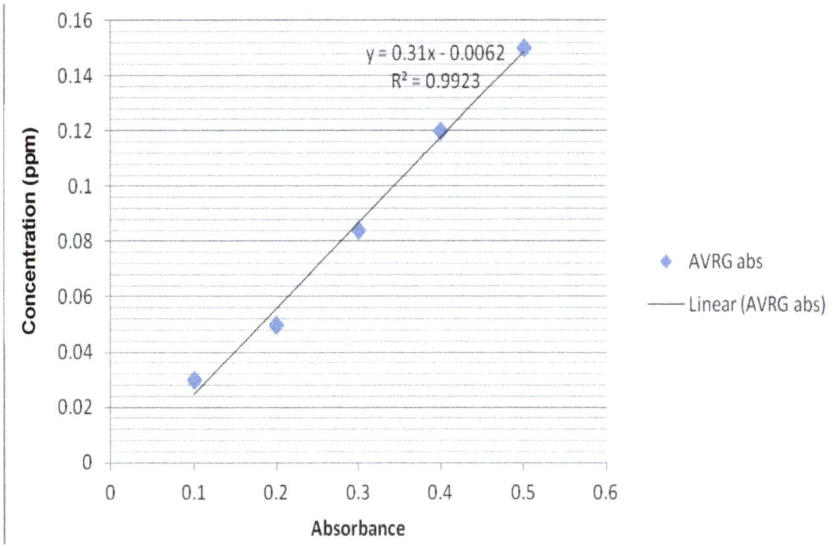

Figure 1. Graphical representation of nitrate standards for concentration (ppm) against absorbance taken at 420 nm.

rapid multiplication of bacterial population. These two factors can elevate the levels of nitrates in the adjacent water bodies through aerobic and anaerobic bacteria activities.

The nitrate level for both river and well samples from Kingwal-Sitatunga area was quite high with 0.41 and 0.68 mg/L, respectively. This area is inhabited by the largest population of a rare antelope species called Sitatunga in the country. The animals graze in the swamp and live there, their wastes go directly to the river; this could explain the higher level of nitrate concentration in this area. Gray (1994), Roeve et al. (1994), Calley et al. (1997) and Chapman et al. (1992) qualified the usual claim that excessive use of nitrate fertilizers and animal manure constitute most of the elevated levels of nitrates in water bodies.

Tulon source, Kimondi Cataracts and Samoo Bridge showed relatively low levels of nitrate anion concentrations. Tulon source had the lowest levels of nitrate concentration both for the river and well samples which according to the research is attributed to the fact that it is the source of the river under study and according to European Environmental Agency (1999), the natural nitrate level in water source and groundwater are generally low, but their concentration grow due to human activities, such as agriculture, industry, domestic effluent and emission from combustion engines.

From the readings, it was evident that there is slightly higher concentration at Kimondi Bridge (0.66), Kingwal Bridge (0.36), which according to our study could be attributed to the chemical present in run-off water from the surrounding agricultural farm and vehicle depositions of the exhaust fumes. The sampling area is a rich agricultural place, where there is large scale farming of tea, maize and horticulture where famers use inorganic fertilizers in planting and in top dressing. The presence of nitrates in groundwater could be anthropogenic or as a result of irrational use of nitrogenous fertilizers.

The increasing use of artificial fertilizers, the disposal of wastes (particularly from animal farming), and changes in land use are the main factors responsible for the progressive increase in nitrate levels in river water and supplies over the last 20 years (Akinsola, 2005). Generally, the nitrate concentration in the water samples for both the river points and the wells was far below the International Nitrate Standard Threshold; however, there is evidence that water is getting enriched with nutrients due to the intense growth of papyrus plants along the river Kimondi. The WHO has guideline values for nitrate from agricultural activities that are of health significance in drinking water at 50 mg/L (WHO, 2006). The low levels of nitrate in the river waters, that is, along the river points is not surprising because plants use it up and bacteria catalysts decomposes it too.

Well waters contained appreciable nitrate level because the soil contains nitrate-rich rock minerals, which can dissolve gradually. Leaching from the soil surface can also contribute to its presence. However, the results indicate that the NO_3 levels of all the samples were below the limits. Natural nitrate levels in groundwater are generally very low (typically less than 10 mg/L NO_3), but nitrate concentrations grow due to human activities, such as agriculture, industry, domestic effluents and emissions from combustion engines.

Nitrates generally moves relatively slow in soil and groundwater, there is a lag time of approximately 20 years between the pollution activity and the detection of the pollutant in groundwater. For this reason, it is

Table 2. The mean concentration of nitrates (mg/L), the standard deviation and the variance for the sample analysis.

Sample location	River point sample concentration (mg/L)				Well adjacent to river point sample concentration (mg/L)			
	Nitrate conc. for three readings (mg/L)	Mean conc. (mg/L)	Standard deviation	Variance	Nitrate conc. for three readings (mg/L)	Mean conc. (mg/L)	Standard deviation	Variance
Kimondi cataracts	0.13, 0.16, 0.14	0.14	0.01528	0.00023	0.33, 0.33, 0.35	0.34	0.01155	0.00013
Kimondi Bridge	0.66, 0.66, 0.67	0.66	0.00577	0.00003	0.66, 0.65, 0.68	0.66	0.01528	0.00023
Kipchabo tea factory	0.32, 0.34, 0.36	0.34	0.02	0.0004	0.79, 0.79, 0.77	0.78	0.01155	0.00013
Tulon source	0.17, 0.19, 0.19	0.18	0.01155	0.00013	0.27, 0.26, 0.29	0.27	0.01528	0.00023
Kingwal Bridge	0.36, 0.35, 0.35	0.35	0.00577	0.00003	0.43, 0.47, 0.45	0.45	0.02	0.0004
Samoo Bridge	0.33, 0.35, 0.33	0.33	0.01155	0.00013	0.17, 0.18, 0.16	0.17	0.01	0.0001
Sironoi Bridge	0.30, 0.30, 0.31	0.30	0.00577	0.00003	0.62, 0.64, 0.65	0.64	0.01528	0.00023
Kingwal Sitatunga area	0.40, 0.41, 0.42	0.41	0.01	0.0001	0.67, 0.68, 0.70	0.68	0.01528	0.00023

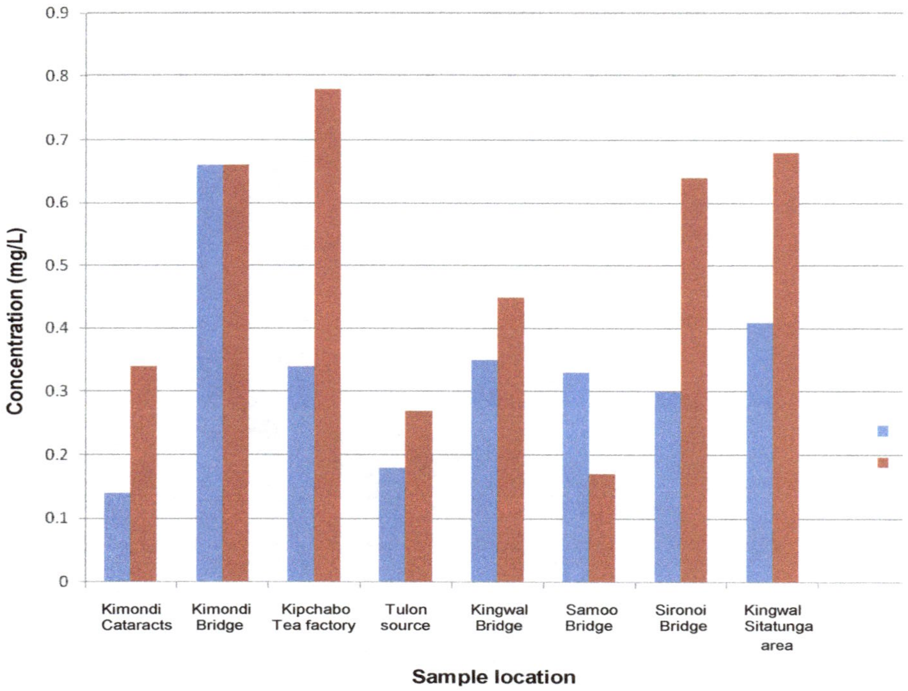

Figure 2. Bar graph showing a comparison between the nitrate anion concentration for both the River points and wells. Key: Red is Well point; Blue is River point.

predicted that current polluting activities will continue to affect nitrate concentrations for several decades (Eaton et al., 1995). Groundwater contains various types of pollutants and several other substances are dissolved in it. Concentration of these pollutants is useful for human body but in a specific limit (Ranjana, 2012).

According to Pulido-Bosch et al. (2000), shallow wells which draw water from intensively cultivated superficial formations, yield waters with a high NO_3 content. When the boreholes are deeper and penetrate low-permeability formations in the superficial layers, the waters contain little NO_3; similar to what happens in areas of recharge where agricultural activities are absent. Due to the detrimental biological effects, treatment and prevention methods must be considered to protect groundwater aquifers from nitrate leaching and high concentrations.

Conclusion

Further water analysis should be carried out periodically to obtain enough data for thorough assessment. Such analysis should include further chemical and microbial investigations. The health authorities and water board should monitor the safety of drinking waters in the communities to avoid all the potential dangers associated with nitrate pollution. Proper agricultural management practices need to be introduced avoiding overuse of nitrogen-based fertilizers. Nutrient pollution coupled with climate change might render clean drinking water scarce and hence bring about more strive.

ACKNOWLEDGEMENT

The authors hereby acknowledge the support of the Staff in Chemistry Department at UEAB and Mr. S. Mule for his tireless guidance during the research.

REFERENCES

Akinsola RO, Godowoli IA (2005). Determination of nitrate content in drinking Water: Asurvey of Bama Local Government Area, Borno State, Nigeria. Chem. Class J. 2:5-9.

APHA (1999). Standard methods for the examination of water and wastewater, American public health association: Washington D.C USA.

Calley AGF, Starfford CF, Stafford DA (1997). Treatment of Industrial Effluent, Holder and Stroughton: London. p. 328.

Chapman D (Ed.) (1992). Water Quality assessment. A guide to the use of biota, Sediments and Water, Chapman and Hall: London. pp. 71-76; 281-282, 412-416.

Clessicens SL, Greenberg EA, 19th ed. American Public Health Association, 1015 Fifteenth Street, NW, Washington, DC, pp. 49-51.

Eaton AD (1995). "Standard methods for the Examination of Water and Waste Water.

Elliot S (2008). "Testing the water", Royal Society of Chemistry, RSC, News Mag. 12(5):12-13.

European Environmental Agency (1999). Ground Water Quality and Quantity in Europe. Fatoki, O.S.S. Afr. J. Sci. 95:237.

Girard J (2005). Principles of Environmental Chemistry. Jones and Bartlett Publishers, Michigan. pp. 243-244.

Gray NF (1994). Drinking water Quality Problem and solution, John Wiley: New York.

Hallberg GR, Keeney DR (1993). Nitrate, Alley, William A., ed., Regional Groundwater Quality, Van Nostrand Reinhold, New York. Pp. 297-322.

Hill MJ (1991). Nitrates and Nitrites in Food and Water, Ellis Horwood: New York. pp. 14-92; 131-187.

Hem JD (1961). Some aspects of Chemical Equilibrium in ground water contamination, Public Health Service Symposium, A. Report Taft Sanitary Engr. Centre, Report WEI-5.

Jagessar RC, Sooknundun L (2011). Determination of nitrate anion in waste water from nine selected areas of coastal Guyana via a spectrophotometric method. Int. J. Res. Rev. Appl. Sci. 7(2):203-212.

Nolan BT, Hitt KJ (2006). Vulnerability of Shallow Groundwater and Drinking-Water Wells to. Nitrate in the United State. J. Environ. Sci. Technol. 40:7834-7840.

Pulido-Bosch AS, Bensi LM, Vallejos JMAC, Pulido-Leboeuf P (2000). Nitrates as indicators of aquifer interconnection. Application to the Campo de Dalia (SE-Spain). Environ. Geol. 39(7).

Ranjana A (2012). Nitrate contamination in ground water samples of Gangapurcity town (SawaiMadhopur District) Rajasthan. J. Chem. Biol. Phys. Sci. 2:511-513.

Roeve RN (1994). Enviromental Analysis (analytical chemistry by open learning), John Wiley: New York. pp. 17-19.

Speiran GK (1996). Geohydrology and Geochemistry near coastal ground-water-discharge areas of the Eastern Shore, Virginia: U.S. Geol. Surv. Water Supply pap. 2479, USGS, Richmond, Virginia.

World Health Organization (2006). Guidelines for Drinking water quality incorporating First addendum. Vol. 1, Recommendations. – 3rd Ed. WHO Publication Centre, Albany, 191 New York.

2

Chromium, an essential nutrient and pollutant: A review

Mandina Shadreck and Tawanda Mugadza

Department of Chemical Technology, Midlands State University, P. Bag 9055, Gweru, Zimbabwe.

The increased use of chromium (Cr) in several anthropogenic activities has led to the subsequent soil, surface water and ground water contamination. It has wide applications in the dyes, stainless steel, leather tanning, electroplating of chrome, and wood preservatives industries. Chromium exists in the environment in several diverse forms such as Cr(0), Cr(III) and Cr(VI) species. Cr toxicity depends on its valence state. Cr(VI) which is regarded as being highly mobile is toxic, while Cr(III) is less mobile and less toxic. Cr(VI) being more mobile in soil, more toxic and a stronger oxidant penetrates more readily into the cell membranes than the trivalent form. Chromium does not bioaccumulate in plants and animals therefore high levels of chromium in the environment are highly toxic to plants and animals. Chromium toxicity in human beings is expressed in liver and kidney damage as well as skin lesions or rashes. Symptoms of chromium toxicity in plants include alterations in the seed germination process, reduced growth of roots, stems and leaves, which results in low total dry matter production and yield. This paper reviews properties of trivalent and hexavalent chromium with respect to their essentiality as micronutrients and their toxic harmful effects. Trivalent chromium is essential to normal carbohydrate, lipid and protein metabolism, by making the action of the hormone, insulin, more effective while the hexavalent chromium is toxic and involved in mutagenicity, carcinogenicity, and teratogenicity.

Key words: Chromium, micronutrient, toxicity, pollutant, metabolism.

INTRODUCTION

Contamination of the environment by toxic heavy metals is a worldwide problem (Mahvi, 2008). The development of industries such as mining, electroplating, fertilizer, pesticides, metals finishing, leather, painting and dying has led to the discharge of metallic species into the environment causing serious environmental pollution to both flora and fauna (Cossich et al., 2002). Heavy metallic species are an environmental concern because of their toxicity, bio-accumulation tendency, persistency in nature and non biodegradability (Garg et al., 2007).

The toxicity of heavy metals is well documented, with past disasters in Minamata tragedy (Japan) due to methyl mercury contamination and "Itai-Itai" due to contamination of cadmium in Jintsu river of Japan (Friberg and Elinder, 1985; Kjellstrom et al., 1977). In order to assess the risk caused by heavy metal pollution, knowledge of the bioavailability of the different chemical species of the metals is essential (Eary and Rai, 1989). Among the toxic metal ions, chromium (VI) is a threat due to its toxicity even at low concentrations (Mahvi, 2008).

CHROMIUM OCCURRENCE AND SOURCES

Chromium (Cr) is the 17th most abundant element in the Earth's mantle and naturally occurs as chromite ($FeCr_2O_4$)

Table 1. Uses of Chromium.

Form	Uses
Cr(O)	Stainless steel production Alloy production Metal and alloy manufacturing
Cr(III)	Metal and alloy manufacturing Brick lining Chrome plating Leather tanning Textiles
Cr(VI)	Copying machine toner Chrome plating Leather tanning Textiles Copying machine toner

in ultramafic and serpentine rocks or complexed with other metals like crocoite ($PbCrO_4$), bentorite $Ca_6(Cr,Al)_2(SO_4)_3$ and tarapacaite (K_2CrO_4), vauquelinite ($CuPb_2CrO_4PO_4OH$), among others (Avudainayagam et al., 2003; Babula et al., 2008). The anthropogenic sources of Cr in the environment stem from the use of Cr in the metallurgy, refractory and chemical industries (Zayed and Terry, 2003). Chromium from anthropogenic sources can be released to soils and sediments indirectly by atmospheric deposition, but releases are more commonly from dumping of Cr-bearing liquid or solid wastes such as chromate by-products ("muds"), ferrochromium slag, or chromium plating wastes. Such wastes can contain any combination of Cr (III) or Cr(VI) with various solubilities (Stanin and Pirnie, 2004).

Chromium exists in the environment in three stable oxidation states, Cr(0), Cr(III) and Cr(VI) which have different toxicities and transport characteristics (Papassipoi et al., 2009). Cr(0) is the metallic form, produced in industry and is a solid with high fusion point usually used for the manufacturing of steel and other alloys. Cr (VI) typically exists as the oxyanion chromate (CrO_4^{2-}), dichromate ($Cr_2O_7^{2-}$) and the trioxide (CrO_3) have a high solubility in soils and water and very mobile in the environment. Additionally, Cr(VI) is considered the most toxic form of chromium, as it presents high oxidizing potential, high solubility, and mobility across the membranes in living organisms and in the environment (Becquer et al., 2003). The trivalent chromium, Cr(III), in contrast forms oxides, hydroxides and sulphates and is less toxic due to its insolubility in water. It presents lower mobility and is mainly bound to organic matter in soil and aquatic environments (Oliveira, 2012). Cr(III) has a tendency of forming hydroxide precipitates with Fe at typical ground water pH values. At high concentrations of oxygen or Mn oxides, Cr(III) can be oxidized to Cr (VI) (Peralta-Videa et al., 2009).

USES OF CHROMIUM

Chromium and its compounds are very useful in everyday life, as presented in Table 1. It is used on a large scale in many different industries, including metallurgical, electroplating, production of paints and pigments, tanning, wood preservation, chromium chemicals production, and pulp and paper production (Zayed and Terry, 2003). Chromium is resistant to ordinary corrosive agents at room temperature, which explains its uses as an electroplated, protective coating. It is also used in ferrous and non-ferrous alloys, in refractories, and in chemicals. Ferrous alloys, mainly stainless steels, account for most of the consumption. These steels have a wide range of mechanical properties as well as being corrosion and oxidation resistant (Bielicka et al., 2005).

CHROMIUM CHEMISTRY

Chromium is a polyvalent element, found naturally in the air, soil, water and lithosphere (Vasilatos et al., 2008). Chromium can exist in several chemical forms displaying different oxidation states from zero to six, but in the natural environment, only trivalent and hexavalent chromium are stable (Skovbjerg et al., 2006; Zayed and Terry, 2003). Cr(III) is the most stable state in water with a standard potential of -1.74V as shown in Figure 1, therefore a considerable amount of energy would be needed to convert it to lower or higher oxidation states. The chromium oxidation states of I - III have negative potentials thus oxidation is favored while oxidation states

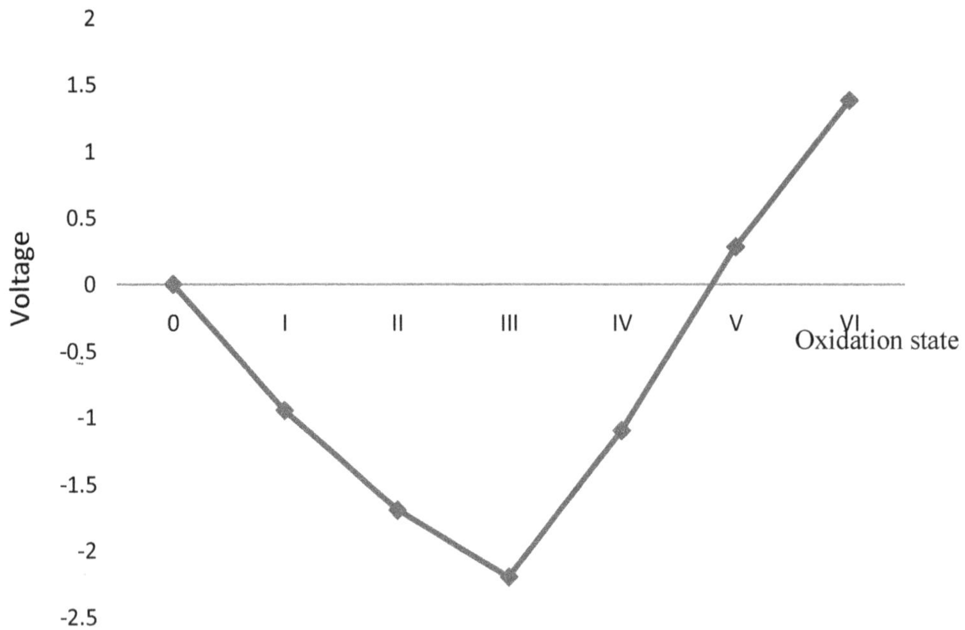

Figure 1. Oxidation state free energy diagram for chromium.

of IV - VI have positive potentials thus reduction is favored.

In acidic solutions, Cr(VI) has a very high positive redox potential (1.38 V) denoting that it is strongly oxidizing and unstable in the presence of electron donors (Figure 1). Under common environmental conditions of pH and Eh, Cr(III) compounds are sparingly soluble in water but Cr(VI) compounds are quite soluble (Swietlik, 1998). The resulting Cr(VI) solutions are powerful oxidizing agents under acidic conditions, but less so under basic conditions. For example, H_2CrO_4 is used for cleaning glassware in chemical laboratories by oxidizing organic residues. Thus, Cr(VI) is much more toxic and mobile in groundwater than the relatively immobile Cr(III). Depending on the concentration and acidity, Cr(VI) can exist either as chromate ion (CrO_4^{2-}) or as dichromate ion ($Cr_2O_7^{2-}$). The common dissolved Cr entities of Cr(VI) are the hydrogen chromate ion ($HCrO_4^-$), CrO_4^{2-} and $Cr_2O_7^{2-}$ (Testa, 2005). The entity that will dominate in a particular environment depends upon the specific conditions like pH, Eh, total concentration of Cr and the aqueous chemistry (Figure 2) (Rai et al., 1987).

At pH > 6.5, CrO_4^{2-} species dominate while at pH < 6.5, $HCrO_4^-$ counterparts dominates at low concentrations (<0.03 mol/L), but at concentrations greater than 0.001 mol/L, $HCrO_4^-$ ions begin to change to $Cr_2O_7^{2-}$ which becomes the dominant entity at concentrations greater than 0.03 mol/L (Palmer and Puls, 1994). $HCrO_4^-$ imparts a yellow colour to the water while $Cr_2O_7^{2-}$ imparts an orange colour (Testa, 2005). In aqueous solution, Cr(III) dominates as soluble Cr^{3+}at pH<3. As pH increases, Cr(III) hydrolyzes to $Cr(OH)^{2+}$, $Cr(OH)_3$ and $Cr(OH)_4^-$ species. In slightly acidic to alkaline conditions, Cr(III) precipitates as amorphous $Cr(OH)_3$ which subsequently crystallize to $Cr(OH)_3{\cdot}H_2O$ (Rai et al., 1987).

The Chromium Eh-pH diagram (Figure 2) provides a generalized depiction of the aqueous species stabilities in different redox conditions. Chromium introduced into or existing naturally in the environment may undergo changes in oxidation state if the redox chemistry of the environment is altered or if chromium that is migrating as aqueous species encounters different redox conditions (Vasilatos et al., 2008).

Chromium(VI) is strongly oxidizing only under high redox potentials (Figure 3), and it reacts rapidly with numerous reducing agents found commonly in the environment. Eary and Rai (1988, 1989) reported that Cr (VI) is reduced in seconds by ferrous ions and in a matter of hours to days by ferrous-iron containing oxide and silicate minerals. Similar reactions were reported to occur in low-pH soils that contain small amounts of ferrous iron in clay minerals (Rai et al., 1988). Reduction might occurred rapidly even in the presence of dissolved oxygen. Chromium(VI) is also reduced by organic matter (Schroeder and Lee, 1975; Bartlett and Kimble, 1976; James and Bartlett, 1983) and by $H_2S(g)$ (Smillie et al., 1981). Ferrous iron and organic matter are ubiquitous in soils and ground waters. Consequently, Cr(VI) is reduced to Cr(III) in many natural environments.

The oxidation potential for transforming the trivalent chromium into the hexavalent one is high, and the probability of transformation into a higher oxidation form in environmental conditions is reduced. Due to the high redox potential of the Cr(VI)/Cr(III), there are few oxidants present in natural systems that are capable of oxidizing Cr(III) to Cr(VI). Common oxidants include

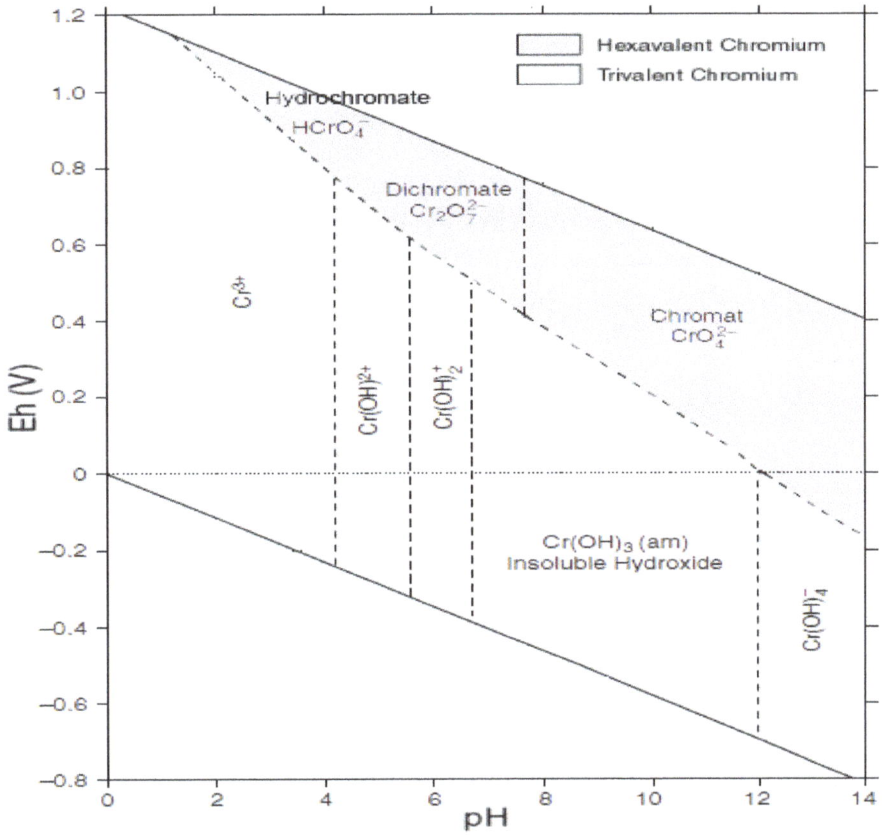

Figure 2. Eh-pH diagram for Chromium (Palmer and Wittbrodt, 1991).

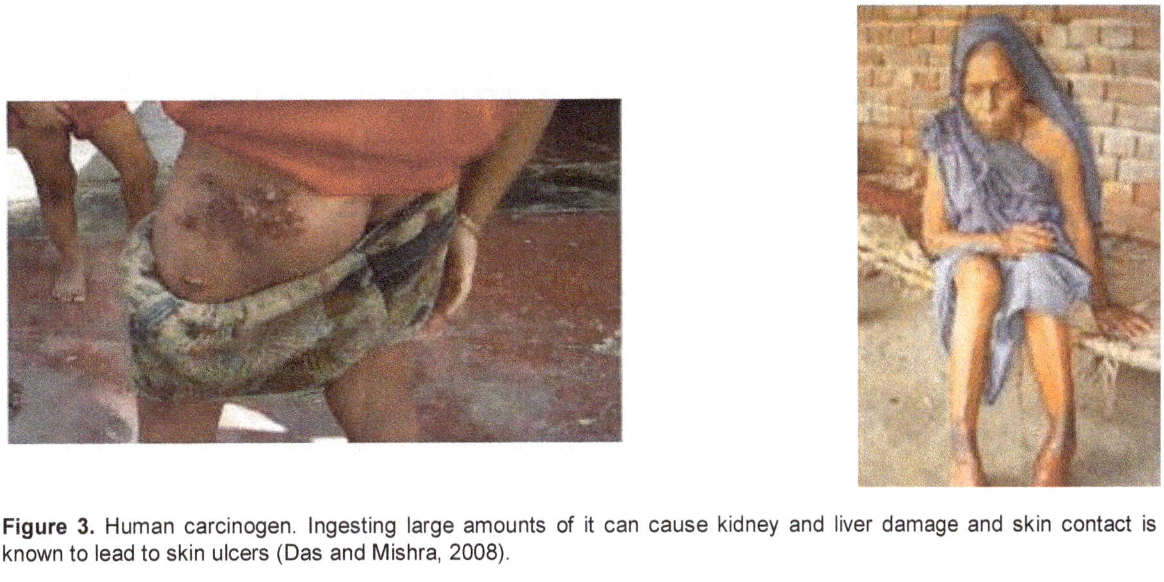

Figure 3. Human carcinogen. Ingesting large amounts of it can cause kidney and liver damage and skin contact is known to lead to skin ulcers (Das and Mishra, 2008).

dissolved oxygen and manganese oxides. However, the oxidation of Cr(III) by dissolved oxygen is reported to be very slow (Schroeder and Lee, 1975; Eary and Rai, 1988), while the oxidation of Cr(III) by manganese oxides is reported to be more rapid. Therefore, manganese oxides are more important oxidants for Cr(III) in groundwater systems (Rai et al., 1989). Despite the wide range of chromium in the soil and plants, the hexavalent chromium is rarely found in natural water above the concentration of the natural background, (1 µg/L). Higher concentrations indicate anthropogenic pollution (Brilly et al., 2003).

Table 2. Effects of Cr (III) and Cr (VI) on humans.

Cr(VI) Toxicant	Cr(III) Bio-element
Dermatitis	glucose metabolism (glucose tolerance factor)
Allergies	cholesterol and triglyceride levels decrease
Cancers	enzymes reactions
Mutations	body fat decrease
Teratogenic impact	muscle mass increase

CHROMIUM TOXICOLOGY

Chromium exists in the environment in several diverse forms such as trivalent (Cr(III)) and hexavalent, of which hexavalent chromium (Cr(VI)) is a carcinogen and a potential soil, surface water and ground water contaminant, while its reduced trivalent form (Cr^{3+}) is much less toxic, insoluble and a vital nutrient for humans. Cr(III) occurs naturally in the environment and is an essential nutrient required by the human body (De Flora, 2000; Flegal et al., 2001; Gibb et al., 2006).

The sources of trivalent chromium include many fresh vegetables and fruits, meat, grains, and yeast (Das and Mishra, 2008). Relatively insoluble, it is the most prevalent form in surface soils where reduction processes (which convert chromium from the hexavalent to trivalent form) are most common. The hexavalent chromium also occurs naturally and most notably in water-saturated (reducing) conditions and is an indicator of human pollution. Inside cells, Cr(III) can complex with organic compounds and this interfers with metallo-enzyme systems at high concentrations (Kotas and Stasicka, 2000).

Oliviera (2012) noted that, of the several chromium oxidation states, only the trivalent and hexavalent forms are of interest. Chromium in its trivalent oxidation state is considered to be essential for human and animal nutrition, while the hexavalent species have been identified as being toxic, carcinogenic and destroy the DNA structure (Oliviera, 2012). Bielicka et al. (2005) also gave the positive effects of Cr(III) as well as the harmful effects of Cr(VI) (Bielicka et al., 2005) (Table 2).

CHROMIUM(VI) TOXICITY

Exposure to Cr(VI) exerts toxic effects on biological systems (Das and Mishra, 2008). Inhalation of Cr(VI) has been shown to cause perforation of the nasal septum, asthma, bronchitis, pneumonitis, inflammation of the larynx and liver and increased incidence of bronchogenic carcinoma while exposure due to dermal contact of Cr(VI) compounds can induce skin allergies, dermatitis, dermal necrosis and dermal corrosion (Bielicka et al., 2005). Cr(VI) is a powerful epithelial irritant and also considered a human carcinogen (IARC, 1990).

Studies of workers in various industries with exposure to chromium compounds (including production of chromate and chromate pigments and chromium plating showed that they are at risk of developing various cancers such as of the nasal or sinonasal cavity, the lung and the stomach (Beaumont et al., 2008). Several chromium(VI) compounds like calcium chromate, lead chromate (and its derived pigments), chromium trioxide and sodium dichromate have been tested for carcinogenicity by several routes in several animal species and strains and were shown to cause various cancerous tumors (IARC, 1990). Other Cr(VI) containing compounds such as sodium dichromate dehydrate (NTP, 2008), potassium chromate (Davidson et al., 2004) have also been shown to be carcinogenic.

Chromium(VI) compounds are known to exert genotoxicity both *in vivo* and *in vitro*. Several studies have shown lymphocytes of workers exposed to dusts of chromium(VI) compounds to have elevated frequencies of DNA strand breaks (Gambelunghe et al., 2003), sister chromatid exchange (Wu et al., 2001), and micronuclei (Benova et al., 2002). DNA single-strand as well as double strand breaks may arise due to the reaction of chromium(VI) with hydrogen peroxide, forming hydroxyl radicals (Quievryn et al., 2003). This induces mismatches during replication, leading to aberrant mismatch repair. Thus chronic exposure to toxic doses of chromium(VI) provokes the selective outgrowth of mismatch-repair-deficient clones with high rates of spontaneous mutagenesis, and thus, genomic instability (Reynolds et al., 2007). The binding of Cr(VI) to double stranded deoxyribonucleic acid (DNA), alters gene replication, repair, and duplication, a proposed mechanism for cancer formation (Kendrick et al., 1992).

Cr(VI) is capable of penetrating cell membranes and get reduced to Cr(III), with the released electrons damaging the membrane (Namieśnik and Rabajczyk, 2012). This reduction ($Cr^{6+} + 3e \rightarrow Cr^{3+}$) contributes to mutagenic changes in the organism. It is also one of the ways in which Cr^{6+} is removed from organisms. Reduction of Cr(VI) to Cr(III) in the immune system forms reactive intermediates, which, in combination with oxidative stress, tissue damage (the result of oxidation), and a cascade of cell collisions, give rise to cytotoxicity, genotoxicity, and carcinogenicity (Shrivastava et al., 2002). It is known that the reduction of Cr(VI) produces the following free radicals: Cr(V), Cr(IV) and Cr(III), which are responsible for the observed toxic and carcinogenic effects (Li et al., 2011; Das and Mishra, 2008).

Based on epidemiologic investigations of workers and of experimental studies with animals, hexavalent chromium compounds were confirmed to be carcinogenic (Das and Mishra, 2008). Epidemiological studies conducted 40 years ago in the USA indicated a 10 to 30 fold- increased risk of lung cancer among workers in the chromate industry compared to the general population (Das and Mishra, 2008). In most studies, a positive correlation between the duration of exposure and lung cancer death was found. Gastrointestinal bleeding, tuberculosis and asthma infertility, birth defects, skin cancers, skin ulcers (Figure 3) and stillbirths have also been recorded among workers exposed to high levels of chromium(VI), (Das and Mishra, 2008). The International Agency for Research on Cancer (IARC) in 1990 concluded that there was sufficient evidence in humans for the carcinogenicity of chromium(VI) compounds as encountered in the chromate production, chromate pigment production, and chromium plating industries for the carcinogenicity of chromium(VI) compounds in humans based on the combined results of epidemiological studies, carcinogenicity studies in experimental animals, and evidence that chromium(VI) ions generated at critical sites in the target cells are responsible for the carcinogenic action observed.

Accidental or intentional ingestion of high doses of chromium (VI) compounds results in acute, potentially fatal, effects in the respiratory, cardiovascular, gastrointestinal, hepatic, renal, and neurological systems (IPCS, 2006; ECB, 2005; ATSDR, 2000). Some of these effects can be attributed to the corrosive nature of the compound (HSE, 2005). Typically, in one case a 17-year-old male died 14 h from respiratory distress with severe hemorrhages after ingesting potassium dichromate (29 mg chromium(VI) kg^{-1}) in an attempted suicide (ATSDR, 2000). Caustic burns in the stomach and duodenum and gastrointestinal hemorrhages were noted (ATSDR, 2000). Several other cases have reported fatalities following ingestion of lower doses of chromium(VI). In one case, a 14-year-old boy suffered gastrointestinal ulceration and severe liver and kidney damage and died 8 days after hospitalization after ingesting potassium dichromate (7.5 mg chromium(VI) kg^{-1}), while in another case, a 44year-old man died of severe gastrointestinal hemorrhage one month after ingesting chromic acid (4.1 mg chromium(VI) kg^{-1}) (ATSDR, 2000).

Developmental toxicity of Cr(VI) has been observed in animals, but to date there is not enough evidence to determine the potential for developmental effects on humans (Banu et al., 2008). Chromium has also been shown to be transferred from mother to young through the placenta and mother's milk and increased birth and developmental defects in children have been informally noted in areas of poorly regulated chromite mining, leather tanning (using Cr) and chrome production (Blacksmith Institute, 2007). However, no scientific studies investigating the potential relationship between these effects and specific chromium exposures in these locations have been conducted. A study on chrome plating workers occupationally exposed to chromic acid (mean 2 to 200 µg m^{-3} chromium(VI) for 8 h a day for 0.2 to 23.6 years) found that at low concentrations (mean <2 µg m^{-3}), chromium(VI)) workers developed smeary, crusty and atrophied septum mucosa and at higher concentrations (2 to 200 µg m^{-3} chromium(VI)) nasal irritation, mucosa ulceration and atrophy and septum perforation were observed (IPCS, 2006). These effects may not have resulted from exposure levels actually measured, but also from earlier exposures. Another study on electroplating workers exposed to chromic acid (>0.1 mg m^{-3} chromium(VI)) for less than 1 year reported frequent incidences of coughing, expectoration, nasal irritation, sneezing, rhinorrhea, nose-bleed, nasal septum ulceration and perforation (ATSDR, 2000).

Further evidence also suggests that exposure to chromium(VI) may induce occupational asthma and chromate sensitive workers acutely exposed to chromium(VI) compounds may develop asthma and other signs of respiratory distress (IPCS, 2006; ATSDR, 2000). For example, a study of 5 individuals with a history of contact dermatitis to chromium, found that exposure via nebulizer to a potassium dichromate aerosol containing 0.035 mg ml^{-1} chromium(VI) resulted in decreased forced expiratory volume, facial erythema, nasopharyngeal pruritis, blocked nose, coughing and wheezing (IPCS, 2006; ATSDR, 2000).

Some studies of workers exposed to airborne chromium(VI) have found increased levels of low-molecular-weight urinary proteins, such as retinol binding protein, β2-microglobulin and tubular antigens, indicative of early kidney changes, for example one such study identified a lowest - observed - adverse - effect - level (LOAEL) of 4 µg m^{-3} chromium(VI) (IPCS, 2006). Work-related cough or dyspnoea, production of phlegm, and shortness of breath was also noted in workers exposed to dust containing chromium oxide at an approximate concentration of 240 to 480 µg m^{-3} chromium(III) (IPCS, 2006).

CHROMIUM(III) AS AN ESSENTIAL NUTRIENT

Trivalent chromium is found in a wide range of foods, including egg yolks, whole-grain products, high-bran breakfast cereals, coffee, nuts, green beans, broccoli, meat, brewer's yeast, and some brands of wine and beer (Cefalu and Hu, 2004). Chromium is also present in many multivitamin/mineral supplements, and there are also specific chromium picolinate (CrP) supplements that contain 200 to 600 µg chromium per tablet (FSA, 2002). The U.S. National Academy of Sciences has established the Recommended Daily Allowances for chromium as 50 to 200 µg/day for adult men and women, which is also the Estimated Safe and Adequate Daily Dietary Intake (ESADDI) for chromium for children aged 7 years to adulthood (Anderson, 1997; 1998). Trivalent chromium is essential for animal and human health at trace levels (Ali and Aboul-Enein, 2006) and plays an essential role in many metabolic processes (García et al., 1999). Chromium(III) is present in the active centers of many enzymes, which is why it is classified as one of the essential elements. It facilitates the transport of glucose from the blood to the cells (Chen et al., 2009). The presence of Cr reduces the demand for insulin (the hormone regulating the level of glucose in the blood), thus Cr(III) helps to regulate blood sugar levels by collaborating with insulin in facilitating the uptake of glucose into cells (Horvath et al., 2008; Sahin et al., 2007; Lai et al. 2006). It also cooperates with insulin in protein synthesis. Cr(III) is an important bioelement, a deficiency of which disturbs glucose tolerance. It has been found, inter alia, that the glucose tolerance factor (GTF) contains Cr(III), which positively affects the action of insulin (Brown et al., 1986).

In the +3 oxidation state, chromium has been shown to reduce the risk of myocardial infarction (heart attack) and the development of atherosclerosis by lowering the concentration of total cholesterol and its low-density lipoprotein (bad cholesterol) fraction and raising the high-density lipoprotein (good cholesterol) level (USA- EPA, 1991b). Its deficiency has been associated with impaired glucose tolerance, fasting hyperglycemia, glucosuria, elevated percent body fat, decreased lean body mass, maturity-onset diabetes, cardiovascular disease, decreased sperm count, and impaired fertility (Valko et al., 2005). On the other hand, Cr(III) has been shown to have a detrimental effect on various components of the immune system, giving rise to immune stimulation or immune inhibition (Shrivastava et al., 2002). Chromium(III) also plays an important function in lipid metabolism thus reducing the risk of atherogenesis (Pechova and Pavlata, 2007), increase in lean body mass (Anderson, 2000), and promotion of weight loss (Morris et al., 1995).

CONCLUSION

This review has highlighted the two sides of Chromium (Cr) as an essential nutrient and a health hazard. Specifically Cr(VI), is considered harmful even in small doses while Cr(III) is considered essential for good health in moderate intake. As a micronutrient, Cr(III) is required for carbohydrate, lipid and protein metabolism while on the other hand the haxavalent form is predominantly involved in mutagenicity, carcinogenicity, and teratogenicity.

REFERENCES

Agency for Toxic Substances and Disease Registry (ATSDR). (2000). Toxicological Profile for Chromium, U.S. Department of Health and Human Services, Public Health Service, ATSDR, September 2000.

Ali I, Aboul-Enein HY (2006). Instrumental methods in metal ion speciation, Chromatographic science series. CRC Taylor Francis Group, 96:305-337.

Anderson RA (2000). Exercise effects on trace element metabolism. Trace Element in Man and Animal 10, in Rousssel A.M., Anderson R.A., Favier A.E. eds, Kluwer Academic/Plenum Publishers. New York, P. 393.

Anderson RA (1998). Chromium, glucose intolerance and diabetes. J. Am. College Nutr. 17:548-555.

Anderson RA (1997). Nutritional factors influencing the glucose/insulin system: chromium. J. Am. College Nutri.16:404-410.

Avudainayagam S, Megharaj M, Owens G, Kookana RS, Chittleborough D, Naidu R (2003). Chemistry of chromium in soils with emphasis on tannery waste sites," Rev. Environ. Contamin. Toxicol. 178:53-91.

Babula P, Adam V, Opatrilova R, Zehnalek J, Havel L, Kizek R (2008). Uncommon heavy metals, metalloids and their plant toxicity: A review. Environ. Chem. Lett. 6(4):189-213.

Banu SK, Samuel JB, Arosh JA, Burghardt RC Aruldhas MM (2008). Lactational exposure to hexavalent chromium delays puberty by impairing ovarian development, steroid genesis and pituitary hormone synthesis in developing Wistar rats. Toxicol. Appl. Pharmacol. 232:180-189.

Bartlett RJ, Kimble JM (1976). Behavior of chromium in soils: II. Hexavalent forms. J. Environ. Quality, 5:383-386.

Beaumont JJ, Sedman RM, Reynolds SD (2008). Cancer mortality in a Chinese population exposed to hexavalent chromium in drinking water. Epidemiol. 19:12-23. doi:10.1097/EDE.0b013e31815cea4c PMID: 18091413.

Becquer T, Quantin C, Sicot M, Boudot JP (2003). Chromium availability in ultramafic soils from New Caledonia. Sci. Total Environ. 301(1-3):251-261

Benova D, Hadjidekova V, Hristova R (2002). Cytogenetic effects of hexavalent chromium in Bulgarian chromium platers. Mutatation Res. 514:29-38. PMID:11815242

Bielicka A, Bojanowska I, Wiśniewski A (2005). Two Faces of Chromium - Pollutant and Bioelement. Polish J. Environ. Stud. 14(1):5-10.

Blacksmith Institute. (2007). Top 10 worst polluted sites. The Blacksmith Institute, New York. Available at: http://www.worstpolluted.org/ Accessed on: October 18 2012.

Brilly M, Jamnic B, Drobne D (2003). Chromium contamination of the Ljubljansko Polje aquifer. RMZ - Materials and Geoenvironment, 50(1):71-74.

Brown DH, Smith WE, El-Shahawi, MS, Wazir MFK (1986). Chromium (III) complexes with sugars. Inorganica Chimica Acta, 124(4):L25-L26.

Cefalu WT, Hu FB (2004). Role of Chromium in Human Health and in Diabetes. Diabetes Care, 27(11):2741-2751.

Chen WY, Chen CJ, Lui CH, Mao FC (2009). Chromium supplementation enchances insulin signaling in skeletal muscle of obese KK/HIJ diabetic mice. Diabetes Obesity and Metabolism, 11:293-303.

Cossich ES, Tavares CRG, Ravagnani TMK (2002). Biosorption of Cr (III) by Sargassum sp. Biomass. Electr. J. Biotechnol. 5(2):133-140.

Das AP, Mishra S (2008) Hexavalent Chromium (VI): Environment Pollutant and Health Hazard. J. Environ. Res. Develop. 2(3):386-392

Davidson T, Kluz T, Burns F (2004). Exposure to chromium (VI) in the drinking water increases susceptibility to UV-induced skin tumors in hairless mice. Toxicol. Appl. Pharmacol. 196:431-437. Doi: 10.1016/j. taap.2004.01.006 PMID: 15094314.

De Flora S (2000). Threshold mechanisms and site specificity in chromium (VI) carcinogenesis. Carcinogenesis 21(4):533-541.

Eary LE, Rai D (1988). Chromate removal from aqueous wastes by reduction with ferrous ion. Environ. Sci. Technol. 22:972-977.

Eary LE, Rai D (1989). Kinetics of chromate reduction by ferrous ions derived from hematite and biotite at 25°C. Am. J. Sci. 289:180-213.

European Chemicals Bureau (ECB) (2005). European Union Risk Assessment Report. Chromium trioxide, sodium chromate, sodium dichromate, ammonium dichromate and potassium dichromate Risk Assessment. EUR Report No. 201508 EN. ECB.

Flegal R, Last J, McConnell E, Schenker M, Witschi H (2001) Scientific Review of Toxicological and Human Health Issues Related to the Development of a Public Health Goal for Chromium(VI), Report Prepared for the Chromate Toxicity Review Committee, P. 6.
Friberg L, Elinder CG (1985). *Encyclopedia of Occupational Health*, third ed. International Labor Organization, Geneva.
Food Standard Agency, FSA (2002) Expert Group on Vitamins and Minerals: Review of chromium.UK.
Gambelunghe A, Piccinini R, Ambrogi M (2003). Primary DNA damage in chrome-plating workers. Toxicology 188(2-3):187-195.
García EM, Cabrera C, Sánchez J, Lorenzo ML, López MC (1999). Chromium levels in potable water, fruit juices and soft drinks: influence on dietary intake. Sci. Total Environ. 241:143-150.
Garg UK, Kaur MP, Garg VK, Sud D (2007). Removal of hexavalent Cr from aqueous solutions by agricultural waste biomass. J. Hazard. Mater. 140:60–68.
Gibb HJ, Lees PS, Pinsky PF. Rooney BC (2006). Lung cancer among workers in chromium chemical production. Am. J. Ind. Med, 38(2):115-126.
Health and Safety Executive (HSE) (2005). EH40/2005 Workplace Exposure Limits.
Horvath EM, Tackett L, McCarthy AM, Raman P, Brozincik JT Elmandorff JS (2008). Antidiabetogenic effects of chromium mitigate hyperinsulinemia-induced cellular resistance via correction of plasma membrane cholesterol imbalance. Mole. Endocrinol. 22:937-950.
International Agency for Research on Cancer, "Chromium, nickel and welding," in IARC Monographs on the Evaluation of Carcinogenic Risks to Humans, vol. 49, The International Agency for Research on Cancer, Scientific Publications, Lyon , France, 1990.
International Programme on Chemical Safety (IPCS) (2006). Inorganic chromium (III) compounds. Draft. Concise International Chemical Assessment Document. WHO. Geneva.
James BR, Bartlett JR (1983). Behavior of chromium in soils: VII. Adsorption and reduction of hexavalent forms. J. Environ. Q. 12:177-181.
Kendrick MJ, May MT, Plishka MJ, Robinson KD (1992). Metals in biological systems, Ellis Horwood Limited, London
Kjellstrom T, Shiroishi K, Erwin PE (1977). Urinary beta. /sub 2/microglobulin excretion among people exposed to cadmium in the general environment. Environ. Res. 13:318-344.
Kotas J, Stasicka Z (2000). "Chromium occurrence in the environment and methods of its speciation," Environ. Pollut. 107:263-283.
Lai MH, Chen YY. Cheng HH. (2006). Chromium yeast supplementation improves fasting plasma glucose and LDL-cholesterol in streptozotozin induced diabetic rats. Int. J. Vitam. Nutr. Res. 76:391-397.
Li Z-H, Li P, Randak T (2011). Evaluating the toxicity of environmental concentrations of waterborne chromium (VI) to a model teleost, oncorhynchus mykiss: a comparative study of in vivo and *in vitro*. Comparative Biochemistry and Physiology, Part C. 153:402-407.
Mahvi AH (2008) Application of agricultural fibers in pollution removal from aqueous solution. Int. J. Environ. Sci. Technol. 5(2):275-285.
Morris GS, Guidry KA, Hegsted M, Hasten, DL (1995). Effects of dietary chromium supplementation on cardiac mass, metabolic enzymes and contractile proteins, Nutr. Res. 15(7):1045.
Namieśnik J, Rabajczyk A (2012). Speciation Analysis of Chromium in Environmental Samples, *Critical* Reviews in Environ. Sci. Technol. 42(4):327-377
NTP (2008). NTP Toxicology and Carcinogenesis Studies of Sodium Dichromate Dihydrate (CAS No. 7789-12-0) in F344/N Rats and B6C3F1 Mice (Drinking Water Studies). National Toxicology Program Technical Report Services, 546:1-192. PMID: 18716633.
Oliveira H (2012). "Chromium as an Environmental Pollutant: Insights on Induced Plant Toxicity. J. Botany. pp.1-8.
Palmer CD, Puls RW (1994). Natural attenuation of hexavalent chromium in groundwater and soils: United States Environmental Protection Agency, EPA Ground Water Issue, EPA/540/5 94/505, P.12
Palmer CD, Wittbrodt PR (1991). Processes affecting the remediation of chromium-contaminated sites. Environ. Health Perspect. 92:25-40.
Papassiopi N, Kontoyianni A, Vaxevanidou K, Xenidis A (2009). Assessment of chromium biostabilization in contaminated soils using standard leaching and sequential extraction techniques. Sci. Total Environ. 407:925-936.
Pechova A, Pavlata L (2007) Chromium as an essential nutrient: a review Veterinarni Medicina, 52(1):1-18
Peralta-Videa, JR, Lopez ML, Narayan M, Saupe G, Gardea-Torresdey J (2009). The biochemistry of environmental heavy metal uptake by plants: implications for the food chain," Int. J. Biochem. Cell Biol. 41(8-9):1665-1677.
Quievryn G, Peterson E, Messer J, Zhitkovich A (2003). Genotoxicity and mutagenicity of chromium (VI)/ ascorbate-generated DNA adducts in human and bacterial cells. *Biochemistry,* 42:1062-1070. Doi: 10.1021/ bi0271547 PMID: 12549927
Rai D, Eary EL, Zachara MJ (1989). Environmental Chemistry of Chromium. Sci. Total Environ. 86:15-23.
Rai D, Eary EL, Zachara MJ, Ainsworth CC, Amonette EJ, Cowan EC, Szelmeczka WR, Resch TC, Schmidt LR, Girvin CD, Smith CS (1988) Chromium Reactions in Geologic Materials. Electric Power Research Institute, Palo Alto, California, EA5741.
Rai D, Sass BM, Moore DA (1987). Chromium (III) hydrolysis constants and solubility of chromium (III) hydroxide, Inorg. Chem. 26:345-349.
Reynolds M, Stoddard L, Bespalov I, Zhitkovich A (2007). Ascorbate acts as a highly potent inducer of chromate mutagenesis and clastogenesis: linkage to DNA breaks in G2 phase by mismatch repair. Nucleic Acids Research, 35:465-476. doi:10.1093/nar/gkl1069 PMID: 17169990
Sahin K, Onderci M, Tuzcu M, Ustandag B, Cikim G, Ozercan I.H, Sriramoju V, Juturu V, Komorowski JR (2007). Effect of chromium on carbohydrate and lipid metabolism in rat model of type 2 diabetes mellitus: the fat-fed streptozotocin-treated rat. Metabolism 56:1233-1240.
Shrivastava R, Upreti RK, Seth PK, Chaturvedi UC (2002). Effects of chromium on the immune system. FEMS Immunol. Med. Microbiol. 34(1):1-7.
Schroeder DC, Lee GF (1975). Potential transformations of chromium in natural waters. Water Air Soil Pollut. 4:355-365.
Skovbjerg LL, Stipp SLS, Utsunomiya S, Ewing RC (2006), The mechanisms of reduction of hexavalent chromium by green rust sodium sulfate: Formation of Cr-goethite, Geochimica et Cosmochimica Acta. 70:3582-3592.
Smillie RH, Hunter K, Loutit M (1981). Reduction of chromium (VI) by bacterially produced hydrogen sulfide in a marine environment. Water Res. 15:1351-1354.
Stanin FT, Pirnie M (2004). The Transport and Fate of Cr(VI) in the Environment in Guertin et al Eds. Chromium(VI) Handbook, (pp 165-214), CRC Press; Florida
Swietlik R (1998). Speciation analysis of chromium in waters. Polish J. Environ. Stud. 7:257-266.
Testa SM (2005). Sources of Chromium Contamination in Soil and Groundwater in Guertin et al Eds. Chromium(VI) Handbook, pp.143-164, CRC Press; Florida .
U.S. Environmental Protection Agency (1991b). Chromium (III).Integrated risk information system. Cincinnati, OH: Environmental Criteria and Assessment Office.
Vasilatos C, Megremi I, Economou-Eliopoulos M, Mitsis I (2008) Hexavalent chromium and other toxic elements in natural waters in the Thiva - Tanagra - Malakasa Basin, Greece. Hellenic J. Geosciences, 43:57-66.
Valko M, Morris H, Cronin MTD (2005). Metals, Toxicity and Oxidative Stress, Current Med. Chem. 121161-1208.
Wu FY, Wu WY, Kuo HW, Liu CS, Wang RY, Lai JS (2001). Effect of genotoxic exposure to chromium among electroplating workers in Taiwan. Sci. Total Environ. 279:21-28. doi:10.1016/ S0048-9697(01)00685-4 PMID:11712598.
Zayed AM, Terry N (2003). Chromium in the environment: factors affecting biological remediation. Plant and Soil 49(1):139-156.

Physicochemical study of acemannan polysaccharide in *Aloe species* under the influence of soil reaction (pH) and moisture application

Jyoti Nema[1]*, S. K. Shrivastava[1] and N. G. Mitra[2]

[1]Department of Applied Chemistry, Government Engineering College, Jabalpur-482011, (MP), India.
[2]Department of Soil Science and Agriculture Chemistry, Jawahar Lal Nehru Krishi Vishwa Vidyalaya, Jabalpur (MP), India.

Two Aloe plant species were subjected to different treatment of pH (6.0, 6.5, 7.0, and 7.5) and moisture supplements at crop coefficient (kc 0.2, 0.3, 0.4, 0.5) under cultivation practices. Aloe polysaccharide (acemannan) was found to be a major component in the leaves of Aloe plant. Acemannan is considered a major and the main active ingredient in Aloe gel. Therefore, knowledge of its chemical composition and physical properties are quite necessary for preparation of medicinal drugs. Results of experiments revealed that for both the plant species, lower soil pH (6.0), and moisture supplements (k_c > 0.3) were not suitable as their acemannan exhibit lower density, less solubility in the main solvent water, poor thermal stability, and less viscosity when compared to the control. But acemannan of higher pH (7.5) and moderate moisture (k_c 0.3) showed statically better physical properties than the other one. Acemannan of *Aloe ferox* showed higher density and viscosity, higher solubility in water and thermal stability than *Aloe vera* acemannan. In hydrolytes composition of Aloe polysaccharide, mannose saccharide was found to be the major concentration in *A. vera* acemannan and glucose in *A. ferox* acemannan. They were found statically and marginally different in their concentration under treatment of pH and moisture supplements.

Key words: Aloe polysaccharide, acemannan, physicochemical properties, Aloe gel.

INTRODUCTION

Many compounds with diverse structures have been isolated from both the central parenchyma tissue of Aloe leaves and the exudates arising from the cells adjacent to the vascular bundles. Aloe is made up of a vast range of compounds which can be divided into two groups for the convenience of this study, viz., minor composition and major composition. The group of major composition includes complex sugars in Aloe leaf gel exhibiting immune stimulating action. Acemannan stands out as a significant component in the fraction of major components (Bassetti and Sala, 2005). The saccharide group makes up most of the organic substances existing in the world; it is the significant component of nutrition, and serves as energy-producing base. It is also known as carbohydrates. In plants, its synthesis is associated with photosynthesis mediated by chlorophyll. Aloe has two categories of sugars, glucose and mannose as monosaccharides (simple molecules), and acemannan and cellulose as polysaccharides (complex molecules).

In Aloe gel, the precursor beta-(1,4)-acetyl-polymannose is better known as the acemannan. In glucomannan it is made up of 97% water and 0.7% solids, a mix of simple sugars and polysaccharides with varied chain lengths and of varying molecular weight. The longer chain polysaccharides which range from 10,000 to 20,000 monomer units of glucose and mannose are called mucopolysaccharides. Because of their distinct characteristics, the mucopolysaccharides are water

*Corresponding author. E-mail: Jyoti1221@rediffmail.com.

bound in nature and develop a viscous form similar to mucilage. Upon immersion in water, acemannan becomes viscous in the same way as mucilage. The chemical nomenclature of acemannan is beta-(1,4)-acetylpolymannose, because it contains a long chain polymer made up of glucose and mannose and reaches a molecular weight of about 18,000 to 20,000 units of molecular mass.

From the previous mentioned study, acemannan is considered to be the major and the main active ingredient in Aloe gel. Therefore, knowledge of its chemical composition and physical properties are quite necessary for preparation of medicinal drugs.

Physical properties of acemannan like density, solubility, viscosity and thermal stability are quite essential in the process of making drugs. These properties were found to be greatly affected by different soil environment such as the physical and chemical properties of soil which are accountable for extraneous control, like soil pH and moisture status.

MATERIALS AND METHODS

Two Aloe plant species (*A. vera* and *A. ferox*) were cultivated under different treatment of pH (6.0, 6.5, 7.0, 7.5) and moisture application at crop coefficient (k_c 0.2, 0.3, 0.4, 0.5) with one set of control (no irrigation nor control of soil pH).

During treatment, soil desiccation were maintain up to a given crop coefficient level (k_c) through irrigation of the required water to the *A. ferox* plants and soil pH were maintain up to given stresses through addition of HCl/NaOH. Required irrigation for maintaining moisture level up to crop coefficient was calculated by using following equation given by Hellman (2004).

$$\text{Required water} = \frac{\text{Evapotranspiration (loss of water ml/cm}^2\text{/day)}}{\text{Soil water holding capacity}} \times \text{Crop coefficient } (k_c)$$

After one year experiment of assessing identity, purity, quality and quantity of the Aloe polysaccharide (acemannan). Acemannan, the acetylated mannans in the gel of polysaccharides was extracted and isolated by alcohol precipitation method (McAnalley, 1990). Aloe gel was homogenized with ethanol (5 : 95 w/v) and stirred for 20 to 30 min. The alcoholic Aloe gel mixture was then allowed to stand for four hours. The clear supernatant liquid was decanted or siphoned off without disturbing the precipitate at the bottom of the container. The solution was then placed centrifuge buckets and centrifuged at 2000 g for 10 min. The precipitates was collected and washed with fresh ethanol. This fraction was then freeze-dried for 30 min. The dried pellets of polysaccharide were weighed. For qualitative and quantitative estimation of acemannan (%), the polysaccharide obtained from each samples' treatments with control sets of Aloe plant species was hydrolyzed to monosaccharide (Morrison, 1988), derivatized into alditol acetate (Hoebbler et al., 1989), and finally quantified by Gas liquid chromatography technique which is recent and conventional method (t'Hart et al., 1989). The monosaccharide standard was derivatized as the alditol acetate adopting the same procedure and the chromatogramme obtained with gas-liquid chromatography (GLC) was used to identify retention time and RF values. Shimadzu class GC-17 Gas Chromatograph with Capillary Glass column [DB 100 (30 m x 0.25 mm)], Flame ionised detector and He as carrier gas of flow rate 0.4 ml/min was used for determination of hydrolytes compositions of acemannan polysaccharides. Temperature setup was controlled at 270, 220 and 250°C of injection port, oven and detector respectively.

Isolated and purified acemannan of each treatment with control Aloe plant species were used for analysis of physical properties. The density was determined as mass (grams) per volume (milliliters) of the dry acemannan powder (AOAC, 1980). Solubility's in water and other media may differ. The method used for measurement was AOAC (1990). The solvents used were water, acetone, propylene glycol and 0.9% sodium chloride. 0.04 grams of acemannan powder was weighed and added into each of the sample tubes. Ten milliliters (10 ml) of the solvent was added producing a 0.4% (wt./volume) mixture. The mixture was subjected to agitation for 5 h at room temperature. The suspension was transferred to centrifuge tubes and centrifuged for 50 min at 1,500 rpm. The solvent blanks were similarly treated. The solutions were decanted from the solids. Both the solutions and the solids (deposits) were dried in an oven at low heat temperature (< 50°C) until it was completely dried. The dried solids were weighed; the solvent flasks that showed deposits were also weighed. This weight was subtracted from the weight of the samples.

Viscosity measurements were made using a Cannon-Fenske type viscometer (Model 150) at 40°C. The aqueous acemanna (0.2%) solutions were loaded into a viscometer and were allowed to warm for ten minutes in a water bath. The solution flow was then observed to the nearest 0.1 s according to the manufacturer's directions. The calibration factor was 0.04197 centistokes/s at 40°C.

Thermo gravimetric analysis (TGA) technique measures change in loss of weight with temperature. Weight loss was calculated using the following formula and the end of the maximum limit temperature (780°C) ash content was also calculated.

$$\text{Weight loss of acemannan and ash (\%)} = \frac{\text{Initial weight } (W_1) - \text{weight after 200°C (W2, W3, W4)}}{\text{Initial weight } (W_1)} \times 100$$

In order to find test of significance, the data of all the experimental parameters were statistical analyzed using ANOVA. The level of probability p = 0.05 was included to compute critical difference ($CD_{5\%}$) at the respective error degree of freedom (SE_m ±).

RESULTS AND DISCUSSION

The results of physicochemical properties of acemannan of plant species viz., *A. vera* and *A. ferox* under different treatment of pH and moisture were represented in Tables 1 and 2.

Density of acemannan polysaccharide was recorded at 0.53 and 0.88 g/l respectively, and at low pH 6.0, to moderate the moisture supplement k_c 0.3 in *A. vera*. While in *A. ferox*, acemannan was relatively denser exhibiting the density ranging from 0.63 to 1.03 gm/l at

Table 1. Physicochemical properties of acemannan in *Aloe species* under stress conditions of soil reaction (pH) and soil moisture (k_c).

Treatment	Acemannan in *A. vera species*						Acemannan in *A. ferox species*					
	Density (g/ml)	Viscosity (centistokes/sec)	Solubility (%)				Density (g/ml)	Viscosity (centistokes/sec)	Solubility (%)			
			Water	0.9% NaCl	Acetone	Propylene glycol			Water	0.9% NaCl	Acetone	Propylene glycol
k_c 0.2	0.79	1.18	24.65	85.60	0.01	0.13	0.92	1.57	29.77	90.50	0.08	0.09
k_c 0.3	0.88	1.23	26.04	87.10	0.03	0.16	0.97	1.54	28.41	93.37	0.01	0.07
k_c 0.4	0.73	1.21	26.85	86.61	0.08	0.16	0.86	1.47	30.07	88.40	0.07	0.09
k_c 0.5	0.56	1.17	22.43	83.97	0.02	0.09	0.73	1.45	30.19	87.63	0.04	0.08
pH 6.0	0.53	1.14	21.35	77.77	0.02	0.09	0.63	1.42	29.75	84.47	0.02	0.06
pH 6.5	0.59	1.09	24.47	81.47	0.06	0.18	0.68	1.43	29.56	90.43	0.07	0.10
pH 7.0	0.74	1.24	26.53	84.57	0.02	0.11	0.88	1.51	31.48	91.53	0.09	0.13
pH 7.5	0.83	1.29	27.85	87.70	0.03	0.13	1.03	1.59	33.73	94.50	0.08	0.11
control	0.66	1.19	23.46	84.23	0.08	0.12	0.77	1.47	30.55	88.43	0.04	0.09
$CD_{5\%}$ Plant x Treatment	0.031	0.029	0.36	1.77	0.014	0.016	0.031	0.029	0.36	1.77	0.014	0.016
SE_m ±	0.01111	0.01000	0.12522	0.61511	0.00510	0.00510	0.01111	0.01000	0.12522	0.61511	0.00510	0.00510

pH of 6.0 to higher pH of 7.5, respectively. For both plant species, lower soil pH (6.0) and moisture supplements (k_c > 0.4) were not suitable, as their acemannan exhibited lower density compared to the control one. The variation in density of acemannan depends on the content of inorganic salt which co-precipitate with the acemannan and also with the rate of hydration. The higher density of acemannan powder, the more rapid will be the lyophilization and the more ease in drugs formation.

The viscosity of the aqueous solution of acemannan is depicted in Table 1. Acemannan in *A. ferox* was more viscous ranging from 1.42 to 1.59 centistokes/s in comparison with that of *A. vera* which range from 1.09 to 1.29 centistokes/s. The viscosity of acemannan significantly varied with plant species and the soil environment during cultivation. This studied ascribed *A. ferox* polysaccharide as more useful for commercial production of herbal products.

Solubility of acemannan signifies penetrating capacity of the component in a medium which is conducive attribute in making of drugs. Therefore, pharmaceutical industries use the medium as a suitable solvent. Acemannan, by visual observation, is a white to off white amorphous powder. The data on solubility of acemannan are presented in Table 1. Acemannan powder dissolving in pure water will produce a highly viscous solution. Solubility of acemannan was significantly but not drastically affected with various source of production. McAnalley (1990) also reported that acemannan powder varied in character of solubility that depends slightly on the source of Aloe leaf but largely on the degree of processing such as filtration, ethanol precipitation and drying. From the aforementioned study, it was concluded that the solubility of acemannan was significantly affected by plant species but marginally affected with soil treatments during cultivation. Acemannan in *A. vera* at pH 7.5 was soluble in water to a great extent (27.85 to 21.35%), while in *A. ferox*, the solubility varied from 33.73 to 28.41% at pH 7.5 and pH 6.0 treatment.

The data in Table 1 reveals that the various solvents attained different solubility ratio of acemannan polysaccharide for both the plant species. Acemannan was particularly insoluble in common organic solvents such as acetone and propylene glycol, but fully soluble in inorganic solvent such as 0.9% NaCl.

Table 1 presents the solubility of acemannan in 0.9% NaCl solution as 88% in the case of *A. vera* and 94% in case of *A. ferox*. Water exhibited medium range of solubility. Therefore, acemannan formed a thick gel in propylene glycol and used in pharmaceutical industries.

Thermo gravimetric analysis gives an idea in characteristic weight loss profile with temperature. Effect of temperature on acemannan weight loss is presented in Table 2. The data revealed that

Table 2. Physico-chemical as thermal stability properties of acemannan (% of weight loss) in Aloe species under stress conditions of soil reaction (pH) and soil moisture (k_c).

Treatment	*A. vera*								*A. ferox*							
	Thermal stability (% of weight loss)				Chemical composition (%)				Thermal stability (% of weight loss)				Chemical composition (%)			
	200°C	400°C	600°C	>600°C (residue)	Mannos	Glucose	Galactose	Xylose	200°C	400°C	600°C	>600°C (residue)	Mannos	Glucose	Galactose	Xylose
k_c 0.2	40.30	45.23	78.70	25.17	60.9	15.9	3.9	1.9	38.08	42.48	72.23	30.63	25.4	50.3	5.1	6.1
k_c 0.3	40.41	45.38	78.36	26.71	64.2	22.7	6.6	3.0	38.38	42.66	71.76	32.79	31.0	71.2	8.2	5.6
k_c 0.4	40.50	45.67	77.93	26.34	73.7	24.0	6.0	1.2	38.53	43.06	72.26	32.45	34.9	82.0	6.3	4.0
k_c 0.5	40.39	45.54	78.25	23.15	56.0	19.1	3.2	0.6	38.43	42.95	71.84	30.66	29.7	45.3	5.3	2.6
pH 6.0	39.74	44.81	78.16	21.67	33.1	6.6	2.0	0.5	38.38	43.15	72.09	28.79	14.8	48.7	2.7	2.8
pH 6.5	40.77	44.58	77.88	22.58	56.8	10.8	1.8	1.6	38.44	43.27	72.46	29.22	18.3	58.8	3.5	3.6
pH 7.0	41.81	45.22	78.27	24.99	66.7	20.4	2.9	2.2	38.44	42.77	72.35	31.67	21.4	75.5	4.5	4.2
pH 7.5	39.99	45.08	78.14	25.71	85.4	32.5	5.0	2.4	38.32	42.36	71.65	32.93	23.4	79.9	8.9	4.7
Control	38.20	44.84	78.36	22.34	49.0	20.6	2.3	1.9	38.09	43.16	71.55	29.46	14.4	56.5	3.2	3.8
$CD_{5\%}$ Plant x Treatment	0.080	0.087	0.087	0.085	3.10	2.50	0.78	0.67	0.080	0.087	0.087	0.085	2.59	4.04	0.86	0.71
SE_m ±	0.02811	0.03000	0.03000	0.03000	1.064	0.857	0.268	0.228	0.02811	0.03000	0.03000	0.03000	0.887	1.385	0.296	0.242

acemannan decomposition pattern was different from those of other polysaccharides. Under identical operating conditions, the pattern of thermal decomposition was different from those of cellulose, dextran or amylan. Significant weight loss of acemannan was identified at temperature ranging from 200 to 600°C. The acemannan fractionation was controlled by these two temperature levels. However, the various treatments of soil pH and moisture supplements during cultivation of *A. species* marginally affected (not dramatic change) the thermal stability of acemannan.

Data in Table 2 represent the mean hydrolyte composition of acemannan polysaccharide under influence of treatment of pH and moisture supplements. The data revealed that several monosaccharides like mannose, glucose, gulactose, xylose and arabinose in trace amount were found in acemannan polysaccharides of Aloe plants (Figure 1a and b). The quantity was significantly affected by the different treatments of soil pH and moisture supplements of individually *Aloe species.*

Glucose was found to be maximum (79.9%) in the hydrolyte of *A. ferox* species while mannose (85.4%) was in *A. vera*. Other monosaccharide released often hydrolysis include arabinose, galactose and xylose in lesser contents (Table 2). Moreira and Filho (2008) also reported similar composition of polysaccharides.

In the *A. vera* gel, a fraction of polysaccharide acemannan was seen as being composed of different constituting monosaccharides were mannose (33.1 to 85.4%), glucose (6.6 to 32.5%), galactose (1.8 to 6.6%), xylose (0.5 to 3.0%) and arabinose remained un-detected.

In *A. ferox* species, a fraction of acemannan contained 45.3 to 79.9% glucose, 14.4 to 34.9% mannose, 2.7 to 8.9% galactose, 2.6 to 6.1% xylose and arabinose un-identified. The variation in the constituting monosaccharides is the significant result of the various treatment effects of soil pH and moisture supplements during cultivation of the two Aloe species. Table 2 shows the effect of species difference of Aloe plants as being the most significant one. It was explained by the fact that the mannosyl residues are contained in a reserve polysaccharide within the parenchyma cells which is significantly influenced by seasons and cultivation practices. Femenia et al. (1999) drew similar reasoning for the quantitative variation of acemannan composition in Aloe plants.

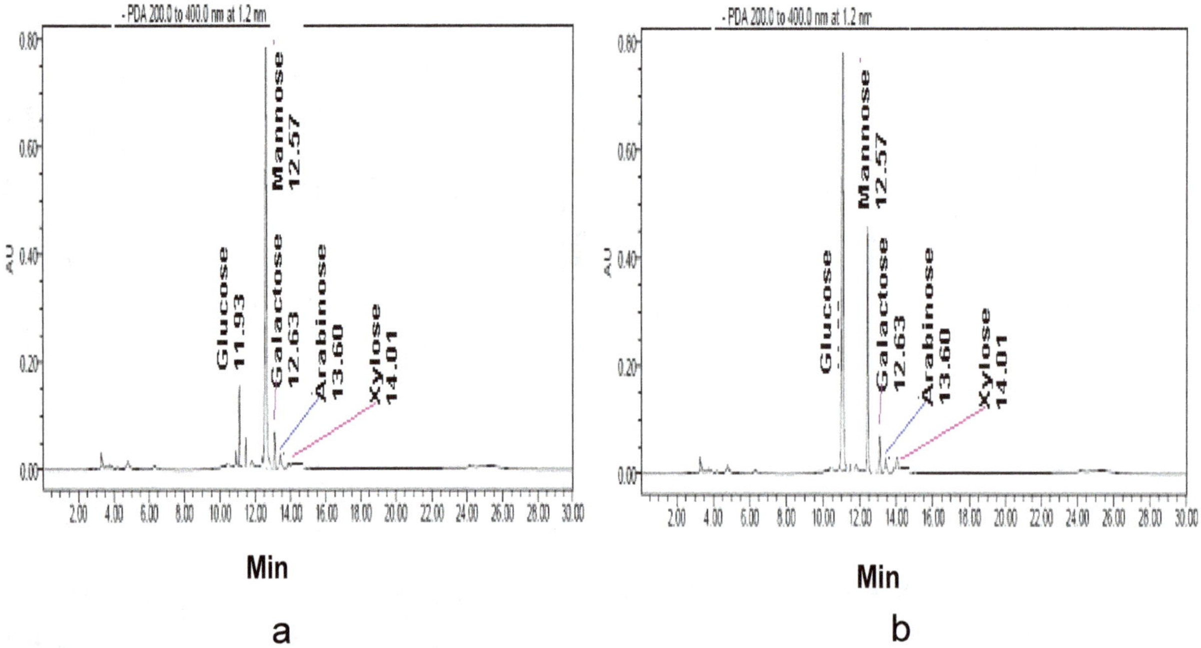

Figure 1. GLC chromatogram of various nutritive contents in gel hydrolyte of acemannan (a) *A. vera* and (b) *A. ferox* of treated the samples (pH 7.5).

REFERENCES

AOAC (1980). Official and Tentative Methods of the American Oil Chemists' Society, I and II, 3rd Ed.

AOAC (1990). Official methods of analyses. Association of Official Analytical Chemists: Washington, DC.

Bassetti A, Stefano S (2005). The Great Aloe Book, USA Edition Ist Edition, Zuccari Editions: 47-51.

Femenia A, Sanchez ES, Simal S, Rosello C (1999). Compositional features of polysaccharides from *Aloe vera* (*Aloe barbadensis* Miler) plant tissue. Carbohydr. Poly., 39: 109-117.

Hellman Ed (2004). Irrigation scheduling of grapevines with evapotranspiration data (http:// winegrapes.tamu.edu).

McAnalley BH (1990). Processes for preparation of *Aloe* products. United States Patent, 4,959,214.

Moreira LRS, Filho EXF (2008). An overview of mannan structure and mannan-degrading enzyme systems. Appl. Microbiol. Biotechnol., 79: 165-178.

Evaluation on the application of immobilized konkoli (*Maesopsis eminii*) immobilized leaves for the removal of Pb^{2+}, Cd^{2+} and Zn^{2+} in aqueous solution

S. A. Osemeahon and D. I. Esenowo*

Department of Chemistry, Federal University of Technology, Yola, Nigeria.

Calcium alginate immobilized konkoli leaves (KIL) was studied as a bio-sorbent for the sorption of Pb^{2+}, Cd^{2+} and Zn^{2+} in aqueous solution. The sorption capacities for Pb^{2+}, Cd^{2+} and Zn^{2+} were found to be 98.52, 90.1 and 81.34% respectively. The sorption behavior with respect to contact time, initial pH, ionic strength and initial metal ion concentration were investigated in order to determine the optimum sorption conditions. The result showed that the metal ion uptake by KIL decreases with increasing ionic strength while increase in initial metal ion concentration led to an increase in metal ion uptake. Maximum sorption was found to occur at pH 6 for Pb^{2+} and pH 5 for both Cd^{2+} and Zn^{2+}. The sorption capacity of KIL is encouraging and therefore may be used in the bio-remediation of wastewaters.

Key words: Konkoli leaves, immobilization, sodium alginate, sorption, metal ions.

INTRODUCTION

Fresh water quality and availability remain one of the most critical environmental and sustainability issues of the twenty first century (Adebo and Adetonyinbo, 2009). Water will continue to be a major issue with definite and profound importance on our live and that of the planet earth. The availability of safe and reliable source of water is an essential pre-requisite for sustainable development.

However, with the ever increasing scientific and industrial developments, man is continually exposed to some health hazards from heavy metal through water intake among other sources. Heavy metals are presently in the soil, natural water and air, in various forms and may become contaminant in food and drinking water. Some of them are constituent of pesticide, paints, fertilizers etc. The concentration of these metals needs to be reduced to meet the ever changing legislative standards. Wuyep et al. (2007) reported that the metals of most immediate concern are cadmium, cobalt, lead, nickel, mercury and zinc. The presence of such metals (>5 $mgcm^{-3}$) (Wuyep et al., 2007) in aquatic environment, causes severe damage to aquatic life, killing micro-organisms during biological water purification process. Moreover, these metals have exacting consequences on human such as brain damage, reproductive failures, nervous system failures, tumor formation etc (Wuyep et al., 2007).

Conventional processes for removal of metals from industrial waste include chemical precipitation, oxidation-reduction, filtration, electrochemical technique and other sophisticated separation procedure using membranes. These processes are expensive and time consuming and have inherent limitations (Akporhonor and Egwaikhide, 2007; Adebo and Adetonyinbo, 2009; Osemeahon et al., 2007). The need for economical and effective method of removing heavy metals from waste water has resulted in the search for unconventional materials that may be useful in reducing the levels of accumulation of heavy metals in the environment. Biological methods such as bio-sorption or bio-accumulation strategies for the removal of metal ions may provide an attractive alternative to existing technologies (Akporhonor and Egwaikhide, 2007).

Konkoli tree is one of the trees that grow wild in the south, east and middle belt of Nigeria. The seed gum is widely used as a thickener in soup and other traditionally

*Corresponding author. E-mail: davidesenowo@gmail.com.

baked products, primarily because of high viscosity, binding and swelling propensity (Barminas and Eremosele, 2002; Osemeahon et al., 2007).

In our earlier experiments (Osemeahon et al., 2007, 2008, 2011), we reported the potential use of konkoli seed gum for various industrial products. However, konkoli seed is produce relatively in small quantity and only once a year for both man and animals to compete for. Whereas konkoli leaves are abundantly produced and are available all year round. To the best of our knowledge, no report is presently available on the industrial application of konkoli leaves which are presently wasting away in the wild. In this experiment, we set out to evaluate the application of immobilized konkoli leaves as a bio-sorbent for the removal of lead, cadmium and zinc metal ion from aqueous solution.

MATERIALS AND METHODS

Sodium alginate, calcium chloride, sodium chloride, sodium hydroxide, hydrochloric acid, cadmium chloride, zinc chloride, lead chloride, were product from the British Drug House (BDH), while konkoli leaves were purchased from Taraba State of Nigeria. All the materials were used as supplied.

Preparation of konkoli leaves

The konkoli (*Maesopsis eminii*) leaves were sun-dried, pounded in a mortar into powder and sieved through 100 µm sieve screen to produce a fine powder. The sieved material was packed in a polythene bag for further use. 4.0 g of the leave was mixed with 100 cm^3 of distilled water to obtained solution A.

Preparation of sodium alginate and calcium chloride

Sodium alginate was made by weighing 4.00 g and making it up to 100 cm^3 mark with distilled water in a volumetric flask and left overnight for complete dissolution to give solution B (Toti et al., 2002). 0.12 M of calcium chloride was prepared according to a standard procedure described by (Wuyep et al., 2007).

Immobilization of the konkoli leaves

25 cm^3 of solution A and 25 cm^3 of solution B were mixed and stirred vigorously in 250 cm^3 beaker, to obtain a homogenous mixture. The mixture was subsequently poured into another beaker containing of 0.12 M calcium chloride solution. A retention time of 1 h was allowed for the reaction to obtain complete precipitation of the immobilized konkoli leave. The precipitated blend solid was removed and allowed to dry at room temperature (30°C). The dried solid mass was stored in a polythene bag for further use (Wuyep et al., 2007).

Determination of metal ion in solutions

The metal ion chosen for this study were Pb^{2+}, Zn^{2+} and Cd^{2+}. A concentration of 200 ppm of each of the metal ion was prepared. From aforementioned concentration, 50 cm of solution of the metal ion was taken into the conical flask; 0.2 g of dried KIL was added and then shaken vigorously for 2 h by using flask shaker (Stuart Scientific, SF1). The solution was then filtered and the residual metal ion concentration determined using Atomic Absorption Spectrophotometer (AAS) (210 VGP Buck Scientific) (Osemeahon et al., 2007).

KIL sorbent sorption capacity

For this equilibrium studies, 0.2 g of the sorbent was shaken with 50 cm^3 of Pb^{2+} ion solution at room temperature for 24 h. The synthetic waste water sample was filtered and analyzed for residual metal ion concentration using AAS (Charamathy et al., 2001). This process was repeated for all the metal ions studied.

Effect of pH on sorption capacity

The sorption characteristics of the KIL at different pH values (1.0 to 6.0) were investigated at 30°C. 1.0 M hydrochloric acid and 1.0 M sodium hydroxide was used to adjust the solution pH as the case may be. The residual metal ion was measured as stated previously (Charamathy et al., 2001).

Effect of the ionic strength on sorption capacity

Useful information regarding the effect of the ionic strength was obtained by measuring the sorption of KIL in various concentrations of NaCl solution (0.1 to 2.0; % w/w). 0.2 g of the sample was added to 50 cm^3 of the prepared NaCl solution and the equilibrium concentration of the residual metal ion determined (Wuyep et al., 2007).

Effects of the contact time on sorption capacity

To determine the kinetics of the sorption for the various metal ions, several set of samples consisting of 0.2 g of the dried sorbent and 50 cm^3 of the metal ion solution for each of the ions was prepared. As the samples were undergoing agitation (with the flask shaker), they were removed one after the other at a predetermined time interval ranging from 0.5 to 24 h for analysis. The solution was filtered and analyzed for residual metal ion. This was done for all metal ions at 30°C (Osemeahon et al., 2007).

Effect of the initial metal ion concentration on sorption capacity

The effect of the initial metal ion concentration on the sorption capacity of different samples consisting of 50 cm^3 each of different metal ion concentrations ranging from 5 to 100 ppm, but each containing 0.2 g of the dried KIL was prepared and shaken until an equilibrium was achieved at 30°C. The synthetic waste water was filtered and analyzed for residual ion concentration (Osemeahon et al., 2007).

RESULTS AND DISCUSSION

KIL sorption capacity

The equilibrium sorption for Pb^{2+}, Zn^{2+} and Cd^{2+} of KIL are shown in Figure 1. It was observed that KIL can take up a significant quantity of Pb^{2+}, and relatively, the lowest amount is from Zn^{2+}. The values of 98.52, 91.7 and 81.34% for Pb^{2+}, Cd^{2+} and Zn^{2+} respectively were found.

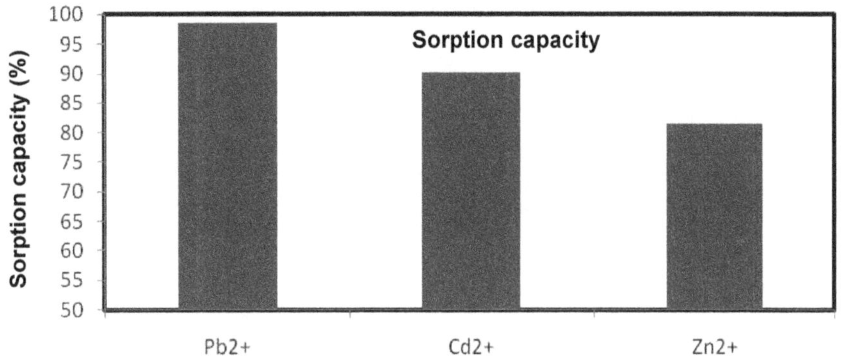

Figure 1. Sorption capacity for Pb^{2+}, Cd^{2+} and Zn^{2+} by KIL.

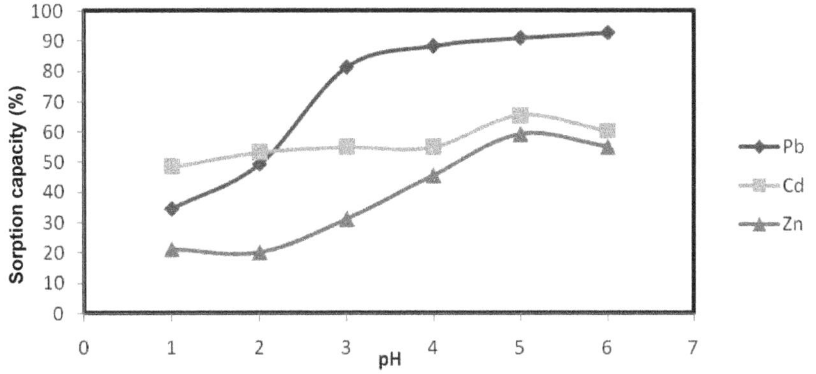

Figure 2. Effect of pH on sorption of Pb^{2+}, Cd^{2+} and Zn^{2+} by KIL.

The report from the present study are comparable with those reported for konkoli seed gum and other similar studies (Osemeahon et al., 2007; Lee and Lin, 2000; Charamathy et al., 2001). Although, the sorption capacity recorded for lead in this work (98.52%) is relatively lower than that of konkoli seed gum (99.8%) reported by Osemeahon et al. (2007), by a very small margin. The differences observed in the sorption capacities for the different metal ions can be explained in terms of the differences in hydration free energy, the ability of metal to form covalent bond with ligand, the metal polymeric cations within the membrane structure and the nature of the surface sites available (Cooper et al., 2002).

Effect of pH

pH plays an important role in the sorption of the metal ions by different sorbents. Figure 2 depicts the effect of pH on removal of metal ions. It was seen at low pH values that the sorption is low, but as the pH of the solution increases, the metal ions uptake increases, however, at different level depending on the metal. The trend of result obtained was similar to the ones reported by other workers (Souag and Buayada, 2009; Osemeahon et al., 2007; Wuyep et al., 2007; Chen and Wu, 2000).

The optimal pH of sorption for the different metal ions is 6 for Pb^{2+} and 5 for Cd^{2+} and Zn^{2+}. The pH of the sorption media is related to the metal sorption - mechanism of the surfaces from the water and reflected the nature of the physiochemical interaction of the ions in solution and nature of the sorption sites (Osemeahon et al., 2007). Therefore, it can be suggested that the differences in the pH optima for different metal ions could probably be due to the differences solution chemistry of the metal ions (Mcox et al., 2002).

The pH influences the equilibrium of metal ion uptake in aqueous solution. The counter reaction of the competiting hydrogen ion as well as the chemistry of the active binding sites of the sorbent is responsible for the depression in sorption capacity. Increasing the pH leads to precipitation of insoluble hydroxide or hydrated oxide, thereby lowering the metal ion availability for sorption. On the other hand, a decrease in pH results in the increase in the hydrogen ion concentration and hence possible competition for binding sites (Yen-Peng and Sung, 2000).

Effect of time on sorption

The effect of contact time on sorption capacity is shown

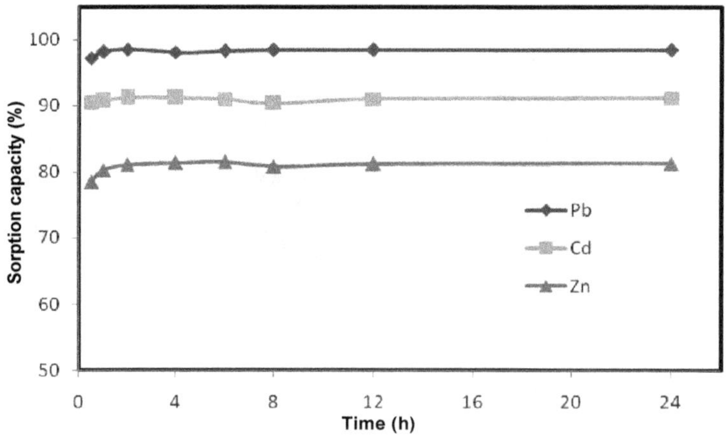

Figure 3. Effect of contact time on sorption of Pb^{2+}, Cd^{2+} and Zn^{2+} by KIL.

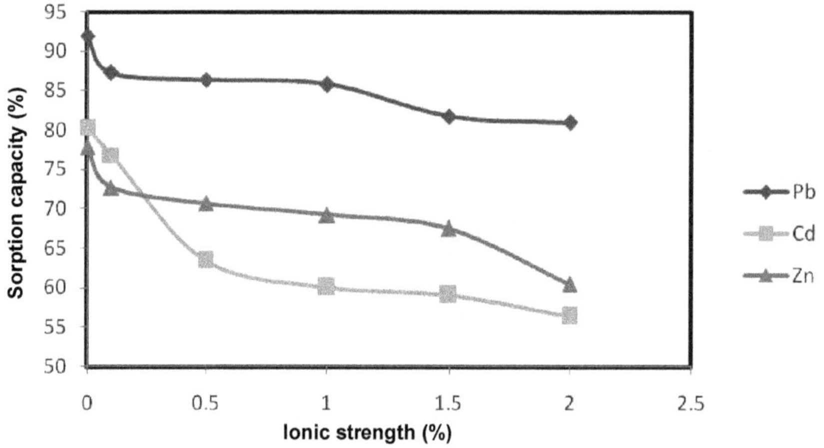

Figure 4. Effect of ionic strength on sorption of Pb^{2+}, Cd^{2+} and Zn^{2+} by KIL.

in Figure 3. Saturation was reached after 2 h of contact time for all the metal ions, but high sorption rate were observed during the first 30 min. Equilibrium time of 30 min was recorded for Pb^{2+} and 2 h for the rest of the metal ions.

The rapid sorption of metal ions is attributed to highly porous structure of the sorbent which provide readily access and large surface area for the sorption of the metal ions to the binding site (Aiabuk et al., 2007). This result demonstrates that the sorbent (KIL) can compete favorably with other industrial sorbents in terms of kinetics.

Effect of ionic strength on sorption

Industrial waste and natural water often contains dissolved salt which competes with heavy metal ions for binding sites on the sorbent carbon chain (Osemeahon et al., 2008; Cox et al., 2000), hence, there is need to investigate the level of influence of this factor on the sorption capacity of KIL.

The effect of the ionic concentration on the removal of the heavy metal ions from the waste water by KIL is represented in Figure 4. It was observed that the sorbent showed a decrease in sorption capacity with increase in initial ion concentration for all the metal ions studied.

The decrease in sorption capacity with increasing ionic strength is primarily attributed to the difference in the ionic osmotic pressure between the sorbent and the external solution (Lee and Lin, 2000; Osemeahon et al., 2008). The difference in osmotic pressure between the sorbent and the external solution decreases as the ionic strength of external solution increases. Therefore, the sorption of metal ion decreases when the ionic strength of external solution increases (Lee and Lin, 2000). From this experiment, the coexistence of metal ions in same solution caused the interference with the performance of KIL sorbent. Therefore a pretreatment of the effluent will remove the interfering ion and this appears to improve

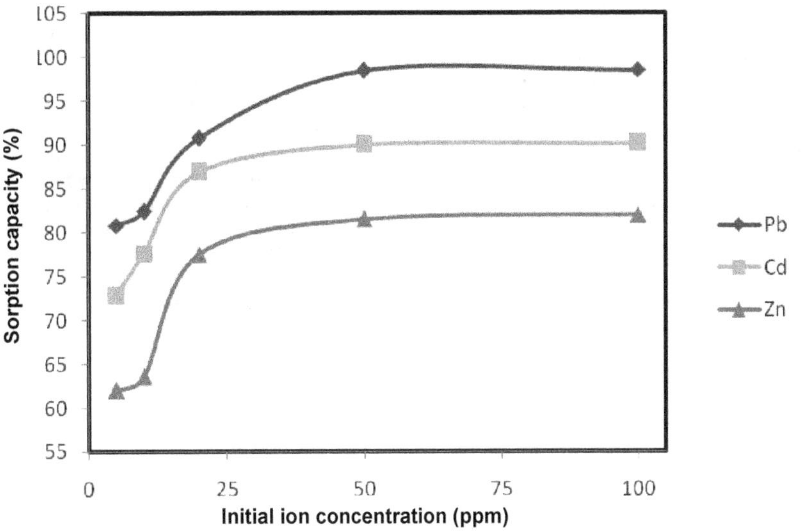

Figure 5. Effect of initial ion concentration on Pb^{2+}, Cd^{2+} and Zn^{2+} by KIL.

the efficiency of the sorbent (Osemeahon et al., 2008).

Effect of initial ion concentration

The result of metal ions sorption by the KIL as a function of initial metal ion concentration is shown in Figure 5. It was observed that the sorption efficiency of the metal ions by the sorbent increases with initial ion concentration but leveled up at certain point of the initial ion concentration. This behavior can be explained in term of increase in flux of the metal ion. The flux of the metal ion varies directly with the metal ion concentration, and hence, there should be an increase in flux with increase in initial concentration as reported by Iulia et al. (2007). The constant regime observed for all metal ions is explained as the point of saturation.

One of the problems of using conventional methods for the removal of the metal ions from the aqueous systems is effectiveness of the sorbent when the metal ion is present at a very low concentration (Asiagwu et al., 2009). The behavior of KIL even at low concentration of 5 ppm (removal of 80.8, 72.8 and 62% for Pb^{2+}, Cd^{2+} and Zn^{2+} respectively) shows that KIL sorbent can even be used at low metal ion concentration. This may reduce the present problem associated with most sorbents performance at low metal ion concentration.

Conclusion

KIL were immobilized and the sorption behavior in aqueous solution was studied. Sorption capacities of 98.52, 90.1 and 81.34% for Pb^{2+}, Cd^{2+} and Zn^{2+} respectively were obtained. It was observed that the uptake of metal ion by KIL decreases as the ionic strength of the external solution increases. While on the other hand, an increase in the initial metal ion concentration gave rise to increase metal ion uptake. The study revealed that KIL has potential for bioremediation of wastewater.

REFERENCES

Adebo BA, Adetoyinbo AA, (2009). Assessment of groundwater quality in unconsolidated sedimentary coastal aquiver in Lagos State, Nigeria. Sci. Res. Essay, 4(4): 314-319.

Ãiabuk A, Akar S, Tumali S, Gedikli (2007). Biosorption of Pb (II) by industrial strain of *Saccharomyces cerevisiae* immobilized on the biomatrix of cone biomass of *Pinus nigra*: Equilibrium and mechanism analysis. J. Chem. Eng., 131: 293-300.

Akporhonor EE, Egwaikhide AS (2007). Removal of the selected metal ion from aqueous solution by adsorption onto the chemically modified maize cobs. Sci. Res. Essay, 2(4): 132-134.

Asiagwu AK, Okoye PAC, Ifeoma O, Oniuku PE (2009). Sorption Potentials of waste tire for some heavy metals (Cd, Pb) in aqueous solution. Leonard J. Sci., 6: 227-234.

Barminas JT, Eremosele IC (2002). Rheological properties and potential industrial application of konkoli(*maesopsis eminii*) seed gum. In: Williams PA, Philip EO (editors) Gum and Stability for the Food Industry. Royal Soc. Chem. UK., pp. 306-311.

Charmarthy S, Chung W, Marshall WE (2001). Adsorption of Selected Toxic Metals by Modified Pea nut shells. J. Chem. Technol. Biotechnol., 75: 791-797.

Chen JP, Wu S (2000). EDTA study on EDTA chelated copper adsorption by granulated activated carbon. Society for Chemical Industry, J. Chem. Technol. Biotechnol., 75: 791-797.

Cooper C, Jiang JQ, Ouki S (2002). Preliminary evaluation of polymeric Fe and Al modified clay as adsorbent for heavy metal removal in water treatment. J. Chem. Technol. Biotechnol., 77: 546-551.

Cox M, Elshafey E, Pichugin AA, Appleton Q (2000) Removal of mercury (II) ions from aqueous solution on a carbonaceous sorbent prepared from flax shives. J. Chem. Technol., 75: 427-435.

Iulia N, Iorgulescu M, Spiriou MF, Ghiurea M, Petcu C, Cinteza O (2007). Adsorption of Heavy Metal Ions on Calcium Alginate Microparticles, Ann. Bucharest - Chime, Year (new series), 1: 59-67.

Lee W, Lin G, (2000). Super-sorbent polymeric material viii: swelling behavior of polymeric cross-linked poly (sodium acrylate-co-trimethyl-metaacryloli oxyethyl ammonium iodide) in aqueous solution. J. Appl. Polym. Sci., 79: 1165-1674.

Mcox E, Pichugin AA, Appleton Q (2002). Application of carbon sorbent for the removal of cadmium and other heavy metals from the aqueous salt solution. J. Chem. Technol. Biotechnol., 77: 429-436.

Osemeahon SA, Barminas JT, Aliyu BA, Nkafamiya II (2008). Development of sodium alginate and konkoli gum grafted polyacryamide blend membrane: Optimizing of grafting conditions. Afr. J. Biotechnol., 7(9): 1309-1313.

Osemeahon, SA, Barminas JT, Aliyu BA, Maina HM (2007). Preliminary evaluation on the application of grafted membrane for sorption of copper and lead ions in aqueous solution. Int. J. Phys. Sci., 2(11): 294-9.

Osemeahon SA, Barminas JT, Nkafamiya II, Esenowo DI (2011). Immobilization of konkoli (*Maesopsis eminii*) leaves with calcium alginate and study of its swelling behavior. Int. Res. J. Plant Sci., 2(6): 186-190.

Souag R, Touaibia D, Buayada B (2009). Adsorption of heavy metals (Pd, Zn, Cd) Using Keratin powder prepared form Algerian Sheep Hoofs. Eur. J. Sci. Res., 3: 416-425.

Toti SU, Mahadevappa Y, Kumaresh K, Soppimath S, Aminabhavi TM (2002). Pervaporation separation of water and acetic acid mixture through blend membrane od sodium alginate and guar gum grafted polyacrylamide. J. Appl. Polym. Sci., 83: 259-277.

Wuyep PA, Chuma, AG, Awodi, S. Nok AJ (2007). Biosorption of Cr, Mn, Fe,Ni, Cu, and Pb metals from refinery effluent by calcium alginate immobilized mycelia of *polyporus squamosunolos*; Sci. Res. Essay, 2(7): 217-221.

Yen-peng T, Sung G,(2000). Use of polyvinyl alcohol as a cell immobilized matrix of copper bio-sorption by yeast cell. Chem. Technol. Biotechnol., 75: 541-546.

Evaluation of phytochemical and antimicrobial potentials of roots, stem-bark and leaves extracts of *Eucalyptus camaldulensis*

M. H. Shagal[1]*, D. Kubmarawa[1], K. Tadzabia[2] and K. I. Dennis[1]

[1]Department of Chemistry, Modibbo Adama University of Technology, P.M.B. 2076, Yola, Nigeria.
[2]Department of Chemistry, Umar Suleiman College of Education, P.M.B. 2 Gashua, Yobe State, Nigeria.

The extracts for phytochemical screening and antimicrobial activity were carried out on dried powdered leaves, stem-bark and roots of *Eucalyptus camaldulensis* using ethanol and water as solvent. The results of the phytochemical screening showed that the plant parts contained saponins, tannins, phenols and glycosides. The disc diffusion method was adopted for the antimicrobial activity of the plant extracts. The antimicrobial activity test of the plant extracts on *Escherichia coli*, *Staphylococcus aureus*, *Salmonella typhi* and *Bacillus subtilis* showed that both water and ethanol extracts had inhibitory activity on all the tested organisms except the roots and stem-bark extracts of ethanol which showed no measurable zone of inhibition. The results obtained from this study revealed that extracts of *E. camaldulensis* possess antimicrobial activities against some microorganism that causes diseases.

Key words: *Eucalyptus camaldulensis*, phytochemical, medicinal, ethanol, diseases.

INTRODUCTION

Antimicrobial agents are substances that interfere with the growth and metabolism of microbes. In common usage, the term denotes inhibition of growth and with reference to specific groups of organisms, terms as antibacterial, antifungal, antiviral and antiprotozoa are frequently employed. Antimicrobial agents may either kill microorganisms or inhibit their growth. Those that inhibit growth are called bacteria static. These agents depend on the normal host defenses to kill or eliminate the pathogens after its growth has been inhibited. For example, sulfa drugs, which are frequently prescribed for urinary infections, inhibit the growth of bacteria in the bladder until they are eliminated during the normal process of urination. Antimicrobial agents that kill are bactericidal. These antimicrobial agents are particular useful in situations in which the normal host defenses cannot be relied on to remove or destroy pathogens. A given antimicrobial can be bactericidal in one situation, yet bacteria static in another, depending on the concentration of the drug and the growth stage of the microorganism (Nester et al., 2004).

Some antimicrobial agents are used to treat infection and they are called chemotherapeutic agents. Chemotherapeutic agents are chemicals substance used for the treatment of infectious diseases or disease caused by the proliferation of malignant cells. These substances are prepared in the chemical laboratory or obtained from microorganisms and some plants and animals in general, naturally occurring substances are distinguished from synthetic compounds by the name antibiotics. Some antibiotics are prepared synthetically, but most of them are prepared commercially by microbial biosynthesis.

Chemotherapeutic agents must have selective toxicity for the parasite, which means a low toxicity for host cells and high toxicity for the parasite (Pelezer et al., 1993).

In Nigeria, application of medicinal plants especially in traditional medicine is currently well acknowledged and established as a viable profession (Kafaru, 1994).

*Corresponding author. E-mail: shagal2006@yahoo.com.

Table 1. Results of phytochemical screening of *E. camaldulensis*

Bioactive compounds	Water extracts			Ethanol extracts		
	Roots	Stem-bark	Leaves	Roots	Stem-bark	Leaves
Tannins	+	+	-	+	+	+
Saponins	+	+	-	+	-	+
Phenols	+	+	+	+	-	+
Glycosides	+	+	+	+	+	+
Alkaloids	-	-	-	-	-	-

+ = Present, - = Absent.

Extraction of bioactive compounds from medicinal plants permits the demonstration of their physiological activity. It also facilitate pharmacology studies leading to synthesis of a more portend drugs with reduced toxicity (Ebana et al., 1991).

Presently, in the developing countries, synthetic drugs are not only expensive and inadequate for the treatment of diseases, but are also often with adulterations and side effects (Shariff, 2001) and the increasing resistance of most synthetically derived antimicrobial agents are of utmost concern (Adeniyi and Ayepola, 2008). Therefore, there is need to search for suitable plants of medicinal value to be effectives in the treatment of diseases, which must be harmless to human tissue.

Eucalyptus species have been planted on many parts of the world. It belongs to the order myrtales and myrtaceae. It is a large genus of aromatic trees indigenous to Australia, Tasmania and the neighboring Island, but today, it can be found growing in subtropical regions of the world. The genus consists of about 700 species of evergreen trees and shrubs (Adeniyi et al., 2006).

Eucalyptus oil is readily steam distilled from the leaves and can be used for cleaning, deodorizing and in very small quantities in food supplements, especially sweets, cough drops and decongestants (Schulz et al., 1998). It may also provide antiseptic properties (Chao and Young, 1998). The leaf extract of some species (*Eucalyptus globulus, Eucalyptus maculate and Eucalyptus viminalis*) have been reported to inhibit some Gram-positive bacteria (Takahashi et al., 2004). Fungicide activity has also been reported (Essien and Akpan, 2004; Mehraban et al., 2005).

The aim of this study was to investigate the phytochemical composition and antimicrobial activities of the crude leaves, stem-bark and root extracts of *Eucalyptus camaldulensis* against some pathogenic microorganisms.

MATERIALS AND METHODS

Sampling and sample preparation

Fresh sample of roots, stem-bark and leaves of the *E. camaldulensis* were collected from the premises of Modibbo Adama University of Technology, Yola and was identified by Dr. Jatau, D.F. The identified plant parts were washed with tap water and air-dried. The dried parts were chopped into pieces, milled into fine powder by pounding manually with a clean and sterile pestle and mortar. The dried powdered samples were each collected into sterile cellophane bags and labelled to prevent mix up. The samples were kept in cool dry place till further use. The dried powdered samples were used for extraction purposes.

Extraction procedure

250 ml of water was added to 50 g each of the dried powdered samples in a flask. Each of the soaked sample was stirred, sealed with aluminium foil and allowed to stand for 72 h. The content was then filtered with Whatman No. 1 filter paper. The filtrates were concentrated using rotary evaporator at 40°C. The extracts were stored in a universal bottle and refrigerated at 4°C prior to use (Mann et al., 2008).

The same procedure was followed with dried powdered samples as described earlier except that ethanol was used as solvent in place of distilled water.

Phytochemical analysis

The preliminary phytochemical analysis of the extracts were carried out to determined the presence of tannins, saponins, alkaloids, phenols and glycosides as described (Harborne, 1998; Evans, 1998; Abulude, 2007; Fadeyi, 1987). Results are as shown in Table 1.

Test for saponins

5 ml of the extract was vigorously shaken with 10 ml of water in a test tube. Frothing which persisted was taken as an evidence for the presence of saponins.

Test for tannins

Extract plus 4 ml of water and drops of ferric chloride were mixed. Immediate green precipitate was taken as evidence for the presence of tannins.

Test for alkaloids

2 ml of the extract plus picric acid were mixed; an orange coloration was taken as evidence for the presence of alkaloids.

Table 2. Results of antimicrobial activity of roots, stem-bark and leaves of *E. camaldulensis.*

Test organism	Water extracts			Ethanol extracts			Gentamicin (Control)
	Roots	Stem-bark	Leaves	Roots	Stem-bark	Leaves	
E. coli	17	12	13	-	-	18	26
S. aureus	18	13	15	19	20	22	23
S. typhi	16	13	15	16	14	13	28
B. subtilis	17	12	13	22	17	24	22

- = No measurable zone of inhibition.

Test for phenols

Equal volume of the extract was added to equal volume of ferric chloride, a deep bluish green solution was taken as a positive test for the presence of phenols.

Test for glycosides

5 ml of extract plus 25 ml of dilute sulphuric acid were poured into a test tube. The mixture was boiled for 15 min, cooled and neutralized with 10% sodium hydroxide and 5 ml of Fehling A and B was added. Brick red precipitate is a positive test for the presence of glycosides.

Test organisms

The organisms used in this study include, *Escherichia coli, Staphylococcus aureus, Salmonella typhi* and *Bacillus subtilis.*

Collection of test organisms

The organisms were collected from the microbiology laboratory of the Peace Hospital, Jimeta, Yola with the help of the laboratory staff. *E. coli* was collected in peptone water and labeled S_1. The same procedure was repeated for the collection of *S. aureus, S. typhi* and *B. subtilis.* These were labeled S_2, S_3 and S_4, respectively.

Preparation of nutrient agar/nutrient broth

This was carried out as described by Monica (2000). 28 g of the nutrient agar powder was dissolved in 1000 ml of distilled water in a conical flask and was autoclaved for l5 min at 121°C, and then, it was allowed to cool to 47°C and dispensed into plates or slants. The slants were used in culturing and sensitivity test of the organisms.

28 g of nutrients broth powder was dissolved in 1000 ml of distilled water in a conical flask and was autoclaved at 121°C for 15 min; it was then allowed to cool to 47°C and was dispensed into test tubes.

Antimicrobial investigation

The stocks were maintained on nutrient agar slant and sub-cultured in nutrient both for incubation at 37°C prior to each antimicrobial testing. Inoculation of the test organisms on nutrient agar prepared plates was achieved by flaming a wire loop on a spirit lamp, cooling the wire loop (air cooling) and fetching the test organisms. The discs were prepared using a Whatman filter paper, kept in vials-bottles and sterilized in an oven at 150°C for 15 min. Prepared discs containing the various extracts were carefully placed on the inoculated plates using a sterilized forceps in each case (Fatope, 1993). The plates were then turned upside-down and inoculate at 37°C for 24 h in an incubator. After incubation, the inoculated plates were observed for zones of inhibition (in mm diameter). The result was taken by considering the zone of growth and inhibition of the organisms by the test fractions (Mackie and McCartney, 1989). Activity and inactivity were observed in accordance with the standard and acceptable method. Results are as shown in Table 2.

RESULTS AND DISCUSSION

The phytochemical analysis (Table 1) revealed the presence of tannins, saponins, phenols and glycosides in the roots and stem-bark of water extracts. Only phenols and glycosides were present in the water extract of the leaves. In the ethanol extracts, all the bioactive compounds tested were present in the roots and leaves except alkaloids which are absent. Tannins and glycosides were present in the stem-bark of ethanol extract, but saponins, phenols and alkaloids were absent. It is interesting to note that in both the water and ethanol extracts of the roots, stem-bark and leaves of *E. camaldulensis,* alkaloids was completely absent. Phytochemical screening of the extracts varies from one plant part to another as revealed in the results. It could also vary from place to place due to geographical location, climatic conditions and soil condition of a particular area. This may explain why it could be possible to have differences in chemical composition of the same plant of study in other areas.

The crude extracts of medicinal plant studied were found to contain the following phytochemical compounds, saponins, tannins and phenols. Other investigators reported the presence of these components, cardiac glycoside and volatile oils (Ahmad et al., 1998).

The antimicrobial activities of the test organisms are as shown in Table 2. Antimicrobial susceptibility of the extracts against the test organisms showed that both extract has activities on the entire test organism except *E. coli* which shows no measurable zone of inhibition with ethanol extract from roots and stem-bark of the plant.

The inhibitory effects of these medicinal plants on the microorganisms may therefore, be due to the presence of the aforementioned phytochemical components. The results of the present study showed that the crude extracts of *E. camaldulensis* inhibit the growth of *S. typhi, E. coli*, *B. subtilis* and *S. aureus*. Differences in polarity among various solvents have been reported to be accountable for the differences in solubility of plant active principles, hence, variation in degree of activities. Results also showed that activities of all the extracts were concentration dependent. Highest activity was demonstrated by the standard antibiotic gentamicin (control). This is because the antibiotic is in it pure state and has refined processes that have established it as a standard antibiotic (Prescott et al., 2002). Results of this study therefore have shown that *E. camaldulensis* is a potential source of antibiotic substances for drug development for use against this group of organisms that causes diseases. Thus, both extracts had antimicrobial activities and confirmed the historical use of *E. camaldulensis* oil as an antimicrobial agent (Kumar, 1988).

Conclusion

This study has shown that the extracts of *E. camaldulensis* possess antimicrobial potentials found to be effective against pathogenic microorganisms involved in wounds infections, urinary tract infections, gastrointestinal tract infections and typhoid fever. Therefore, the results of this study provide a rationale for the use of the plant parts in traditional medicine practice in Nigeria. The activities of *E. camaldulensis* should further be investigated against wide range of microorganisms. Also, purification and toxicological studies should be carried out with a view of sourcing antimicrobial agents for drug development.

REFERENCES

Abulude FO (2007). "Phytochemical Screening and Mineral Contents of Leaves of some Nigerian Woody Plants." Res. J. Phytochem., 1(1): 33-39.

Adeniyi BA, Ayepola OO (2008). Phytochemical screening and antimicrobial activity of leaf extracts of *Eucalyptus camaldulensis* and *Eucalyptus torelliana* (Myrtaceae). J. Med. Plants Res., 2: 34-38.

Adeniyi BA, Odufowoke RO, Olaleye SB (2006). Antimicrobial and gastro protective properties of *Eucalyptus torelliana* (Myrtaceae) crude extracts. Int. J. Pharmacol., 2: 362-365.

Ahmed I, Mehmood A, Mohammad F (1998). Screening of Some Indian Medicinal Plants for their Antimicrobial Properties. J. Ethnopharmacol., 62: 183-193.

Chao SC, Young DG (1998). Effect of a diffused essential oil blend on bacterial bioaerosis. J. Essent. Oil Res., 10: 517-523.

Ebana RUB, Madunagu BE, Ekpe ED, Otung IN (1991). Microbiological Exploitation of Cardiac glycosides and alkaloids from *Garcinia kola, Barreria ocymoides, kola nitida* and *Citrus aurantifolia.* J. Appl. Biotechnol., 71: 398-401.

Essien JP, Akpan EJ (2004). Antifulgal activity of ethanolic leaf extracts of *E. camaldulensis* against ringworm pathogen. Glob. J. Pure Appl. Sci., 10: 37-41.

Evans WC (1998). Trease and Evans Pharmacognosy. 14th edition, WB Saunders Company Limited, pp. 15-16.

Harborne JB (1998). Phytochemical Methods: A Guide to Modern Techniques of Plant Analysis. 3rd edition, Chapman and Hall London, pp. 1-198.

Fatope MO, Adoum OA (1993). "Bioactivity of some savannah plants in the brine shrimp lethality test and *in-vitro* anti-microbial assays." Int. J. Pharmacog., 35(5): 334-337.

Fadeyi MO, Adeoye AO, Olowokudejo ID (1987). Epidermal and phytochemical studies with genus of *Bohervia* (Nyctanginaceae) in Nigeria. Int. J. Crude Drug Res., 27: 178-184.

Kafaru E (1994). Immense Help from Nature's Workshop. Elika Health Services Ltd. Academic Press Plc, Lagos, Nigeria, pp. 1-27.

Kumar A (1988). Antibacterial properties of some *Eucalyptus* oils. Fitoterapia. 59: 141-144.

Mackie R, McCartney C (1989). Practical Medicinal Microbiology. 3rd edition, Churchill Livingstone publishers, London and New York, 2: 100–106, 121, 141, 163-167, 303, 432-491.

Mann A, Yahaya AY, Banso A, Ajayi GO (2008). Phytochemical and antibacterial screening of *Anogeissus leiocarpus* against some microorganisms associated with infectious wounds. Afr. J. Microb. Res., 2: 60-62

Mehraban F, Tabrizib NO, Jahaniani F (2005). Antidermotophyte activities of *E. camaldulensis* in comparison with griseofulvin. Iran. J. Pharmacol. Ther., 4: 80-83.

Monica C (2002). District Laboratory Practice in Tropical Countries part 2, Cambridge, low price edition. Cambridge University Press, p. 200.

Nester EW, Robert CE, Pearsall NN, Anderson DG, Nester MT (2004). Microbiology: A Human Perspective. 4th Edition, McGraw Hill Inc., pp. 507-508.

Pelezer MJ, Chan ECS, Noel RK (1993). Microbiology Concepts and Applications. McGraw Hill International USA, pp. 507-508.

Prescott MI, Harley PS, Klein AD (2002). Microbiology. 5th Edition, McGraw Hill Inc., p. 39.

Schulz V, Hansel R, Tyler VE (1998). Rational Phytotherapy. 3rd edition, Springer-Berlin, Germany, pp. 146-147.

Shariff LU (2001). Modern Herbal Therapy for Common Ailments. Nature Pharmacy Series (Volume 1). Spectrum Books Limited, Ibadan, Nigeria in Association with Safan Books (Export) Limited, United Kingdom, pp. 9–84.

Takhashi T, Kokubo R, Sakaino M (2004). Antimicrobial activities of *Eucalyptus* leaf extracts and flavonoids from *Eucalyptus maculate.* Lett. Appl. Microbial., 39: 60-64.

Heavy metal levels in selected green leafy vegetables obtained from Katsina central market, Katsina, North-western Nigeria

Shuaibu I. K.[1] , Yahaya M.[1], and Abdullahi U. K.[2]

[1]Department of Pharmacy, School of Health Technology, Kankia, Katsina State College of Health Sciences, Katsina, Nigeria.
[2]Department of Pure and Applied Chemistry, Umaru Musa Yar' Adua University, Katsina, Nigeria.

The levels of Lead (Pb), Cadmium (Cd), Iron (Fe), Zinc (Zn) and Copper (Cu) were determined in four different samples of vegetables purchased from Katsina central market using atomic absorption spectrometer. The mean concentrations of metals ranged from 0.071 mg/kg Pb to 0.632 mg/kg Cu. The relative abundance of metals in vegetables followed the sequence Cu (0.483 mg/kg) > Zn (0.268 mg/kg) > Fe (0.260 mg/kg) > Pb (0.095 mg/kg). The levels of Pb and other metals were below the FAO/WHO recommended limits for metals in vegetables. Low concentrations of Pb and absence of Cd in all the samples are indications that these plants contribute less toxic effects of metals. The results showed that, these vegetables are the main sources of essential trace elements. The daily human intakes of metals have also been computed and were observed below recommended values by the FAO/WHO.

Key words: Heavy metals, vegetables, Katsina, spectrometry.

INTRODUCTION

The term vegetable applies to edible part of the plant that stores food in roots, stems, or leaves. Vegetables are green and leafy-like in appearance bearing edible stems or leaves and roots of plants (Sharma, 2004). Vegetables constitute essential diet components by contributing carbohydrates, proteins, vitamins, iron, calcium and other nutrients that are in short supply.

Vegetables also contain both essential and toxic elements over a wide range of concentrations. Metals in vegetables pose a direct threat to human health. Plants and vegetables take up elements by absorbing them from contaminated soils and waste water used for irrigating them as well as from deposits on different parts of the plants exposed to the air from polluted environment (Funtua et al., 2008).

Vegetables, especially those of leafy vegetables grown in heavy metals contaminated soils, accumulate higher amounts of metals than those grown in uncontaminated soils because of the fact that they absorb these metals through their roots (Muhammad et al., 2008). Vegetables accumulate heavy metals in their edible and non-edible parts. Absorption capacity of heavy metals depends upon the nature of vegetables and some of them have a greater potential to accumulate higher concentrations of heavy metals than others (Akan et al., 2009). Atayese et al. (2009) investigated heavy metal contamination of Amaranthus grown along major highways in Lagos, Nigeria. Ladipo and Doherty (2011) studied heavy metal levels in vegetables from selected markets in Lagos, Nigeria.

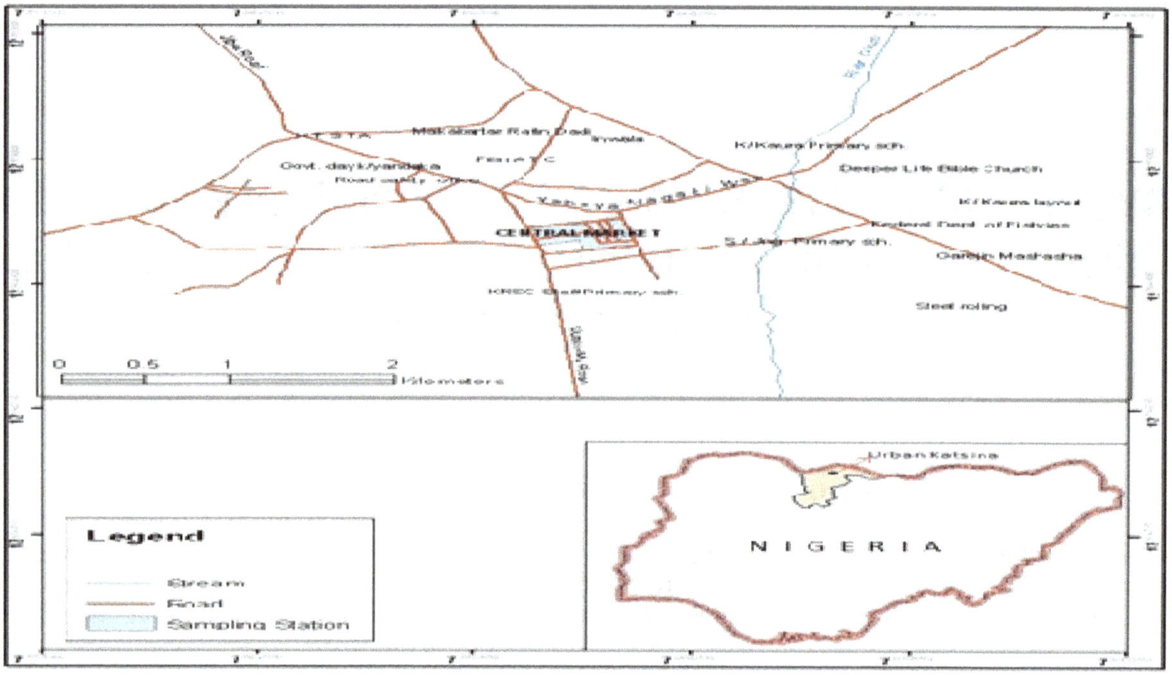

Figure 1. Map of the studied area showing sampling station (Katsina central market).

The use of green leafy vegetables for the preparation of soups cuts across different cultures in Nigeria and other parts of West Africa (Ladipo and Doherty, 2011; Akan et al., 2009). Intake of vegetables is an important path of heavy metal toxicity to human being and based on persistent nature and cumulative behaviour as well as the probability of potential toxicity effects of heavy metals as a result of consumption of leafy vegetables, this study was carried out to determine the dry matter levels of Pb, Cd, Fe, Zn, and Cu in some selected vegetables that are consumed regularly by inhabitants in Katsina metropolis. These vegetables serve as food sources and thus offer rapid and ideal means of providing adequate vitamins, mineral salts, trace essential elements and fibre as suggested by Ihekeronye and Ngoddy (1995).

MATERIALS AND METHODS

Chemicals of analytical grade purity and distilled deionized water were used. All glass wares and plastic containers used were washed with detergent solution followed by (20% v/v) nitric acid and then rinsed with tap water and finally with distilled deionized water. Also, standard solutions of the metal salts and other reagents were prepared.

Study area

The study area covered in this research was Katsina, a city in North-western Nigeria. It is the capital of Katsina state, one of Nigeria's 36 states. It is also the headquarters of Katsina Local Government Area. The city is located on the latitude 12° 59′ N and longitude 7° 36′ E with an average area of 142 km^2 and a population of 318,459 as of 2006 (Figure 1).

Sample and sampling

A total of 120 samples of four different vegetables that is, Foetid cassia (*Cassia tora*), Kenaf (*Hibiscus cannabinus*), Tossa jute (*Corchorus olitorius*) and Wild jute (*Corchorus tridens*) were purchased from Katsina Central Market.

Sample treatment

The vegetables were washed with 20% (v/v) nitric acid and then rinsed with distilled deionized water. These samples were cut into pieces with knife and were air-dried in the laboratory for 4 days before oven-dried at 105°C for about 24 h. The samples were crushed into powder in a mortar with a pestle. The samples were then sieved through a 2 mm nylon sieve and transfer into a labeled polyethylene container for analysis.

Sample digestion

The procedure according to Awofolu (2005) was used for digestion of plant sample. 0.5 g of sieved leaf samples were then weighed into 100 cm^3 beaker. A mixture of 5 cm^3 concentrated HNO_3 and 2 cm^3 $HClO_4$ were added to dissolve the sample. The beaker was heated at moderate temperature of 110°C on a hot plate for 1 h in a fume hood until the content was about 2 cm^3.

The digest was allowed to cool, filtered into 50 cm^3 standard volumetric flask and mad e up to the mark with distilled deionized water.

Table 1. Concentrations of the heavy metals (mg/kg) in green leafy vegetables.

Plant sample	Pb	Cd	Fe	Zn	Cu
Foetid Cassia	0.118 ± 0.071	ND	0.242 ± 0.050	0.227 ± 0.060	0.333 ± 0.034
Kenaf	0.095 ± 0.000	ND	0.250 ± 0.000	0.250 ± 0.020	0.467 ± 0.000
Tossa jute	0.071 ± 0.022	ND	0.334 ± 0.014	0.221 ± 0.031	0.500 ± 0.017
Wild jute	0.095 ± 0.010	ND	0.214 ± 0.050	0.375 ± 0.015	0.632 ± 0.022
Mean ± SD	0.095 ± 0.026		0.260 ± 0.029	0.268 ± 0.032	0.483 ± 0.018
Range	0.071 - 0.118		0.214 - 0.334	0.221 - 0.375	0.333 - 0.632
FAO/WHO Safe limit (2001) [a]	0.300	0.2	425.00	99.40	73.00

ND = Not detected; SD = standard deviation; source: *a* = Adu et al. (2012).

Sample analysis

A serial dilution method was used to prepared the working standards and the concentrations of the metals in each sample digest were determined using Atomic Absorption Spectrophotometer (Buck Model 210 VGP) equipped with a digital readout system.

Data analysis

Data obtained were analyzed using Microsoft Excel and results were expressed as mean ± standard deviation.

Daily intake of heavy metals from vegetables

The daily intake of heavy metals through the consumption of vegetables tested was calculated according to the equation (Cui et al., 2004):

Daily intake of metals (DIM) = DVC × VMC

DVC = daily vegetable consumption; VMC = mean vegetable metal concentrations (mg/day, fresh weight).

Where daily vegetable consumption was taken as 98 g of vegetables per person per day as set by the FAO/WHO (1999), for heavy metal intake based on body weight for an average adult (60 kg body weight).

RESULTS AND DISCUSSION

The metal contents obtained from each vegetable from the central market site are listed in Table 1. Among the 4 different vegetables examined, Wild jute recorded highest level of Cu and Zn which are 0.632 mg/kg and 0.375 mg/kg, respectively. Tossa jute recorded highest level of Fe which is 0.334 mg/kg and Foetid cassia recorded highest level of Pb which is 0.118 mg/kg. Similarly, Tossa jute recorded the lowest mean levels for lead (0.071 mg/kg) and zinc (0.221 mg/kg).

The mean levels of the metals examined in Foetid cassia, Kenaf and Tossa jute samples were found to be in the order: Cu > Fe > Zn > Pb. These trends suggest that Foetid cassia, Kenaf and Tossa jute samples have a high retention capacity for Cu followed by Fe and then Zn and Pb. However, the mean levels of the metals in Wild jute are found to be in the order: Cu > Zn > Fe > Pb. This trend indicates that Wild jute has a high retention capacity for Cu followed by Zn and then Fe and Pb. However, it could be observed that all vegetables have higher retention capacities for essential metals (Cu, Zn, and Fe) than the toxic ones (Pb and Cd).

Pb is a toxic element that can be harmful to plants, although plants usually show ability to accumulate large amounts of lead without visible changes in their appearance or yield. In many plants, Pb accumulation can exceed several hundred times the threshold of maximum level permissible for human consumption (Muhammad et al., 2008). The high levels of Pb in some plants may probably be attributed to pollutants in irrigation water, farm soil or due to pollution from the highways traffic (Qui et al., 2000). Wong et al. (1996) reported that Chinese cabbage picks up Pb more readily compared to other heavy metals such as Cd, Cu, Ni, and Zn. The level of Pb in this study is found highest in the leaf of Foetid cassia (0.118 mg/kg) and lowest in the leaf of Tossa jute (0.071 mg/kg). Both Kenaf and Wild jute have recorded the same Pb content of 0.095 mg/kg in their leaves. The levels of Pb reported in this study are higher when compared to that reported in the leaves of lettuce (0.01 mg/kg) by Adu et al. (2012). The Pb contents of the plants in this study are lower when compared to the FAO/WHO (2001) safe limit of 0.3 mg/kg. The study showed that, in the plant, Pb contents are within the permissible limit. Thus, the Pb level in the leafy parts of the vegetables examined seems not to be alarming except in a case of excessive consumption.

Cd is a non-essential in foods and natural waters and it accumulates principally in the kidney and liver (Divrikli et al., 2006). Various sources of environmental contamination have been reported for its presence in foods and various values have been reported for leafy vegetables which include 0.090 mg/kg for fluted pumpkin by Sobukola et al. (2010), 0.049 mg/kg for lettuce by Muhammad et al. (2008). In contrast, no Cd was detected in all vegetable samples examined in this study and hence are within safe limit, regular monitoring is required

Table 2. Estimation of heavy metal intake through consumption of vegetables in Katsina Metropolis.

Heavy metal	Mean conc. (mg/kg)	Daily intake (μg/day)	WHO/FAO limit (μg)[b]
Pb	0.095	9.31	214
Cd	0.000	0.00	60
Fe	0.260	25.48	-
Zn	0.268	26.26	60000 (60 mg)
Cu	0.483	47.33	3000 (3 mg)

Source: b = Elbagermi et al. (2012).

over a long period as the vegetables are transported from different sources.

Fe is essential for the synthesis of chlorophyll and activates a number of respiratory enzymes in plants. The deficiency of Fe results in severe chlorosis of leaves in plants. High levels of exposure to iron dust may cause respiratory diseases such as chronic bronchitis and ventilation difficulties. Fe content is found highest in the leaves of Tossa jute (0.334 mg/kg) and lowest in the leaves of Wild jute (0.214 mg/kg). Substantial amounts were recorded in the leaves of Foetid cassia and Kenaf which are 0.242 and 0.250mg/kg respectively. These values are far below the mean value of 12.873mg/kg in the leaves of lettuce as reported by Adu et al. (2012). The Fe contents of these plants are lower than the FAO/WHO (2001) safe limit of 425.00 mg/kg. These vegetables could be good supplement for Fe.

Zn is the least toxic and an essential element in human diet as it is required to maintain the functioning of the immune system. Zn deficiency in the diet may be highly detrimental to human health than too much Zn in the diet. The recommended dietary allowance for Zn is 15 mg/day for men and 12 mg/day for women Agency for Toxic Substances and Disease Registry (ATSDR, 1994), but high concentration of Zn in vegetables may cause vomiting, renal damage, cramps etc. Wild jute recorded the highest level of Zn which is 0.375 mg/kg and the least concentration was recorded by Tossa jute which is 0.221 mg/kg. Both Foetid cassia and Kenaf recorded moderate amounts of zinc which are 0.227 and 0.250 mg/kg respectively. These values are higher when compared to those reported in available literature. Sobukola et al. (2010) have reported Zn levels of 0.011, 0.070 and 0.050 mg/kg in the leaves of bitter leaf, water leaf and cabbage, respectively. The contents of Zn in all the plants examined are generally lower than the permissible levels by the FAO/WHO in vegetables as shown in Table 1. Regular consumption of these four vegetables may assist in preventing the adverse effect of zinc deficiency which results in retarded growth and delayed sexual maturation because of its role in nucleic acid metabolism and protein synthesis (Barminas et al., 1998).

Cu is an essential micronutrient which functions as a biocatalyst, required for body pigmentation in addition to Fe, maintains a healthy central nervous system, prevents anaemia and interrelated with the functions of Zn and Fe in the body (Akinyele and Osibanjo, 1982). However, most plants contain the amount of Cu which is inadequate for normal growth which is usually ensured through artificial or organic fertilizers (Itanna, 2002). Among all heavy metals, Cu is the most abundant element, which recorded highest concentration of 0.632 mg/kg in the leaves of Wild jute. The least concentration of 0.333 mg/kg was recorded in the leaves of Foetid cassia. Moderate amounts of Cu were recorded by Kenaf and Tossa jute which are 0.467 and 0.500 mg/kg, respectively. The results obtained here were observed to be lower compared to other published results. Elbagermi et al. (2012) reported values of 5.00, 5.75 and 5.32 mg/kg for the concentration of Cu in carrot, cucumber and spinach. The contents of Cu in this study and from other published works are within the permissible level of 73.00 mg/kg by the FAO/WHO in vegetables.

The exposure of consumers and the related health risks are usually expressed in terms of the provisional tolerable daily intake .The FAO/WHO (1999) have set a limit for the heavy metal intake based on the body weight for an average adult, namely, 60 kg body weight. The average diet per person per day of vegetables is 98 g. If the mean levels of Pb (0.095 mg/kg), Cd (0.000 mg/kg), Fe (0.260 mg/kg), Zn (0.268 mg/kg) and Cu (0.483 mg/kg) found here are consumed daily, the contribution of heavy metal intake for an average human being from the vegetable diets were calculated and presented as shown in Table 2. It can therefore be concluded that our estimated daily intakes for heavy metals studied here are below those reported by the FAO/WHO, which had set a PTDI limit for heavy metal intake based on body weight for an average adult (60 kg body weight) for Pb, Cd, Zn, and Cu as shown in Table 2.

Conclusion

The results reported here confirm that the vegetables obtained from the central market site contained substantial amounts of the metals with exception of Cd which recorded zero concentrations or Cd concentrations were found to be below the detection limit of the machine. Levels of the metals are found to be within the safe limits prescribed by the FAO/WHO. This is an important result as human health is directly affected by consumption of vegetables. The monitoring of heavy metals in vegetables needs to be continued; because these are the main sources of food for humans in many parts of the world and are considered as bioindicators of environmental pollution.

ACKNOWLEDGEMENT

The authors appreciate the assistance of Mal. Mustapha Kano of the Soil Science Department, Faculty of Agriculture, Bayero University, Kano for carrying out AAS analysis on the samples.

REFERENCES

Adu AA, Aderinola OJ, Kusemiju V (2012). Heavy metals concentration in Garden lettuce (*Lactuca Sativa* L.) grown along Badagry expressway, Lagos, Transnat. J. Sci. Technol. 2(7):115-130.

Agency for Toxic Substances and Disease Registry, ATSDR (1994). Toxicological profile for Zinc and Cobalt. US Department of Health and Human Services, Public Health Serv. 205-88-0608.

Akan JC, Abdulrahman FI, Ogugbuaja VO, Ayodele JT (2009). Heavy metals and anion levels in some samples of vegetables grown within the vicinity of Challawa indust. Area, Kano state, Nigeria. Am. J. Appl. Sci.6(3):534-542.

Akinyele IO, Osibanjo O (1982).Levels of trace elements in hospital diet. Food Chem. 8:247-251.

Atayese MO, Eigbadon AJ, Oluwa, KA, Adesodun, JK (2009). Heavy metal contamination of Amaranthus grown along major highways in Lagos, Niger. Afr. Crop Sci. J. 16(4):225-235.

Awofolu OR (2005). A survey of trace metals in vegetation, soil and lower animals along some selected major roads in metropolitan city of Lagos, Environ. Monit. Assess. 105:431-447.

Barminas JT, Charles M, Emmanuel D (1998). Mineral composition of non-conventional leafy vegetables. Plant foods for Hum. Nutr. 53:29-36.

Cui YJ, Zhu YG, Zhai RH, Chen DY, Huang YZ, Qiu Y, Liang JZ (2004). Transfer of metals from soil to vegetables in area near a smelter in Nanning, China. Environ. Int. 30(6):785-791.

Divrikli U, Horzum N, Soylak M, Elci L (2006). Trace heavy metal contents of some spices and herbal plants from western Anatolia, Turkey. Int. J. Food Sci. Technol. 41:712-716.

Elbagermi MA, Edwards HGM, Alajtal AI (2012). Monitoring of heavy metal content in fruits and vegetables collected from population and market sites in the Misurata area of Libya. Intl. Scholarly Res. Network. 10:1-5.

FAO/WHO (1999). Joint Expert Committee on Food Additives, "Summary and Conclusions", in proceedings of the 53rd Meeting of Joint FAO/WHO Expert Committee on Food Additives, Rome, Italy.

FAO/WHO, Codex Alimentarius Commission (2001). Food Additives and Contaminants. Joint FAO/WHO Food Standards programme, ALINORM 01/12A:1-289.

Funtua MA, Agbaji FB, Ajibola VO (2008). Assessment of the heavy metal contents of spinach and lettuce grown along the bank of river Getsi, Kano. J. Chem. Soc. Niger. 5(1):11-14.

Ihekeronye AI, Ngoddy PO (1995). Integral Food Science and Technology for the Tropics, 2nd edition. Macmillan Education Ltd. Oxford and London, P. 293.

Ladipo MK, Doherty VF (2011). Heavy metal levels in vegetables from selected Markets in Lagos, Nigeria. Afr. J. Food Sci. Technol. 2(1):18-21.

Muhammad F, Farooq A, Umar R (2008). Appraisal of heavy metal contents in different vegetables grown in the vicinity of an industrial area. Pak. J. Bot. 40(5):2099-2106.

Qui XX, Huang DF, Cai SX, Chen F, Ren ZG, Cai YC (2000). Investigation on Vegetables pollution and pollution sources and its control in Fuzhou, Fujian Province. Fujian J. Agric. Sci. 15:16-21.

Sharma OP (2004). Hills Economy. A textbook of Botany, 2nd editon. Arish press Dhaka, Bangladesh, pp. 18-21.

Sobukola OP, Adeniran OM; Odedairo AA, Kajihausa OE (2010). Heavy metal levels of some fruits and leafy vegetables from selected markets in Lagos, Nigeria. Afr. J. Food Sci. 4(2):389-393.

Wong JW, Li GX, Wong MH (1996). The growth of Brassica Chinensis in heavy metal contaminated sludge compost from Hong Kong. Bioresour. Technol. 58:209-313.

Effect of particle size on loading capacity and water quality in water defluoridation with 200°C calcined bauxite, gypsum, magnesite and their composite filter

Bernard Thole[1]*, Felix Mtalo[2] and Wellington Masamba[3]

[1]Ngurdoto Defluoridation Research Station, P. O. Box 3020, Usa River, Arusha, Tanzania.
[2]College of Engineering and Technology, University of Dar Es Salaam, P. O. Box 35131, DSM, Tanzania.
[3]Harry Oppenheimer Okavango Research institute, University of Botswana, P/Bag 285, Maun, Botswana.

Defluoridation was carried out with bauxite, gypsum, magnesite and their composite calcined at 200°C to determine relationship between particle size, capacity and water quality, in an attempt to formulate a composite filter in the ratio of loading capacities. Particles of 0.5 to 1.0, 1.0 to 1.4, 1.4 to 2.0 and 2.0 to 3.0 mm diameter were employed in batch defluoridation. Color, hardness, pH, $F^-, Cl^-, Al^{3+}, Fe^{2+}$ were determined hourly until fluoride equilibrium was obtained. Results showed that capacities were inversely correlated to particle size obtaining polynomial, linear and logarithmic relations of capacity (mg/g) to particle size (mm); $C_s = -\phi(P_z)^2 - \delta P_z$ (composite, gypsum), $C_s = -\phi P_z - \delta$ (bauxite) and $C_s = -\phi \ln P_z + \delta$ (magnesite). Bauxite, gypsum and composite decreased but magnesite increased pH. Particle size and pH changes were positively correlated, highly for bauxite (r^2 = 0.915), fairly for composite and magnesite (r^2 = 0.855 and 0.853, respectively) and lowly correlated for gypsum (r^2 = 0.562). Smaller particles obtained higher apparent colour, hardness, alkalinity and sulphate. There was no relationship between particle size and Cl^-, Al^{3+}, Fe^{2+}. Sorption capacities of these materials increase with decrease in particle size, however decrease in particle size enhances colour, hardness and sulphate levels. In optimizing the materials for defluoridation particle size can be decreased while monitoring colour, hardness, alkalinity and sulphate levels. Particles of diameter 2 mm were optimum in the present experiments.

Key words: Bauxite, composite, defluoridation, gypsum, magnesite.

INTRODUCTION

Water defluoridation research has been going on world over in an attempt to mitigate dental and skeletal fluorosis. The effects of high fluoride ingestion on teeth and bone are well documented (Varol et al., 2010; Smittakorn et al., 2010; Castillo et al., 2007). Fluoride is known to cause browning and mottling of teeth and weakening of the bones leading to crippling conditions (Vasudevan et al., 2009; Feenstra et al., 2007). A number of technologies have been researched on among which are use of bone char (Smittakorn et al., 2010), precipitation with alum and lime (Yadav et al., 2010; Nath and Dutta, 2010), employment of clay materials (Sujana et al., 2009), use of synthetic adsorbents (Maliyekkal et al., 2010), and, other advanced techniques such as electrolysis, electrodialyis (Amer et al., 2001), reverse osmosis (Fawell et al., 2006), nanofiltration (Feenstra et al., 2007), Crystalactor (Giesen, 1998), solar dew collector system (Solar, 2007), Memstil technology (Hanemaaijer et al., 2007), and The water pyramid (Aqua-Aero, 2007). Bauxite, gypsum and magnesite have been employed in experimental water defluoridation

*Corresponding author. E-mail: tholebernard@hotmail.com.

(Thole, 2005; Sajidu et al., 2008; Singano, 2000).

The results were promising with fairly good sorption capacities obtained for bauxite (12.31 mg/g) and gypsum (9.28 mg/g) (Sajidu et al., 2008; Peter, 2009). Low sorption capacities were recorded for magnesite (3.45 mg/g) (Singano, 2000). Despite the fluoride sorption potential of the three materials the quality of treated water appeared to change. Singano (2000) reported an increase in water pH during defluoridation with magnesite and in defluoridation with bauxite and gypsum apparent colour, turbidity and hardness were reported to increase (Thole, 2005; Sajidu et al., 2008). It is common to optimize adsorbent media through calcination at various temperatures and at times decreasing grain size among other optimisation techniques (Feenstra et al., 2007). The present research investigated similar optimisation techniques such as calcination, particle size reduction, hybridization of materials and changes in design flow rate. However among these only particle size experimenttation is reported here, the other results are reported elsewhere (Thole et al., 2011; Thole, 2011). Research has shown the effect of grain size in defluoridation with activated alumina (Shoeman and Macleod, 1987), synthetic hydroxyapatite, bauxite and gypsum (Thole, 2005) and Magnesite (Singano, 2000).

The reported results show that decreasing grain size has a positive impact on loading capacities. The same reports indicate negative quality changes in water treated with these materials. Activated alumina, for instance, is known to contribute towards residual concentrations of aluminium whereas gypsum increases sulphate concentrations in the water. However the relationship between grain size and residual concentrations of contaminants; such as aluminium, iron, sulphate, chlorides and calcium, in defluoridation with bauxite, gypsum and magnesite has not been widely reported in literature. The current research aimed at determining effects of particle size on loading capacities and quality of the treated water with respect to pH, alkalinity, residual colour, hardness and concentrations of $Cl^-, Al^{3+}, Fe^{2+}, SO_4^{2-}$ in water defluoridation with bauxite, gypsum, magnesite and their composite. This work contributes to research on optimizing the three materials for water defluoridation through calcination, particle size reduction and hybrid filter design.

MATERIALS AND METHODS

Bauxite, gypsum and magnesite were locally sourced within Tanzania, crushed and sieved to grain sizes of ranges 0.5 – <1.0, >1.0 – <1.4, >1.4 – <2.0 and >2.0 – <3.0 mm diameter. These are referred throughout the text as 1.0, 1.4, 2.0 and 3.0 mm, respectively. The segregated materials were then heat treated at 200°C in an open air muffle furnace for 2 h. Preliminary results had shown optimum defluoridation conditions with the materials when calcined at 200°C for all tested calcine temperatures between 150 and 500°C (Thole et al., 2011). Naturally occurring ground water was obtained and tested for colour, hardness, pH, alkalinity and concentrations of $Cl^-, Al^{3+}, Fe^{2+}, SO_4^{2-}$ prior to use in defluoridation experiments. One gram of each grain-size of bauxite, gypsum and magnesite was placed in one litre of the ground water. The media was agitated at 125 revolutions per minute (rpm). Colour, hardness, pH, alkalinity and concentrations of $F^-, Cl^-, Al^{3+}, Fe^{2+}, SO_4^{2-}$ were determined hourly until equilibrium fluoride concentrations were obtained. A mass balance Equation 1 was employed to calculate loading capacities on the adsorbent materials.

$$q_e = \frac{(C_o - C_e)V}{m} \quad (1)$$

In Equation 1 q_e is the amount of adsorbed fluoride at equilibrium (mg g^{-1}); V is the volume of the solution (L); C_o and C_e are the initial fluoride concentration and fluoride concentration at equilibrium (mg l^{-1}) respectively, and, m is mass of adsorbent (g). A composite of bauxite, gypsum and magnesite was prepared in the ratio of their mean loading capacities for each grain size. Batch defluoridation experiments were carried out with the composite, as described for the materials, employing Equation1 for calculating loading capacities. In analyzing for possible relationships particle size was regressed separately with; loading capacities, apparent colour, pH, alkalinity, hardness and concentrations of $Cl^-, Al^{3+}, Fe^{2+}, SO_4^{2-}$. Standard analytical methods were employed in all determinations (HACH, 1997; APHA, 1989). Fluoride was analyzed by the ion selective electrode (ISE) technique. In this method fluoride ions were selectively absorbed by the ISE membrane establishing a potential (voltage) whose magnitude was proportional to fluoride concentration of sample.

This potential was compared to the constant potential of a reference electrode. Aluminon method was employed for determining aluminium and the calmagate spectrophotometric method was used for calcium and magnesium. Sulphates concentrations were determined through SulfaVer 4 turbidimetry and Argentometric method was employed for chloride. To determine alkalinity samples were titrated with acid to a colorimetric end point corresponding to a specific pH. Phenol alkalinity was determined by titration to a pH 8.3 end point, as evidenced by phenolphthalein colour change, and indicates the total hydroxide and one half the carbonate present. Methyl orange or total alkalinity was determined by titration to a pH between 3.7 and 5.1 and includes all carbonates, bicarbonates and hydroxides. HACH one combination pH electrode method was employed to determine pH. The APHA (1989) Platinum-Cobalt standard method was employed to test for colour. Total iron was analyzed through use of FerroVer. FerroVer reagent reacts with all soluble and most insoluble forms of iron in sample to produce soluble ferrous iron. This reacts with the 1, 10-phenanthroline indicators in the reagent to form an orange colour that is proportional to the concentration of iron.

RESULTS AND DISCUSSION

Loading capacities

The loading capacities for bauxite, gypsum, magnesite and their composite increased with decrease in particle size. Plots of loading capacity versus particle size depict the capacity-particle size relationships for bauxite, gypsum, magnesite and the composite, Figures 1 to 4.

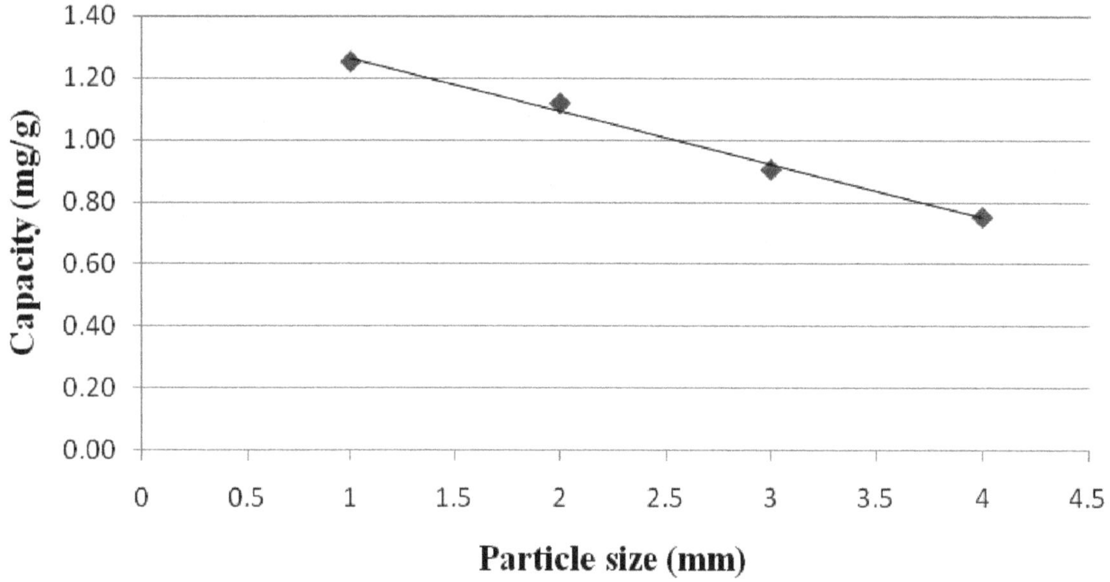

Figure 1. Loading capacity against particle size in defluoridation with bauxite *(r^2 = 0.993)*.

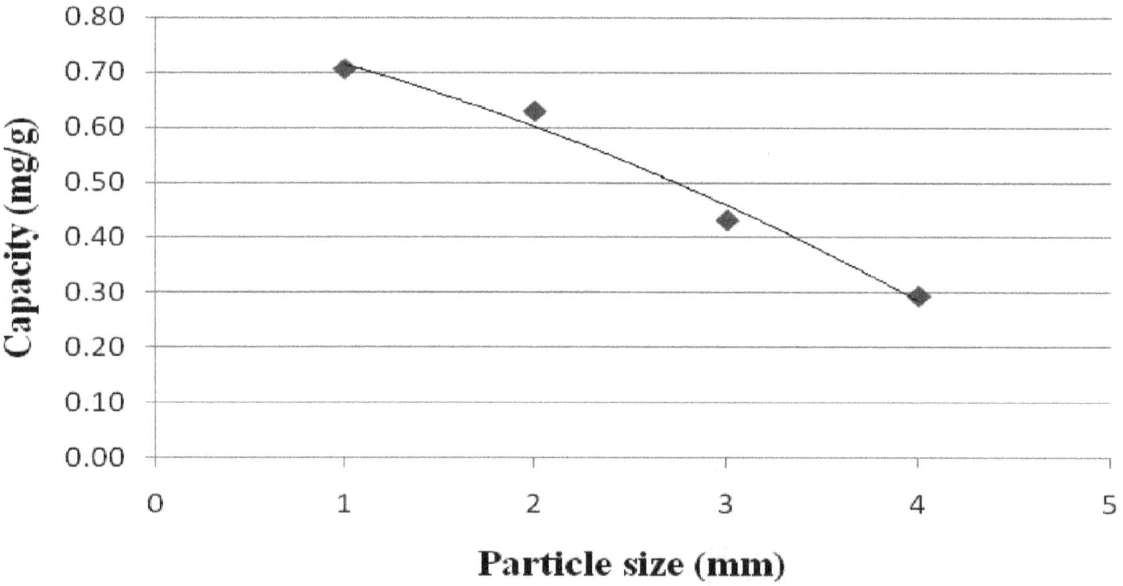

Figure 2. Loading capacity against particle size in defluoridation with gypsum *(r^2 = 0.985)*.

The relation between loading capacity, (C_s, mg/g), to particle size (P_z, mm) was best represented by a linear equation for bauxite (Equation 2), polynomial equations for gypsum and the composite (Equations 3 and 5, respectively), and, natural logarithm for magnesite (Equation 4). The relation equations were;

$$C_s = -0.1718P_z + 1.4367 \quad (r^2 = 0.993) \quad (2)$$

$$C_s = -0.015P_z^2 - 0.0677P_z + 0.7967 \quad (r^2 = 0.985) \quad (3)$$

$$C_s = -0.189\ln P_z + 0.4082 \quad (r^2 = 0.959) \quad (4)$$

$$C_s = -0.0513P_z^2 + 0.0688P_z + 1.0396 \quad (r^2 = 0.9969) \quad (5)$$

Respective best fit regression types were selected for each trend line based on correlation coefficient values. Exponential, linear, logarithmic, polynomial and power trend lines were tried for fitness in each plot and one with highest r^2 value was selected as best representing the relationship between C_s and P_z. In general loading

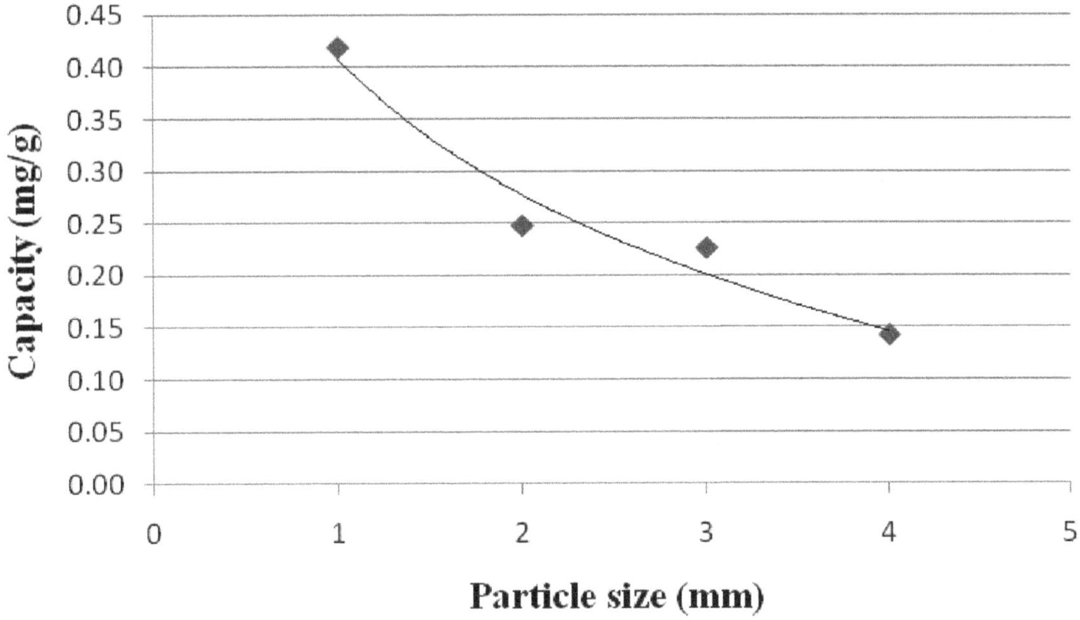

Figure 3. Loading capacity against particle size in defluoridation with magnesite *(r^2 = 0.959)*.

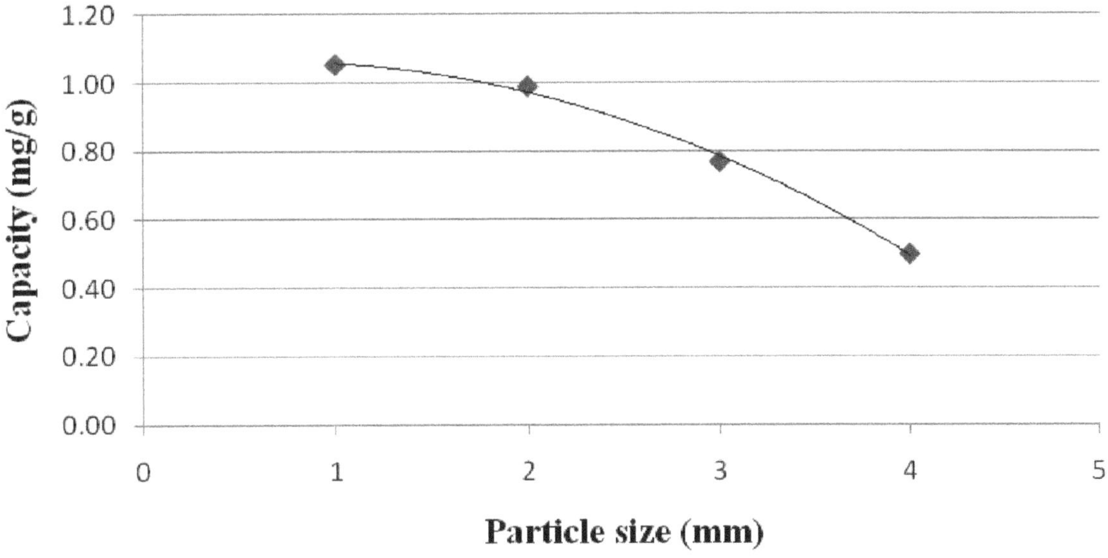

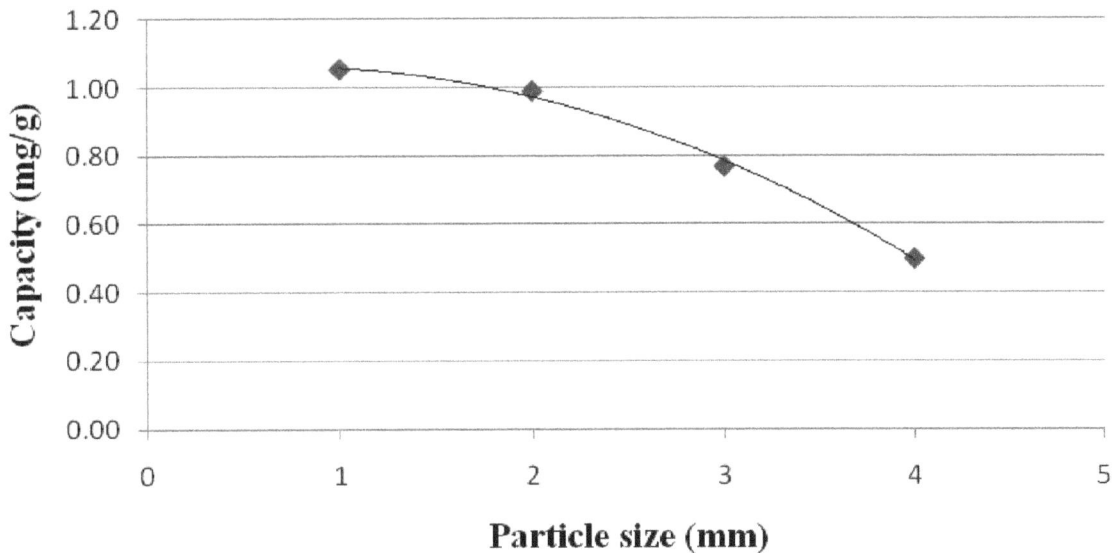

Figure 4. Loading capacity against particle size in defluoridation with composite *(r^2 = 0.997)*.

capacity increased with decrease in particle size.

This result could be attributed to increase in surface area available for fluoride sorption with decrease in particle size. Similar results were obtained elsewhere albeit with bone char (Ko et al., 1999), activated alumina (Shoeman and Macleod, 1987) and other adsorbents (Biswas et al., 2007; Turner et al., 2005). The results appear to suggest that for these three materials and their composite particle size may be decreased to enhance fluoride sorption. However, it must be noted that lower particle sizes obtained greater changes in apparent colour, pH, alkalinity, residual sulphate and hardness presented in water quality. The mean capacities obtained were not comparable to those obtained in earlier experiments. The mean sorption capacities were 1.05, 0.57, 0.28 and 0.82 mg/g for bauxite, gypsum, magnesite and the composite filter, respectively. Higher sorption capacities are reported in earlier experiments for bauxite (12.31 mg/g), gypsum (9.28 mg/g) (Sajidu et al., 2008; Peter, 2009) and magnesite (3.45 mg/g) (Singano, 2000).

Table 1. Raw groundwater quality prior to defluoridation.

Concentrations in raw water (mg/l)					Alkalinity (mg/l)				pH	Colour (TCU)
Hardness as $CaCO_3$	Al^{3+}	Fe^{2+}	SO_4^{2-}	Cl^-	Phenol	OH^-	CO_3^{2-}	HCO_3^{2-}		
14.5	0.00	0.02	20	20	10	0	20	70	8.72	0.0
WHO recommended upper limits / ranges										
Hardness as $CaCO_3$	Al^{3+}	Fe^{2+}	SO_4^{2-}	Cl^-	pH			Colour (TCU)		
500	0.2	0.3	400	250	6.5 – 8.5			5 – 50		

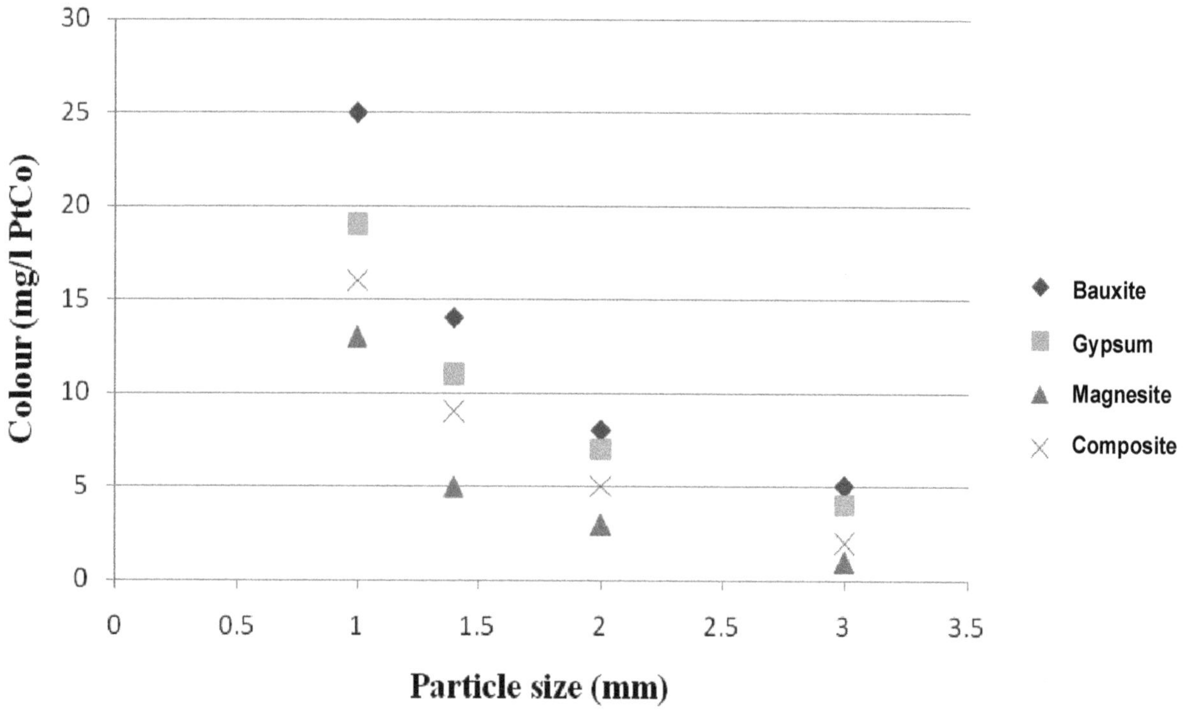

Figure 5. Plot of apparent colour against particle size at equilibrium.

The low sorption capacities obtained in the present experiments could be attributed to low initial fluoride concentrations, 2.5 mg/l, in the solutions employed that resulted in low fluoride ion availability per unit surface area of media material.

Water quality

Table 1 presents quality of the raw water before defluoridation experiments. The raw water quality was within world health organization (WHO) recommended limits in hardness, pH, colour and concentrations of aluminium, iron, sulphate, and chloride. Figure 5 depicts apparent colour at equilibrium with respect to particle size. Apparent colour was undetectable in the raw water as depicted in Table 1 as such the apparent colour obtained at equilibrium indicated colour introduced in the water by the materials. Figure 5 shows that the apparent colour was more intense in water treated with lower particle sized materials. It was also observed that, except for magnesite, all the materials with grain sizes of 1.0 mm caused colouration greater than 15 True Colour Unit (TCU) in water which is the WHO recommended upper aesthetic limit. Note that 1.0 mg/l platinum as chloroplatinate ion is equal to 1 TCU (HACH, 1997). Particles larger than 1.0 mm appeared to colour the water to intensities within this WHO limit.

This result could be attributed to the fact that larger particles settle more quickly than smaller ones as such larger sized particles, >1 mm diameter, stayed less longer in suspension enhancing early decolouration of

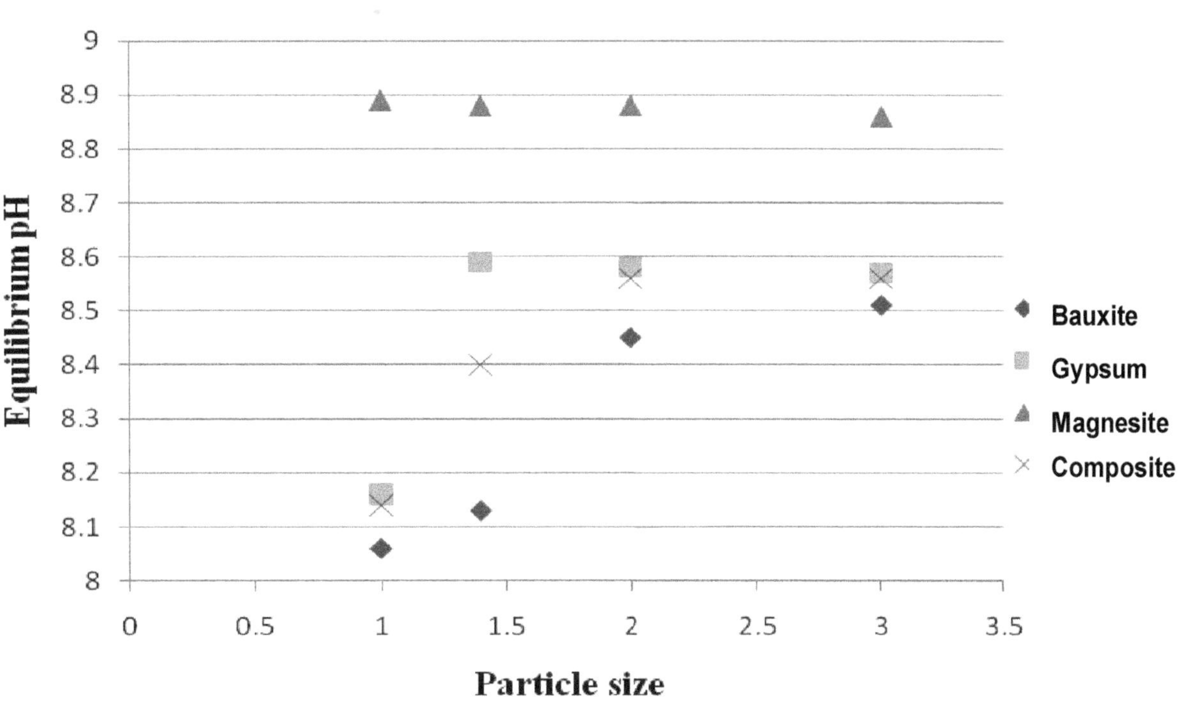

Figure 6. Plot of equilibrium pH versus particle size (mm).

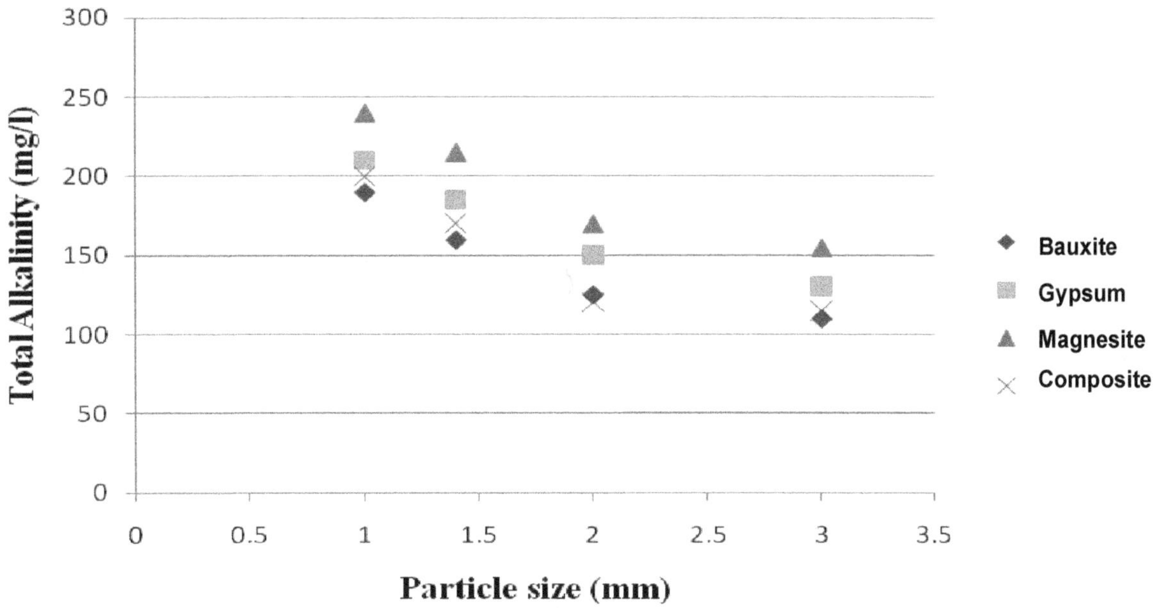

Figure 7. Alkalinity regressed with particle size.

the treated water. Similar results were obtained for pH, alkalinity, residual sulphates and hardness, Figures 6 to 9. The pH of the water before treatment was 8.72. Bauxite and the composite filters decreased pH as shown by the respective plots in Figure 6. The decrease in pH seemed to be enhanced with lower particle sizes. Particles below 1.5 mm diameter lowered the pH from the initial 8.72 to as low as 8.05, below the WHO recommended upper limit for pH of 8.50. Magnesite, on the other hand, increased the water pH to range of 8.85 to 8.91 for all particle sizes. pH changes were very minimal in defluoridation with gypsum for particles larger than 1 mm in diameter. The results showed that the combination of the three; bauxite, gypsum and

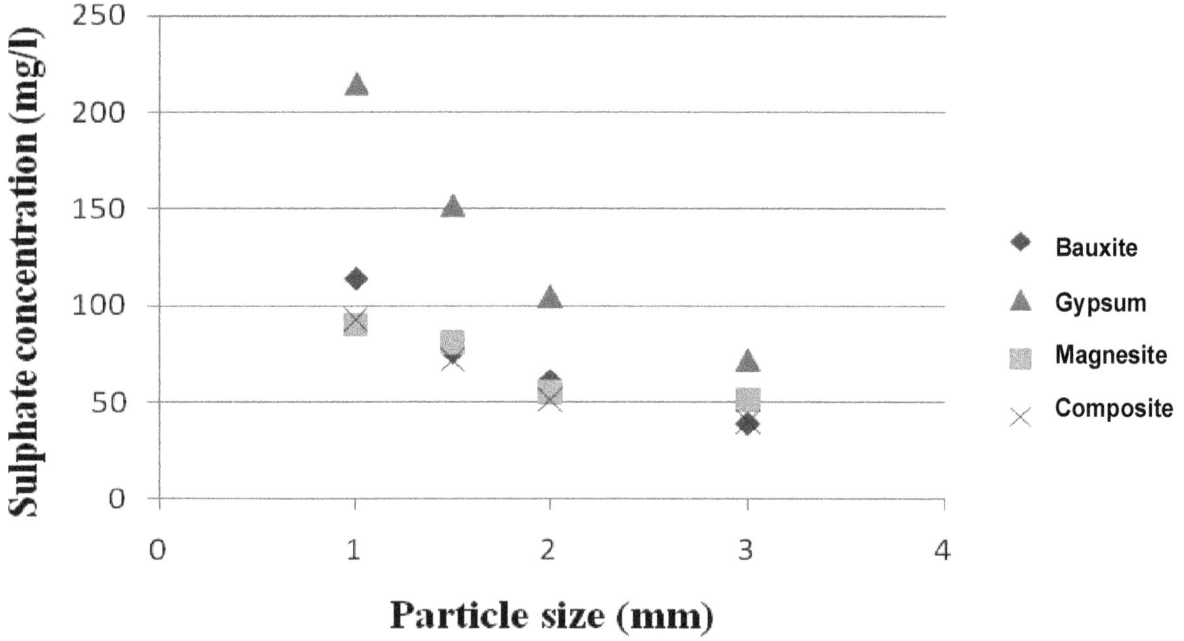

Figure 8. Residual sulphate concentrations against particle size.

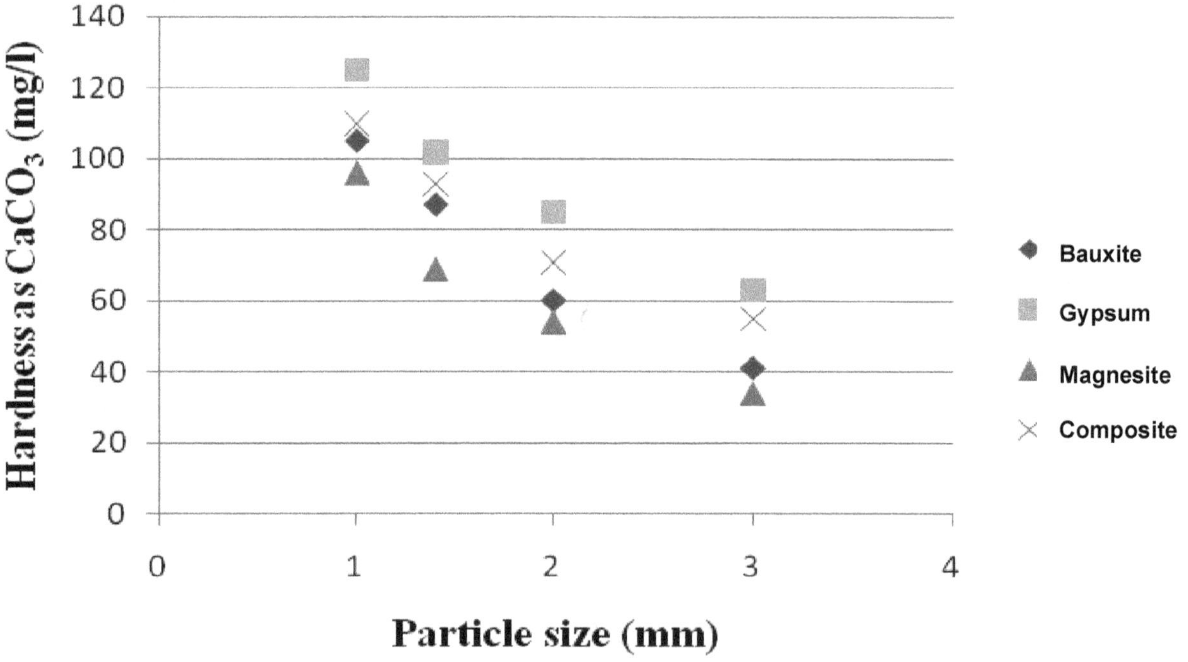

Figure 9. Hardness plotted against particle size.

magnesite, into a composite based on their loading capacities lowers pH. This may be taken advantage of where the initial water pH is higher than 8.50, a typical status in high fluoride water (Fawell et al., 2006). Total alkalinity appeared to increase with decrease in particle size with all values ranging from 100 to 250 mg/l for all the materials and their composite.

Total alkalinity being a measure that includes carbonate, bicarbonate and hydroxide alkalinities, it may be summed up that the ions $CO_3^{2-}, HCO_3^{2-}, OH^-$ are more available for dissolution when particles of these materials are smaller. The concentration of sulphates in the untreated water was 20 mg/l and the concentrations went up to

Table 2. Mean residual concentrations of Cl^-, Al^{3+}, Fe^{2+} (mg/l) in water after defluoridation with the materials of different particle sizes.

Material	Ion	1.0 mm	1.4 mm	2.0 mm	3.0 mm	*R^2
	Cl^-	71.5	57.9	68.3	73.1	0.42
Bauxite	Al^{3+}	0.5	0.3	0.4	0.4	-0.19
	Fe^{2+}	0.1	0.1	0.4	0.2	0.43
	Cl^-	69.5	72.1	54.3	64.7	-0.44
Gypsum	Al^{3+}	0.1	0.0	0.2	0.1	0.28
	Fe^{2+}	0.2	0.3	0.1	0.3	0.22
	Cl^-	145	127	151	147	0.41
Magnesite	Al^{3+}	0.2	0.4	0.3	0.3	0.19
	Fe^{2+}	0.1	0.4	0.2	0.3	0.33
	Cl^-	70.5	64.6	67.3	71.2	0.37
Composite	Al^{3+}	0.1	0.2	0.2	0.1	-0.20
	Fe^{2+}	0.2	0.3	0.1	0.3	0.22

*R^2 = correlation coefficients between ion concentration (mg/l) and particle size (mm).

about 200 mg/l with gypsum and to about 100 mg/l with bauxite, magnesite and the composite for materials of particle size range 0.5 to < 1.0 mm diameter. Gypsum introduced greatest quantities of sulphates in the water compared to the other materials. Generally, lower particle sized materials contributed more residual sulphate concentrations. The WHO upper limit for sulphate concentration is 400 mg/l. Despite that none of the concentrations went to 400 mg/l, it is still plausible that at large scale application the fine grained materials may introduce much more sulphates as to transcend beyond the upper limit. Water with sulphates beyond 400 mg/l being not safe for drinking, it may be important to identify sulphate sorption techniques to be employed alongside this defluoridation technology.

Bauxite, gypsum and magnesite changed the quality of treated water. Similar results were obtained in separate experiments. Singano (2000) reported an increase in water pH during defluoridation with magnesite and in defluoridation with bauxite and gypsum apparent colour, turbidity and hardness were reported to increase (Thole, 2005; Sajidu et al., 2008). The composite filter appeared to adjust pH downwards from 8.72 to within 8.15 to 8.43 that falls within WHO standard limits of 6.5 to 8.5. The water treated with 2 mm grains of the composite filter adhered to WHO standards in apparent colour, turbidity, hardness and residual concentrations of $Cl^-, Al^{3+}, Fe^{2+}, SO_4^{2-}$. However this does not necessarily imply safety for human consumption because other quality parameters of naturally occurring substances such arsenic, cadmium, lead and mercury; that have very low WHO guideline values of 0.05 mg/l and below, were not determined. Hardness determinations revealed that lower sized particles increased water hardness to a greater extent compared to large particles, Figure 9. This could be attributed again to availability of greater surface area hence greater solid-liquid surface interaction.

This effect could result into more dissolution of the cations Ca^{2+}, Mg^{2+}. The initial water hardness before defluoridation was very low, 14.5 mg/l as $CaCO_3$, entailing that all the materials increased hardness. Gypsum had the greatest impact on hardness seconded by the composite. Magnesite had the least effect on hardness. The major component in magnesite was MgO whereas in gypsum it was $CaSO_4$. It is probable that lower solubility of MgO compared to $CaSO_4$ resulted in less availability of total quantities of Ca^{2+}, Mg^{2+} ions from magnesite as compared to gypsum. Residual concentrations of chloride, aluminium and iron appeared not correlated to particle size.

The results are presented in Table 2 with correlation coefficients of the regressions between mean ion concentration and particle size. The correlation coefficients between ion concentration and particle size were below 0.5 indicating absence of linearity in behaviour of ion concentration with respect to particle size. The preliminary results suggest that the dissolution of the ions; Cl^-, Al^{3+} and Fe^{2+} from these materials into aqueous medium, is independent of particle size of the solid surface.

Conclusion

Loading capacities increased with decrease in particles size in defluoridation with bauxite, gypsum, magnesite and their composite filter. Particle size reduction may thus be employed in optimizing defluoridation performance of these materials. Apparent colour, alkalinity, hardness and residual sulphate concentration appeared to increase, where as pH decreased with decrease in particle size as such these parameters limit the extent to which particle size may be decreased. It is therefore apparent that during particle size reduction for optimisation of these materials, apparent colour, pH, alkalinity, hardness and residual sulphate concentration need to be monitored. Residual concentrations of chloride, aluminium and iron seemed independent of particle size.

ACKNOWLEDGEMENT

The authors are grateful for the research funds offered by the Malawi Government that facilitated the research activities from which the results reported herein were extracted.

REFERENCES

Amer Z, Bariou B, Mameri N, Taky M, Nicolas S, Elmidaoui A (2001). Fluoride removal from brackish water by electrodialysis. Desalination. 133: 215-223.

APHA (1989). Standard Methods for Examination of Water and Wastewater, 17th ed. American Public Health Association, Washington DC.

Aqua-Aero Water systems (2007). WaterPyramid. http://www.waterpyramid.nl, retrieved on 15th July, 2011.

Biswas K, Saha SK, Ghosh UC (2007). Adsorption of fluoride from aqueous solution by a synthetic iron (III) - aluminium (III) mixed oxide. Ind. Eng. Chem. Res., 46: 5346-5356.

Castillo NAM, Ramos RL, Perez RO, de la Cruz RFG, Pina AA, Rosales JMM, Coronado GRM, Rubio LF (2007). Adsorption of fluoride from water solution on bone char. Ind. Eng. Chem. Res., 4: 9205-9212.

Fawell J, Bailey K, Chilton J, Dahi E, Fewtrell L, Magara Y (2006). Fluoride in drinking water. IWA Publishing, World Health Organisation, London.

Feenstra L, Vasak L, Griffioen J (2007). Fluoride in groundwater: Overview and evaluation. Report SP 2007-1. International groundwater resources assessment centre, Utrecht.

Giesen A (1998). Fluoride removal at low costs. European semiconductor, 20: 103-105.

HACH (1997). Water analysis handbook. Loveland, Colorado.

Hanemaaijer JH, van Medevoort J, Jansen A, van Sonsbeek E Hylkema H. Biemans R, Nelemans B, Stikker A (2007). Memsti Membrane Distillation: A near future technology for sea wate desalination: Paper presented at the International Desalinatior Conference, June 2007, Aruba.

Ko DCK, Porter JF, McKay G (1999). Correlation-based approach to optimization of fixed-bed sorption units. Ind. Eng. Chem. Res., 38 4868-4877.

Maliyekkal SM, Anshup, Antony KR, Pradeep T (2010). High yielc combustion synthesis of nanomagnesia and its application for fluoride removal. Sci. Total Environ., 408: 2273–2282.

Nath SK, Dutta RK (2010). Fluoride removal from water using crushec limestone. Indian J. Chem. Technol., 17: 120–125.

Peter KH (2009). Defluoridation of high fluoride waters from natura water sources by using soils rich in bauxite and kaolinite. J. Eng Appl. Sci., 4: 240-246.

Sajidu SMI, Masamba WRL, Thole B, Mwatseteza JF (2008). Grounc water fluoride levels in villages of Southern Malawi and remova studies using bauxite. Int. J. Phys. Sci., 3: 001-011.

Schoeman JJ, Macleod H (1987). The effect of particle size anc interfering ions on fluoride removal by activated alumina. Water SA. 13: 229-234.

Singano JJ (2000). Investigation of the mechanisms of defluoridation o drinking water using locally available magnesite. PhD dissertation University of Dar es Salaam, Dar Es Salaam, Tanzania.

Smittakorn S, Jirawongboonrod N, Mongkoinchai-arunya S, Durnford D (2010). Home-made bone charcoal adsorbent for defluoridation o groundwater in Thailand. J. Water Health, 8: 826-836.

Solar D (2007). Solar Dew Collector Sytem http://www.solardew.com/index2.html, Retrieved on 15th July, 2011.

Sujana MG, Pradhan HK, Anand S (2009). Studies on sorption of some geomaterials for fluoride removal from aqueous solutions. J. Hazard Mater., 161: 120-125.

Thole B (2005). Water defluoridation with Malawi bauxite, gypsum and synthetic hydroxyapatite, bone and clay: Effects of pH, temperature sulphate, chloride, phosphate, nitrate, carbonate, sodium, potassium and calcium ions. MSc dissertation, University of Malawi, Zomba, Malawi.

Thole B (2011). Defluoridation kinetics of 200 °C calcined bauxite, gypsum, and magnesite and breakthrough characteristics of their composite filter. J. Fluorine Chem., 132: 529-535.

Thole B, Mtalo FW, Masamba WRL (2011). Water Defluoridation with 150 – 300°C Calcined Bauxite-Gypsum-Magnesite Composite (B-G-Mc) filters. Water Resources Management VI, Wit Transactions on Ecology and Environment, 145: 383-393.

Turner BD, Binning P, Stipp SLS (2005). Fluoride Removal by Calcite: Evidence for fluorite precipitation and surface adsorption. Environ. Sci. Technol., 39: 9561-9568.

Varol E, Akcay S, Ersoy IH, Koroglu BK, Varol S (2010). Impact of chronic fluorosis on left ventricular diastolitic and global functions. Sci. Total Environ., 408: 2295-2298.

Vasudevan S, Lakshmi J, Sozhan G (2009). Studies on Mg-Al-Zn as anode for the removal of fluoride from drinking water in an electrocoagualtion process. Clean: Soil, Air Water, 37: 372-378.

Yadav RN, Singh OP, Yadav R (2010). Study of the aluminum ammonium sulphate as defluoridated agent in drinking water earthenware. Arch. Apll. Sci. Res., 2: 11-22.

Distribution and temporal variation of selected heavy metals in sediment of River Osara mainstream drainage in North Central Nigeria

Olatunji Olatunde Stephen[1]* and Osibanjo Oladele[2]

[1]Department of Chemistry, Faculty of Applied Sciences, Cape Peninsula University of Technology, Bellville, Western Cape, South Africa.
[2]Department of Chemistry, Faculty of Science, University of Ibadan, Ibadan, Oyo State, Nigeria.

The concentration levels of selected iron-ore composite heavy metals were investigated in mainstream sediment of River Osara traversing by Itakpe iron-ore deposit and mining area in North Central Nigeria. The sediment samples collected along the drainage course, up-, mid- and downstream for 24 months were digested using standard methods, and levels of Cd, Cr, Ni, Cu and Pb determined using flame atomic absorption spectrophotometer (AAS). The mean concentrations of Pb, 15.63±2.35-18.98±2.04 mg/kg was the highest followed by Ni, 5.20±1.79-15.83±3.04 mg/kg, relative to other metals. The concentrations of Cu and Cr were 3.18±1.14-10.02±2.89 mg/kg and 1.12±0.27-20.93±4.89 mg/kg, respectively, while Cd, 0.24±0.12-0.40±0.26 mg/kg was the least. There was significant ($p < 0.05$) positive correlation between the concentration of the measured metals with organic carbon level in sediment, with coefficient of correlation γ=0.68, Cd; γ=0.69, Ni; γ=0.54, Cr; γ=0.85, Pb and γ=0.81, Cu implying that metals levels in sediments are partly defined by organic carbon levels. Seasonal fluctuations were observed in sediment concentrations of metals with higher concentration during dry season than in wet season. The concentration of heavy metals in River Osara sediment by Itakpe iron-ore deposit and mining area are low and within natural concentration levels. The result could serve as benchmark required for effective water management and environmental monitoring.

Key words: Heavy metals, concentration, bottom sediment, iron-ore, monitoring, River Osara, Nigeria.

INTRODUCTION

Bottom sediments of aquatic ecosystems are sinks for contaminants, including heavy metals (Yusuf and Osibanjo, 2006). They are able to record and reflect atmospheric signals and nature of the overlying waters (Kakulu and Osibanjo, 1988). Being an integral part of the dynamic hydro-geochemical system involved in the cycling of elements, sediment can accumulate and store, or release heavy metals and other toxic elements to the watershed, depending on conditions defined by the pH, redox potential, organic matter and grain size distribution of the sediment (Davis et al., 1991; Forstner, 2004). As a result, aquatic sediments are used to monitor environmental and natural changes occurring in local, regional or global watersheds and/or changes in aquatic ecosystems forced by anthropogenic factors. This is because of their ability to bear signatures of contaminant inputs and build up from natural and anthropogenic sources including the atmosphere, surface run off, domestic and industrial discharges (Forstner, 1985). The accumulations of elevated levels of heavy metals in sediments of many rivers were reported severally (Amiard et al., 1987; Pardo et al., 1990; Bryan and Langston, 1992; Gupta and Karuppiah, 1996;

*Corresponding author. E-mail: snf_olatunji@ymail.com.

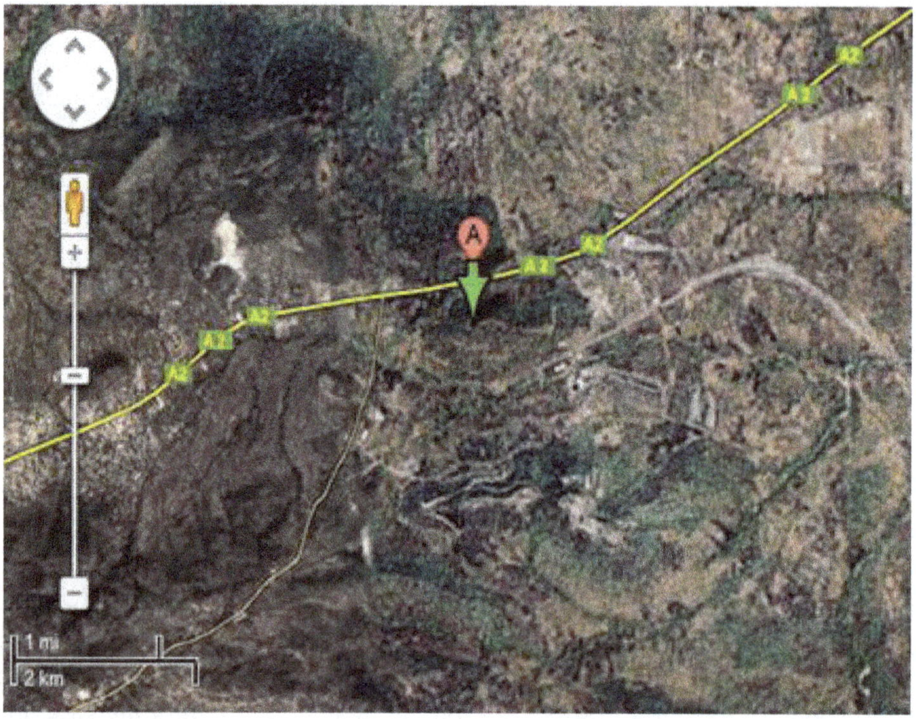

Figure 1. Watershed and drainage pattern of study area (Source: GSN, 1986).

Weng et al., 2008; Inengite et al., 2010).

The concern for heavy metals arises from their non-degradable and persistence nature, tendency for bioaccumulation and ecological toxicity, detrimental effect on aquatic biota and human health. For instance heavy metals such as copper, cobalt and arsenic were reported to be hepatotoxic, causing a liver condition referred to as liver cirrhosis (Hutton, 1987). Cadmium is known to cause kidney damage, hypertension and sperm deformation. Mercury, lead and arsenic toxicity results in neuro-degeneration and senility with possibility of irreversible brain damage. Alzheimer's, Parkinson's disease and dementia are caused by elevation of aluminium in human. Cancer, abdominal pain, skin lesions, speech impairment and autism, loss of hearing, vision and movement are among the potential toxic effects of heavy metals (Hammer and Hammer, 2004; Bakare-Odunola, 2005).

Studies revealed that heavy metals are composite associates of iron ore, occurring in different amounts especially Mn, Cr, Ni, Pb, Cd, Cu, Ti, V, W and Zn (Nriagu, 1989; Thomas, 1992; Kabata et al., 2001). Iron-ore mining operations and mineral beneficiation processes can endanger the integrity of rivers, especially downstream due to particulate depositions from mining, haulage and crushing, acid mine drainage and ore processing wastes. Substantial amount of heavy metals in mine wastes, tailings and low grade ore may also enter into soil and aquatic environments via atmospheric, industrial wastes disposal and improper disposal of sewage sludge. This could lead to metal enrichment in water and sediment body, and consequently ecosystem damage, which can affect adjoining water systems. Although mining activities at Itakpe iron ore deposit, Nigeria is abysmally low due to non-operational; there is a potential risk of contamination of River Osara due to its proximity to the iron-ore deposit, mining area and the ore beneficiation plants by direct and indirect discharge of mine waste, mine tailings, spent liquor and other effluents. Evaluating the baseline sediment geochemistry of the aquatic environment around the mineral deposit vein is crucial to understanding the existing heavy metals status. Information such as this will serve as basis for monitoring environmental changes that may be induced in River Osara either by natural, mining or other anthropogenic forces.

This study therefore investigates the baseline concentration levels of selected iron ore composite associated heavy metals in sediments of River Osara in order to define benchmark for future monitoring.

MATERIALS AND METHODS

Study area

Study location is within the section of River Osara traversing past the Itakpe ferruginous iron-ore deposit, and down beyond the mineral beneficiation plant and industrial area of Nigeria Iron Ore Mining Company in Kogi State, located within the Lokoja – Okene area delimited by latitudes 7° and 8°N and longitudes 6°and 7°E, in North Central Nigeria (Figure 1).

Table 1. Ranges and mean concentrations (mg/kg) of heavy metals, pH and organic carbon in sediments from River Osara at Itakpe.

Sample identity	pH	Organic Carbon (%)	Cd (mg/kg)	Cr (mg/kg)	Ni (mg/kg)	Cu (mg/kg)	Pb (mg/kg)
ITK/SM/D/03-04							
Concentration range	6.52-7.29	3.90-7.30	0.17-0.51	1.78-3.23	9.41-23.52	3.75-10.23	13.67-18.11
Mean (Standard Deviation)	6.90 (0.28)	5.60 (1.22)	0.40 (0.26)	2.37 (0.55)	14.46 (4.09)	6.61 (2.34)	15.75 (1.49)
ITK/SM/D/04-05							
Concentration range	5.90 -7.07	3.20-5.60	0.17-0.32	15.53-27.91	12.34-22.57	6.00-14.50	14.92-19.08
Mean (Standard Deviation)	6.57 (0.40)	4.52 (0.95)	0.24 (0.12)	20.94 (4.89)	15.83 (3.04)	10.02 (2.89)	16.96 (1.45)
ITK/SM/W/03-04							
Concentration range	6.78-7.25	3.60-6.50	0.16-0.32	0.78-1.54	2.86-8.16	1.58-6.36	15.81-22.17
Mean (Standard Deviation)	6.98 (0.14)	4.96 (1.01)	0.25 (0.05)	1.12 (0.27)	5.20 (1.79)	3.18 (1.14)	18.98 (2.04)
ITK/SM/W/04-05							
Concentration range	6.36-7.23	3.30-5.90	0.18-0.42	9.72-15.37	7.28-20.06	1.73-10.76	12.97-19.64
Mean (Standard Deviation)	6.89 (0.28)	4.18 (0.76)	0.30 (0.08)	12.79 (1.93)	12.91 (4.30)	4.88 (2.64)	15.63 (2.35)

Codes: ITK-Itakpe; SM-sediment sample; D-dry season, W-wet season; 03-04 - year 2003-2004 and 04-05-year 2004-2005.

Sediment sampling

Eighty sediment samples were randomly collected from the mainstream depth of River Osara, at the up-stream, mid-stream and down-stream flow portion around Itakpe iron ore deposit area on N 07° 37′ 420″, E 006° 18′ 284″ during dry and wet seasons, spanning a period of twenty-four months between 2004 and 2005.

Samples digestion and analysis

Sediment samples were sorted, air-dried, pulverized and screened through 2 mm mesh nylon sieve. The 2 mm sieved sediment samples were digested according to the methods described by Andersson (1976), Onianwa (2000), and Smejkalova et al. (2003) with slight modifications. Five grams each of sediment samples was weighed into 250 ml teflon beakers and digested with 50 ml, 2 M nitric acid, in water bath. The digests were filtered into 100 mL volumetric flask and quantified for Cd, Cr, Ni, Cu and Pb using flame atomic absorption spectrometer (FAAS) Unicam 969.

Determination of organic carbon in sediment

The method of Walkley and Black, 1934 was used: 5 g of sediment samples were weighed each into 250 ml teflon beakers. Each sample was rapidly subjected to dichromate oxidation by addition of 50 ml potassium dichromate 0.5 M, $K_2Cr_2O_7$ and 2.5 ml concentrated sulphuric acid in 5% $FeSO_4$. Excess $Cr_2O_7^{2-}$ were titrated with 0.25 M ferrous ammonium sulphate $[Fe(NH_4)_2(SO_4)_2.6H_2O]$.

Determination of sediment pH

The pH of the sediment samples were determined according to the method described by McLean, (1982). 1:1 mixtures of sediment samples and distilled water in glass beakers were gently stirred to enhance H^+ release. The resulting mixtures were allowed to stand for 30 min and the sediment pH measured using a pre-calibrated pH meter.

Recovery studies

Analyte recovery was conducted on spiked sediment sample in triplicates; at low, medium and high spike concentrations. The spiked samples were subjected to the same sample treatment and analysis method. The results showed good efficiency in analytes digestion and recoveries. The percentage recoveries and coefficient of variation of the spiked replicate samples were Cd, 90.45-93.81%, C.V 3.87-6.75%; Cr, 79.64-85.92%, C.V 9.16-18.42%; Ni, 89.34-94.68%, C.V 8.58-15.84%; Cu, 85.19-92.55%, C.V 6.34-14.71% and Pb, 82.75-92.10%, C.V 7.29-10.98%, Pb.

RESULTS AND DISCUSSION

Concentrations of heavy metals in sediments

The results showed variations in concentration levels of heavy metals in the analysed sediment (Table 1). Theseasonal concentrations range (mg/kg) of the heavy metals detected in the sediments showed that Pb, 15.63±2.35-18.98±2.04 mg/kg had the highest concentration levels relative to the measured metals. This was followed by Ni, 5.20±1.79-15.83±3.04 mg/kg and then Cu, 3.18±1.14-10.02±2.89 mg/kg. The concentration ranges of other metals in River Osara sediment were Cr, 1.12±0.27-20.94±4.89 mg/kg and Cd, 0.24±0.12-0.40±0.26 mg/kg being the least. The distribution of heavy metals in sediments is in the order Pb > Ni > Cu > Cr > Cd, with the values indicating heterogeneous quantities at different sampling points. This is consistent with the findings of Toma et al. (1981) in fine fractions of continental shelf sediments of Western Nile. The metals retention potential of the sediment may be attributed to the levels of sediment organic carbon (4.18 ±0.76-5.60±1.22%). Heavy metals are usually associated with organic matter and fluidized sediment clay minerals, with variable crustal composition consisting of soluble and insoluble silica, alumina and iron/manganese oxy-hydroxides (Fe_2O_3/MnO) (Rothwell et al., 2007). Ma and Rao (1997) noted that heavy metals are not evenly distributed in their different speciated matrices. The sediment pH values were near circum-neutral with seasonal means ranged 6.57±0.40-6.98±0.14.

The pH, organic carbon, clay minerals and iron/ manganese flux contributes to the total sediment retention capacity in aquatic systems. Material exchange also took place between water column and the sediment system, in which metal partitioning occurs between sediment and water. The mechanisms for mobilization, distribution or accumulation of heavy metals in inland freshwater and marine sediments may include weathering of primary minerals, association and dissolution of metals with iron-manganese oxy-hydroxides and aqueous speciation. The concentrations and variations in distribution of heavy metals in aquatic sediments is a function of metals solubility, the density established when they combine with other contaminants and the effect of organic matter and clay minerals heterogeneity on sorption and desorption of the metals.

Cadmium

The concentration of Cd in sediment ranged 0.17-0.51 mg/kg with mean concentration 0.40±0.26 mg/kg during dry season 2003/2004 and 0.17-0.32 mg/kg with mean 0.24±0.12 mg/kg during dry season 2004/2005. Cadmium concentration during wet season 2003/2004 ranged 0.16-0.32 mg/kg with mean 0.25±0.05, while during wet season 2004/2005 the concentration ranged 0.18-0.42 mg/kg with mean value 0.30±0.08 mg/kg. The relative standard deviation for Cd concentration ranged between 6% during wet season to 26% during dry season. The result showed fairly consistent Cd concentration with low distribution variation through the evaluated column upstream and downstream during the study period. This can be attributed to Cd stabilization, enhanced by less perturbation of the surface sediment. Lovern (2000) reported that stability in Cd concentration may be due to complex assemblage induced by sorption density of sediment reactive site. Seasonal variations in Cd concentrations were low and ranged between ± 0.05 mg/kg, Cd to ± 0.26 mg/kg, Cd, indicating slight homogeneity in sediment concentration levels and distribution.

Lead

The concentrations of Pb were slightly variable between sampling points. Sediment of Pb concentration ranged 13.67-18.11 mg/kg with mean levels 15.75±1.49 mg/kg during dry season 2003/2004 (Table 1), while Pb sediments concentration ranged 14.92-19.08 mg/kg with mean Pb level 16.96±1.45 during dry season 2004-2005. The concentration of Pb during wet season 2003/2004 ranged from 15.81-22.17 mg/kg with mean range of 18.98±2.04 and Pb, 12.97-19.64 mg/kg with mean range of 15.63±2.35 mg/kg during 2004/2005. Contrary to the observed seasonal trend with concentration depletion during wet season, Pb enrichment of about 20.5% was observed during wet season 2003/2004 study period. The reason for this cannot be readily explained, because Pb concentration detected in sediment sample during the subsequent study period 2004/2005 followed seasonal distribution trend. However the seasonal mean concentration levels of Pb were not significantly different ($p > 0.05$). Also the concentration levels detected upstream and downstream were not significantly different ($p > 0.05$). This is consistent with the findings of Violante et al. (2010) who reported that Pb level in sediment is controlled by sediment pH and redox state.

Nickel

Nickel concentration was variable and ranged between 9.41-23.52 mg/kg with mean levels 14.46±4.09 mg/kg during dry season 2003/2004 (Table 1). The concentration was fairly consistent though slightly higher during the subsequent dry season 2004/2005 with sediments Ni concentration ranged 12.34-22.57 mg/kg and mean concentration of 15.83±3.04 mg/kg. There was significant depletion ($p < 0.05$) in Ni levels during wet season 2003/2004, perhaps as a result of increase in the volume of water in the river and flow turbulence. Hence the detected wet season concentration of Ni ranged 2.86-8.16 mg/kg with mean 5.20±1.79 during 2003/2004. Nickel level during subsequent wet season 2004/2005 was hardly depleted (< 0.001%) with Ni concentration ranged 7.28-20.06 mg/kg, and mean 12.91±4.30 mg/kg. The least concentration of Ni, 2.86 mg/kg was observed in mainstream drainage sediment upstream with the highest Ni, 23.52 mg/kg measured in the downstream edges beyond the industrial iron ore beneficiation plant.

Copper

The concentration of Cu ranged 3.75-10.23 mg/kg with mean concentration 6.61±2.34 mg/kg in sediment samples during dry season 2003/2004, and 6.00-14.50 mg/kg with mean 10.02±2.89 mg/kg during dry season 2004/2005. Copper concentration observed during wet season 2003/2004 ranged 1.58-6.36 mg/kg with mean 3.18±1.14, while during wet season 2004/2005 the concentration ranged 1.73-10.76 mg/kg with mean value 4.88±2.64 mg/kg. The concentrations of Cu detected in sediment during dry season 2003/2004 were significantly lower than in the subsequent season 2004/2005 with an enrichment of about 35%. This may be as a result of changes in arable land use which were considered as a major source of Cu in adjoining water column (Kaizer and Osakwe, 2010; Darrell, 2009). Ma and Rao (2007) suggested that the concentration of Cu may be a function of their levels in the parent materials.

Chromium

There was significant Cr enrichment of 8.5-12.5 fold factor during study period 2003/2004 and 2004/2005. This is likely to be as a result of hydro-geochemical processes due to the flow of the river through basement intercalated by sediment, or from land based sources. This is because Cr is known to exist in minor amount in ferruginous quartzite and is reported to be soluble. Chromium concentration ranged from 1.78-3.23 mg/kg with mean concentration of 2.37±0.55 mg/kg during dry season 2003/2004 and 15.53-27.91 mg/kg with mean of 20.94±4.89 mg/kg during dry season 2004/2005. Chromium concentration during wet season 2003/2004 ranged from 0.78-1.54 mg/kg with mean of 1.12±0.27 while during wet season 2004/2005 the concentration ranged from 9.72 - 15.37 mg/kg with mean value of 12.79±1.93 mg/kg.

Sediment pH and organic carbon

The prevailing sediments pH observed during the study ranged from 5.90-7.29 (6.84±0.28), while organic carbon levels ranged 3.20-7.30% with mean value of 4.82±0.99%. According to Lee (1975), the concentration levels of heavy metals found in sediments are dependent on the prevailing pH condition of the hypolimnion water column. There was significant ($p < 0.05$) positive correlation between the concentration of the measured metals with organic carbon level in sediment. The coefficient of correlation were Cd, $\gamma = 0.68$; Ni, $\gamma = 0.69$; Cr, $\gamma = 0.54$; Pb, $\gamma = 0.85$ and Cu, $\gamma = 0.81$. The correlation values are consistent with the observation of Weber et al., (1992) who concluded that the extent of sorption of contaminants such as metals to organic carbon content of sorbents sediments is consistent with the behaviour for partitioning processes. Kile et al. (1999) suggested the existence of an empirical correlation between elemental compositions and natural organic matter, and the extent to which they are sorbed to organic compounds. The concentration of metals detected in River Osara sediments may be attributed to the sorption capacity of sediment and this is defined by the organic carbon content. Sediments also consist of mineral matter flux such as Fe/Mn, onto which metals are sorbed.

Heavy metals distribution in sediment

The variation in the measured heavy metals concentrations ranged from ±0.12 to ±12.68 mg/kg, which indicates that the measured metals are not uniformly distributed. There were variation in concentrations of the heavy metals between sampling points along the river channel, with higher concentration values found in sediments collected from non- tidal area and from entrapped portion of the river bed and other discrete blebs. This showed that heavy metals do not occur as continuous monolayer in the surface sediment. According to Taylor et al. (2002), the distribution of contaminants in river systems cannot be expected to be linear nor simple, as their spatial distribution is controlled by physical processes including stream hydraulics, sediment storage and erosion behaviour. The concentrations of Pb were the highest in the sediment samples which ranged from 12.97-22.17 mg/kg followed by Ni, 2.86-23.52 mg/kg concentrations. The concentrations of Cu and Cr were ranged, 1.58-14.50 and 0.78-27.91 mg/kg, respectively, and Cd, 0.16-0.51 mg/kg being the least. However, Cr had the widest distribution variation followed by Ni, Cu, Pb and Cd in that order.

Seasonal effect on heavy metal concentrations

The seasonal variation of heavy metals concentrations in sediment samples from River Osara at Itakpe are presented in Figure 2.

The concentrations of heavy metals in bottom sediment of River Osara at Itakpe were observed to be lower during wet season than during dry season. This may probably be due to volume increase, increase in flow velocity and flow turbulence which enhances re-dissolution, re-suspension and erosion of metals from the surface of bottom sediments in the water body flow dynamics. Sediment surface are less perturbed under steady flow conditions during dry season, hence metals are sequestered on sediments by organic matter and clay mineral fraction (oxy-hydroxides of Fe/Mn).

Sediments enrichment may also occur during dry season when heavy metals precipitating out of water columns are deposited on bottom sediments. This process enhances the concentration and stabilization of heavy metals. Sediments levels of Ni, Cu and Cr were seasonally changeable, while Pb and Cd levels were fairly stable.

Sediment quality

The concentrations of heavy metals detected in this study is consistent with the findings of Obasohan et al. (2006) in sediment from Ogba River in Benin, Nigeria with Cr, 0.99 mg/kg; Ni, 0.71 mg/kg; Cu, 2.68 mg/kg, and lower concentrations of Pb, 1.80 mg/kg and higher Cd level of 1.03 mg/kg. Kakulu and Osibanjo (1988) reported higher concentration levels of Cu, 23.9 mg/kg; Pb, 32.1 mg/kg

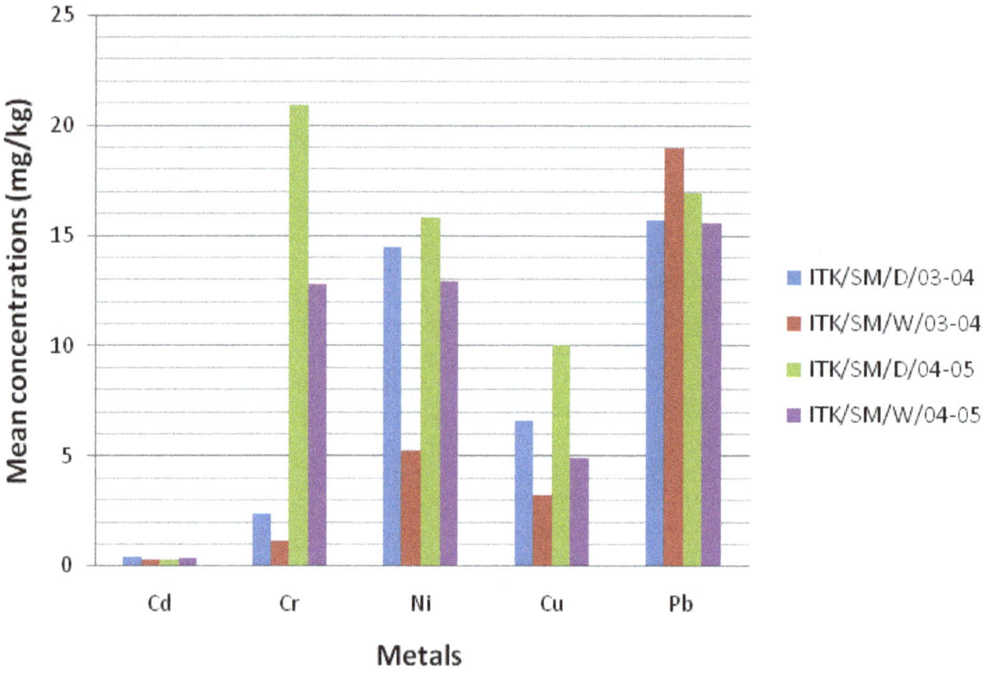

Figure 2. Seasonal variation in concentration levels (mg/kg) of heavy metals in River Osara sediment samples. Code: ITK-Itakpe, SM-sediment, D-dry season, W-wet season, 03-04 - year 2003-2004, 04-05 - year 2004-2005.

and Cd, 0.79 mg/kg in sediment samples from the Niger Delta, Nigeria than levels found in this study, owing to discharge of refinery effluent into the Niger Delta waters. Yusuf and Osibanjo (2006) also reported comparable sediment concentrations of Zn, 35.5 mg/kg; Cu, 11.4 mg/kg; Pb, 11.0 mg/kg and Cd, 0.60 mg/kg with higher level of Mn, 123.4 mg/kg in Ologe Lagoon, South-west Nigeria.

This study's result compares the concentrations of heavy metals found in sediments from rivers studied elsewhere. Biney (1991) reported Cd, <0.20 mg/kg and Pb, 29.30 mg/kg in sediments from Kpong Dam, Ghana, while Cu, 30.30 mg/kg was higher. Watling and Emmerson (1981) reported Cu, 10.5 mg/kg and Pb, 17.8 mg/kg in sediments from Swartkops River South Africa, with higher concentration of Cd, 1.00 mg/kg and Cr, 177.0 mg/kg. The findings of Gabr et al. (2008), in sediments from Suez Canal Egypt with Ni, 14.5±1.5 mg/kg; Cu, 8.8±1.1 mg/kg; and Pb, 13.5±1.5 mg/kg are consistent with these results. Cadmium (1.25±0.2 mg/kg) in sediment of Suez Canal was high compared with 0.16-0.51 mg/kg levels found in River Osara sediment.

Despite the heavy metals seasonal compositional variation in River Osara sediments, the measured concentration values of Pb, Ni, Cu and Cr were low to moderate and below threshold concentrations (Pb, 19.00 mg/kg, Cu, 33.00 mg/kg and Cd, 0.11 mg/kg) recommended by the Joint Group of Experts on the Scientific Aspects of Marine Environment Pollution (GESAMP) for unpolluted sediments, except for Cd (GESAMP 1982). The detected levels may be attributed to geological variations and crustal structural complexities, the impact of human activities and use of the river bodies as well as input from land based sources.

Conclusion

The concentration of heavy metals in sediment River Osara around Itakpe iron-ore deposit and mining area are variable and within natural levels, except for Cd. This suggests the likelihood of source apportionment of Cd in the river. The distribution of heavy metals in sediments is in the order Pb > Ni > Cu > Cr > Cd with the values indicating heterogeneous quantities at different sampling points. Sediment concentrations of the measured heavy metals were seasonally influenced with higher concentrations during dry season, owing to probably less perturbation of sediment state, gradual and consistent sediment enrichment, due to steady flow of water.

REFERENCES

Amiard JC, Amiard-Triquet C, Berthet B, Métayer C (1987). Comparative study of the patterns of bioaccumulation of essential (Cu, Zn) and non-essential (Cd, Pb) trace metals in various estuarine and coastal organisms. J. Exp. Mar. Biol. Ecol. 106:73-89.

Andersson AA (1976). On the determination of ecologically significant fractions of heavy metals in soil. Swed. J. Agric. Res. 6:19-25.

Bakare-Odunola MT (2005). Determination of some metallic impurities present in soft drinks marketed in Nigeria. Nig. J. Pharm. 4(1):51-54.

Biney CA (1991). The distribution of trace metals in the Kyong Headpond and Lower Volta River, Ghana. In Perspectives in Aquatic Ecotoxicity: Shastree NK (Ed). Norendra Publishing House, Delhi, India.

Bryan GW, Langston WJ (1992). Bioavailability, accumulation and effect of heavy metals in sediments with special reference to United Kingdom estuaries: a review. Environ. Pollut. 76:89-131.

Darrell T (Project Lead) (2009). A water quality survey of Nine lakes in the Carleton river watershed area Yarmouth County, Nova Scotia. Water and Wastewater Branch Nova Scotia Environment.

Davis EE, Horel GC, MacDonald RD, Villinger H, Bennett RH, Li H (1991). Pore pressures and permeability's measured in marine sediments with a tethered probe. J. Geophys. Res. 96(B4):0148-0227.

Forstner U (2004). Traceability of sediment analysis. Trends Anal. Chem. 23(3):217-236.

Forstner U (1985). Chemical forms and reactivities of metals in sediments. In R. Lescheber et al. (ed.). Chemical method for accessing bioavailable metals in sludges and soils, Elsevier, London p.190.

Gabr RH, Ali A-F, Gab-Alla A (2008). Effect of transplantation on heavy metal concentration in clams of Lake Timsah, Suez Canal, Egypt. Oceanologia 50(1):83-93.

GESAMP (1982). The health of the oceans. In Salomons W, Forstner U (1984). Metals in the hydrocycle. Berlin, Springer p. 349.

Gupta G, Karuppiah M (1996). Heavy metals in two Chesapeake Bay tributaries Wicomicoa and Pocomoke rivers. J. Haz. Mat. 50:15-29.

Hammer MJ, Hammer MJ (Jr.) (2004). Water Quality: In: Water and Waste Water Technology. 5th ed. New Jersey: Prentice-Hall pp. 139-159.

Hutton M (1987). Human health concern for lead, mercury and arsenic in land: mercury, cadmium, and arsenic in the Environment. Wiley Scope pp. 85-89.

Inengite AK, Oforka NC, Osuji Leo C (2010). Survey of heavy metals in sediments of Kolo creek in the Niger Delta, Nigeria. Afr. J. Environ. Sci. Technol. 4(9):558-566.

Kabata-Pendias A, Pendias H (2001). Trace element in soils and plants. CRC Press, Boca Raton, FL. pp. 1-488.

Kaizer AN, Osakwe SA (2010). Physicochemical Characteristics and Heavy Metal Levels in Water Samples from Five River Systems in Delta State, Nigeria. J. Appl. Sci. Environ. Manag. 14(1):83-87.

Kakulu SE, Osibanjo O (1988). Trace heavy metal and pollutional status in sediments of the Niger Delta area of Nigeria. J. Chem. Soc. Nig. 13:9-15.

Kile OE, Wershaw RL, Chiou CT (1999). Correlation of soil and sediment organic matter polarity to aqueous sorption of non-tonic compounds. Environ. Sci. Technol. 33:2053-2056.

Lee GF (1975). Role of hydrous metal oxides in the transport of heavy metals in the environment session IV transport mechanism 2. Institute for Environmental Sciences, University of Texas, Dallas. Seminar paper p. 18.

Lovern SB (2000). assessing the nonpoint source pollutant removal efficiencies of a two-basin storm water management system in an urbanizing watershed. A thesis submitted to the faculty of the Virginia Polytechnic Institute and State University.

Ma LQ, Rao GN (1997). Chemical fractionation of cadmium, copper, nickel and zinc in contaminated soils. J. Environ. Qual. 26(1):259-264.

Nriagu JO (1989). A global assessment of natural sources of atmospheric trace metals. Nature 338:47-49.

Obasohan EE, Oronsaye JAO, Obano EE (2006). Heavy metal concentrations in *Malapterus electricus* and *Chrysichthys nigrodigitatus* from Ogba River in Benin City, Nigeria. Afr. J. Biotechnol. 5(10):974-982.

Onianwa PC (2000). Roadside topsoil concentration of lead and other heavy metals in Ibadan, Nigeria. Soil Sed. Contam. 10(6):577-591.

Pardo R, Barrado E, Perez L, Vega M (1990). Determination and speciation of heavy metals in sediments of the Pisuerga River. Water Res. 23(3):373-379.

Rothwell JJ, Evans MG, Allott TEH (2007). Lead contamination in fluvial sediments in an eroding blanket peat catchment. Appl. Geochem. 22:446-459.

Smejkalova M, Mikanova L, Boruvka L (2003). Effects of heavy metal concentrations on biological activity of soil micro-organisms. Plant Soil Environ. 49(7):321-326.

Taylor PM, Stuart L, Dustine F (2002). Heavy metal contamination of Mt. Isa–implication for the ecosystem (Isoconferences). Retrieved December, 23, 2005 from www.isoconferences.org.au.

Thomas L (1992). Coal as substance. Handbook of Practical Coal Geology. John Wiley and Sons Limited, England p. 4.

Toma SA, Saad MAH, Salam AMS, Halim Y (1981). The distribution of some absorbed elements on the Nile continental shelf sediments. J. Etudol. Pollut. 5:377-382.

Violante A, Cozzolino V, Perelomov L, Caporale AG, Pigna M (2010). Mobility and bioavailability of heavy metals and metalloids in soil environments. J. Soil. Sci. Plant Nutr. 10(3):268-292.

Walkley A, Black IA (1934). An examination of the Degtjareff method for determining organic carbon in soils: Effect of variations in digestion conditions and inorganic soil constituents. Soil Sci. 63:251-263.

Watling RJ, Emmerson WD (1981). A preliminary pollution survey of the Papenkuils River Port Elizabeth. Water S. Afr. 7:211-215.

Weber WJ, McGinley PM, Katz LE (1992). A distributed reactivity model for sorption by soils and sediments, 1. Conceptual basis and equilibrium assessments. Environ. Sci. Technol. 26:1955-1962.

Weng HX, Zhu YM, Qin YC, Chen JY, Chen XH (2008). Accumulation, discrepancy of heavy metal and organic pollutants. J. Asian Earth Sci. 31:522–532.

Yusuf KA, Osibanjo O (2006). Trace metals in water and sediments from Ologe Lagoon, Southwestern Nigeria. Pak. J. Sci. Ind. Res. 49(2):88-96.

Water quality assessment of Manchar Lake during the month of August and November 2011 by analyzing selected trace elements

Ghulam Murtaza Mastoi[1], Khalid Hussain Lashari[2], Zameer Ali Palh[2], Anila Naz Soomro[2], Zulfiqar Ali Laghari[3], Khalida Faryal Almani[1], Gul- e-Rana Abdul Waheed Mastoi[1] and Abdul Rasool Abbasi[2]

[1]Centre for Environmental Sciences, University of Sindh, Jamshoro, Sindh, Pakistan.
[2]Department of Fresh Water Biology and Fisheries, University of Sindh, Jamshoro, Sindh, Pakistan.
[3]Department of Physiology, University of Sindh, Jamshoro, Sindh, Pakistan.

Manchar Lake is the largest fresh water lake in Pakistan, situated in Jamshoro District. It was created in the 1930 when the Sukkur Barrage was constructed on the river Indus. It is a vast natural depression flanked by Khirthar range in the west, Lakhi hills in south and river Indus in the east. On the north eastern side is the protective embankment. The lake is fed by two canals, the Aral Wah and the Danister from the river Indus. The lake also collects water from numerous small streams in the Khirthar Mountains. The area of the lake fluctuates during the flood season from 350 to 520 km^2. The mean depth of the lake is at present 13 feet. Contamination of Manchar Lake is serious issue of aquatic pollution. Lake gets polluted by different waves although it is a second largest lake of Sindh province. Untreated damping of industrial liquid waste is one of the main causes of its pollution, hence for the justification of its contamination some trace metals, Lead, Copper, Zinic, Cobalt, Chromium and Nickel were analyzed by dual mode of analytical methods flame atomic absorption spectrometry (FAAS) and electro thermal atomic absorption spectrometry (ETAAS) by multi element stranded solution. The highest concentration of Lead, Copper, Zinic, Cobalt, Chromium and Nickel were 54.5 and 58.5 µg/L, 115 and 117.5 µg/L, 3000 and 3200 µg/L, 7 and 9.5 µg/L, 90 and 95.5 µg/L, 90 and 117 µg/L in month of August and November, 2011 respectively (Pb, Cu, Zn, Co, Ni, and Cr) were selected for study.

Key words: Contamination, waste, pollution, environment, fresh water.

INTRODUCTION

The pollution of Manchar Lake is serious issue, various studies has been carried out on Manchar Lake (Arain et al., 2008; Kazi et al., 2009; Arain et al., 2009; Abdul et al., 2009). We know that water is essential for sustaining the life on planet earth. This vital resource makes up more than 60% of the living organisms. Pollutant derived from a growing number of diverse sources (Industrial waste, urban run-off, sewage treatment plants, boating activities, agricultural fungicides, domestic garbage, dumps, anthropogenic activities and mining operations) have increasingly affected aquatic ecosystem (Garrett, 2000; Lashari et al., 2012). Heavy metals aquatic system have attracted considerable attention owing to their toxicity and accumulation in aquatic biota (Elllis 1989; Mason, 1998). The human health is directly concern to the quality of water, sanitation and hygiene. Water required for human consumption must be free from concentration of chemical substances that may be health hazards; as water is highly prone to contamination by bacteria, viruses, protozoa, pathogens, helminthes parasite, and toxic organic, inorganic compounds (Hari et al., 1994; Miranzadeh et al., 2011). Polluted water causes various diseases in consumers such as, diarrhea, dysentery, typhoid, cholera, hepatitis etc (Waheed and Kausar, 1987). The children under the age five have been killed by waterborne diseases and it is reported that the death ratio due to unsafe water is more than all forms of violence including war worldwide annually (World health organization, 1996, 2003). Amongst the water borne diseases diarrhea is considered as one of the major killer among the children. Each year 1.8 million people die from diarrhoeal diseases, and 90% of these deaths are of children under five (World health organization, 2003; WHO/UNICEF, 2004). Moreover, each year 4 billion cases of diarrhea has been reported worldwide and results in 2.2 million deaths occurred mostly of the children under five. This shows that 15% of child deaths every year are attributable to this dreadful disease. It is reported that child is dying every 15 s because of diarrhea worldwide. In India, half million children has dying each year due to diarrhea (UNICEF, WHO, 2009). As, in Pakistan fifty thousands three hundreds children under five year are dying due to diarrhea annually; and Pakistan is ranked sixth among fifteen countries (WHO/UNICEF, 2004). In the source of Industrialization, the quantity of heavy metals and toxic compounds are continuously released into Manchar Lake, hence, the aim of current study was to investigate the heavy metals content from Manchar lake water.

MATERIALS AND METHODS

In the study area of the Manchar lake five sampling stations was selected which includes zero points (waste entrance point into lake), near to zero point (inside the Manchar lake), mid of the Manchar, Danistar Wah (Manchar Lake) and Aral Wah (Manchar Lake) respectively. Sampling was carried out throughout the year; which includes August, 2011, November, 2011. Each sampling was carried out after the interval of three months. Water samples were collected according to the method of in pre-cleaned one liter plastic bottles of scraped caps. The sample bottle was plunged necks downwards about 15 cm below the water surface and about 60 cm from the side of local fisher man boats. The samples were stored in the laboratory at the room temperature on top of cupboards until required for use. All the analyses were conducted according to the standards of American Public Health Association (APHA, 1992). The entire chemicals and study standards solution were obtained from Sigma Fluka.

RESULTS AND DISCUSSION

Natural fresh water contains impurities of trace elements/heavy metals as it dissolves these substances while moving downward as a hydrological cycle (Aamir and Tahir, 2003). Metals can be lethal or harm the organism without killing it directly. Adverse effects on an organism's activity, growth, metabolism, and reproduction are examples of sub lethal effects (Wright and Pamela, 2002). These metals include lead (Pb), Cadmium (Cd), Zinc (Zn), Mercury (Hg), Arsenic (As), Silver (Ag) Chromium (Cr), Copper (Cu) Iron (Fe), and the Platinum group elements (Haijian et al., 2011). Many of these metals are considered essential for human health. Like Iron, Copper, Zinc, Nickel and other trace elements are needed for proper functioning of biological systems, but overexposure can lead to adverse health consequences (Rippey et al., 2008; Ley et al., 2011). However, deficiencies of these trace elements may lead to a number of disorders as well (Prabu, 2009). Generally heavy metal bioaccumulat during metabolic processes (Borgmann et al., 1993; Sharma and Aqrawal, 2005) and yet their biotoxic effects in human biochemistry are hugely unexplored (Duruibe et al., 2007). Their potential accumulation in biosystems and food chain has been documented. Thus a better understanding of heavy metal contamination profiles and their effect in water, soil and plant systems has attracted a lot of research interest (Prabu, 2009). In Pakistan toxic metals in both ground and surface waters, often exceed the maximum admissible concentrations recommended by WHO.

Cobalt

The result shows that maximum concentration of Cobalt was noted up to 7 µg/L in sample Station-1 in the month of August, 2011. However, at sampling Station-5 the minimum value of the cobalt was equal to 3 µg/L. The average value of the samples in the month of August noted 4.6 µg/L, whereas result obtained from the Month of November, 2011 was range as 4.5 to 9.5 µg/L. The average value in November was 7.1 µg/L (Tables 1 to 2). Cobalt content in Keenjhar Lake is Low as compare to that of Manchar Lake. Cobalt concentration were higher than that WHO/UNICEF (2004)) but our result are not in agreement with the result of Lashari et al. (2012) and Korai et al. (2008).

Lead

Child lead poisoning causes delayed brain development

Table 1. Heavy metals of Manchar Lake in the month of August 2011.

Sampling Stations	Co µg/L	Pb µg/L	Cr µg/L	Ni µg/L	Cu µg/L	Zn µg/L
MNVD 1	7	54.5	90	90	115	3000.90
Near zero point 2	5	52.5	25	30	110	3000
Mid of Manchar 3	6	50.5	50	15	100	3000.50
Danister wah 4	2	50	30	57	105.55	2000
Aral wah 5	3	51.5	70	60	90	2500
Average	4.6	51.8	53	50.4	104.11	2700.28
Max:	7	54.5	90	90	110	3000.90
Min:	2	50	25	15	90	2000
WHO Standards µg/L	N/A	0.05X1000= 50	0.05X1000= 50	0.02X1000 =20	1.0X1000 =100	3.0X1000 =3000

Table 2. Heavy Metals of Manchar Lake in the Month of November 2011.

Sampling Stations	Co µg/L	Pb µg/L	Cr µg/L	Ni µg/L	Cu µg/L	Zn µg/L
MNVD 1	9.5	58.5	95.5	92.5	117.5	3200
Near zero point 2	8.5	55.5	30.5	33.5	110.5	3100
Mid of Manchar 3	5.5	52.5	52.5	17.5	105.5	2300
Danister wah 4	7.5	45.5	30.5	63.5	107.55	3100
Aral wah 5	4.5	53.5	72.5	60.5	92.5	2700
Average	7.1	53.1	56.3	53.5	106.71	2880
Max:	9.5	58.5	95.5	92.5	117.5	3200
Min:	4.5	45.5	30.5	17.5	92.5	2700
WHO Standards µg/L	N/A	0.05X1000= 50	0.05X1000= 50	0.02X1000 =20	1.0X1000 =100	3.0X1000 =3000

or acute brain problems (Shi and Zhaoyu, 2008). It also affects the safety of aquatic organisms at a concentration of 0.16 mg/L. The result shows that maximum concentration of lead equal to 58.5 µg/L was found in sampling Station-1 of the Manchar Lake and minimum value equal to 45.5 µg/L in the month of August, 2011, (Tables 1 to 2). Maximum lead concentration of 0.235 µg/L and minimum lead concentration of 0.1 µg/L in Keenjhar Lake but Maximum lead concentration of 58.5 µg/L and minimum lead concentration of 45.5 µg/L in Manchar Lake, but our result are not agreement with the result of Lashari et al. (2012) and Korai et al. (2008).

Chromium

Chromium in humans causes mouth ulcers, nosebleeds, kidney disease, low white blood cell counts and a variety of cancers (Bradl, 2005; David and Pamela, 2002). The most bioavailable and most toxic form of chromium are the hexavalent ion Cr^{+6}. However, low concentrations of hexavalent chromium cause sub lethal toxic effects in aquatic plants and animals. For example, 62 ppb inhibits growth in algae and 16 ppb inhibits growth in chinook salmon (Taub, 2004). As is the case with other metals, chromium toxicity to aquatic organisms increases as water temperature increases and as pH and salinity decrease. Additionally, chromium is more toxic in soft water than in hard water. The concentration of chromium that caused death in 50% of the exposed population was 3 ppm in soft water and 72 ppm in hard water for fathead minnows and 18 ppm in soft water and 133 ppm in hard water for goldfish (Taub, 2004).

In the month of August the maximum concentration of chromium was found at Mid of Manchar sampling Station-1 which is equal to 90 µg/L and the minimum concentration was noted as 25 µg/L at near zero point sampling Stations-2. The average value was 53 µg/L in the August, 2011. Where is during same year in the month of November, samples were collected from the same stations to know the concentration of metals in the Manchar Lake. The results shown that the maximum concentration of Chromium was found up to 95.5 to 30.5 µg/L at sampling station -1. The average value was found equal to 56.3 µg/L. Chromium content in Keenjhar Lake water was maximum 4.9 µg/L and minimum 1.8 µg/L but Chromium content in Manchar Lake was maximum 95 µg/L and minimum 25 µg/L. Chromium level was below in Keenjhar Lake and Chromium level was high in Manchar Lake (WHO/UNICEF, 2004) but our result are not agreement with the result of Lashari et al. (2012), Ansari et al. (2004) and Anazawa et al. (2004).

Nickel

There is evidence suggesting that nickel may be an essential trace element for mammals (Goyer, 1991). Nickel is primarily found combined with oxygen or sulphur as oxides or sulphides that occur naturally in the earth's crust. Nickel combined with other elements is present in all soils, in meteorites, and is emitted from volcanoes. As for most metals, the toxicity of nickel is dependent on the route of exposure and the solubility of the nickel compound (Coogan et al., 1989). The route of exposure may be inhalation, oral, or dermal and can be classified according to systemic, immunologic, neurologic, reproductive, developmental, or carcinogenic effects following acute (01 day), subchronic (10 to 100 days), and chronic (100 days or more) exposure periods. The most common harmful health effect of nickel in humans is an allergic skin reaction in those who are sensitive to nickel. The metal is not only an allergen but also a potential immunomodulatory and immunotoxic agent in humans (Das and Buchner, 2007). Based on studies of nickel workers and laboratory animals, all nickel compounds, except for metallic nickel, have been classified as human carcinogens by the International Agency for Research on Cancer (IARC, 1990) and the U.S. Department of Health and Human Services (DHHS, 1994).

In the month of August, 2011 maximum concentration of nickel was found at Sampling Station-1 which is equal to 90 µg/L and the minimum concentration was noted as 15 µg/L at sampling Station-3. The average value was 50 µg/L in the month of August, 2011. In the month of November the concentration of nickel ranged from 17.5 to 92.5µg/L. The average value was found equal to 53.5 µg/L (Tables 1 to 2). Maximum Nickle in Keenjhar Lake 39.5 µg/L and minimum3.5 µg/L but Nickle in Manchar Lake was maximum 92.5 µg/L and minimum 15 µg/L. Nickle level was below in Keenjhar Lake and Nickle level was high in Manchar Lake (WHO/UNICEF, 2004) but our result are not in agreement with the result of Aamir and Tahir (2003) and Lashari et al. (2012).

Copper

Copper is an essential trace nutrient that is required in small amounts (5 to 20 µg per gram (µg/g) by humans, other mammals, fish and shellfish for carbohydrate metabolism and the functioning of more than 30 enzymes. It is also needed for the formation of haemoglobin and haemocyanin, the oxygen-transporting pigments in the blood of vertebrates and shellfish respectively. Most toxic form of copper is the cupric ion (Cu^{2+}). However, copper concentrations that exceed 20 µg per gram (µg/g) can be toxic, as explained by Bradl (2005) and David and Pamela (2002). Fish and crustaceans are 10 to 100 times more sensitive to the toxic effects of copper than are mammals. As copper, effects on gills, liver and kidneys of the fishes and decrease the sperm count as well respectively. There are many biological effects from copper toxicity at different concentration. Such as, copper concentration of 0.01 mg/L can inhibit water self-purification; at a concentration of 0.0002 mg/L, water starts to become toxic for fish. Copper can remain in soil, so sewage irrigation or sludge fertilizer could easily lead to soil pollution, hinder root development, and inhibit nutrient absorption and growth. Copper toxicity also affects aquatic organisms; in coastal and harbor areas the presence of copper once caused green oyster events. Moreover, an epidemiological survey found that areas where drinking water contains high level of copper are correlated with high rates of deaths from cardiovascular problems. Symptoms of acute copper poisoning include: low blood pressure, vomiting, melena, jaundice, hemolytic anemia, and coma to death (Shi and Zhaoyu, 2008). Copper is used in alloys, tools, coins, jewelry, food and beverage containers, automobile brake pads, electrical wiring and electroplating respectively.

The concentration of copper in the month of August, 2011 ranged from 90 to 105.55 µg/L, whereas the average concentration in the same month was recorded as 104.11. The maximum concentration of Copper was found at Sampling Station-1, 115 µg/L and the minimum concentration was noted at sampling Station-5, 90 µg/L. In the month of November the average concentration of copper was recorded as 106.71 µg/L, the maximum concentration of Copper was 117.5 µg/L at sampling Station-1, whereas the minimum concentration was 92.5 µg/L at sampling Station-5. The concentration of copper is high 7.3 µg/L and low 0.6 µg/L in Keenjhar Lake, but the concentration of copper is high 117.5 µg/L and low 90 µg/L in Manchar Lake. Copper concentration were higher (WHO/UNICEF, 2004).

Zinc

The human body contains 2 to 3 g of zinc with the highest levels in muscles, liver, kidneys, bones and prostate. The recommended daily zinc intake is 12 mg/day for adult women and 15 mg/day for adult men. Daily intake is depending on sex, age and general health status. Like as growing infants, children, adolescents, and women in pregnancy and the elderly have a higher zinc requirement. The uptake of elements by organisms is required in an optimal concentration range. As, when uptake is too low, deficiency occurs this includes reduced sense of taste and smell, skin disorders, mental lethargy and reduced fertility. On the other side, uptake of too much of an essential element can lead to toxicity. Such as, high doses can lead to gastro-intestinal disorders It is found naturally in rocks, air, water and soil. The average natural level of zinc in the earth's crust is 70 mg/kg (dry weight), ranging between 10 and 300 mg/kg (Malle, 1994).

The average concentration of Zinc in the month of August, 2011 was recorded as 2700.28 μg/L. The maximum concentration of Zinc was recorded at Sampling Station-1 which was up to 3000.90 μg/L and the minimum concentration was recorded at sampling Station-4, 2000 μg/L. Whereas, during same year 2011 in the month of November, the average concentration of Zinc was recorded as 2880 μg/L. In the month of November maximum concentration of zinc was recorded as 3200 μg/L sampling Station-1, whereas the minimum concentration of zinc was recorded as 2300 μg/L at sampling Station-3 (Tables 1 to 2). Zinc is high 104 μg/L and low 1.4 μg/L in Keenjhar Lake but zinc is high 3200 μg/L and low 2000 μg/L in Manchar Lake our result are not agreement with the result of Lashari et al. (2012), Ansari et al. (2004) and Anazawa et al. (2004).

Conclusion

From this study it was found that lake is contaminated with trace elements including lead, cobalt, zinc, nickel, copper and chromium. The concentration of these elements was high from the standard limit of World Health Organization. The aquatic organisms were found in danger other hand local human population is also at risk because they are utilizing the water of lake without any treatment. This lake also a major source of fishing and fish cached from lake may be highly contaminated by various toxic species. Hence it is highly recommended that government should take necessary steps to save the precious fresh water lake.

Conflict of Interest

The authors have not declared any conflict of interest.

REFRENCES

Aamir I, Tahir S (2003). Study of trace elements water in the vicinity of Palosi Drain, Peshawar. Pak J. Biol. Sci. 6(1):86-91. http://dx.doi.org/10.3923/pjbs.2003.86.91

Abdul QS, Tasneem GK, Muhammad BA, Muhammad KJ, Hassan IA, Nusrat J, Jamil AB Ghulam AK (2009). Accumulation of arsenic in different fresh water fish species – potential contribution to high arsenic intakes. Food Chem. 112(2):520–524. http://dx.doi.org/10.1016/j.foodchem.2008.05.095.

Anazawa K, Kaida Y, Shinomura Y, Tomlyasu T, Sakamota H (2004). Heavy metals distribution in river water and sediments around a "Firefly village" Japan: application of multivariate analysis. Anal. Sci. 201:79-84. http://dx.doi.org/10.2116/analsci.20.79

Ansari TM, Marr IL, Tariq N (2004). Heavy metals in Marine pollution perspective. A Mini Rev. J. App. Sci. 4(1):1-20. http://dx.doi.org/10.3923/jas.2004.1.20

Arain MB, Kazi TG, Baig JA, Jamali MK, Afridi HI, Shah AQ, Jalbani N, Sarfraz RA (2009). Determination of arsenic levels in lake water, sediment, and foodstuff from selected area of Sindh, Pakistan: Estimation of daily dietary intake, Food. Chem. Toxicol. 47(1):242–248. http://dx.doi.org/10.1016/j.fct.2008.11.009 PMid:19041679

Arain MB, Kazi TG, Jamali MK, Jalbani N, Afridi HI, Shah A (2008). Total dissolved and bioavailable elements in water and sediment samples and their accumulation in Oreochromis mossambicus of polluted Manchar Lake. Chemosphere 70(10):1845–1856. http://dx.doi.org/10.1016/j.chemosphere.2007.08.005 PMid:17889926

Arain MB, Kazi TG, Jamali MK, Jalbani N, Afridi HI, Sarfraz RA, Baig JA, Shah AQ (2009). Assessment of water quality of polluted lake using multivariate statistical techniques, A case study. Ecotox Environ. Safety. 72(2):301–309 http://dx.doi.org/10.1016/j.ecoenv.2008.02.024 PMid:18423587

Borgmann U, Norwood WP, Clarke C (1993). Accumulation, regulation and toxicity of copper, zinc, lead and mercury. Hyalella azteca Hydrobiologia. 259(2):79–89. http://dx.doi.org/10.1007/BF00008374

Bradl H (2005). Heavy metals in the environment: Origin, Interaction and Remediation. Elsevier/Academic Press, London PMCid:PMC1606398

Coogan TP, Latta DM, Snow ET, Costa M (1989). Toxicity and carcinogenicity of nickel compounds, In: McClellan RO, editor. Critical reviews in toxicology. 19, Boca Raton, FL. CRC Press, pp. 341-384 http://dx.doi.org/10.3109/10408448909029327 PMid:2663022

Das KK, Buchner V (2007). Effect of nickel exposure on peripheral tissues: Role of oxidative stress in toxicity and possible protection by ascorbic acid. Rev. Environ. Health. 22:133-49 http://dx.doi.org/10.1515/REVEH.2007.22.2.157

David AW, Pamela W (2002). Environmental Toxicology: Cambridge University Press.

Duruibe JO, Ogwuegbu MOC, Egwurugwu JN (2007). Heavy metal pollution and human biotoxic effects. Int. J. Phys. Sci. 2(5):112-118.

Elllis KN (1989). Surface water pollution and its control: Macmillan press ltd, hound mill, Basingstoke, Hampshire RG 21 London, 3-18:97-208.

Garrett RR (2000). Natural sources of metals to the environment, Hum Ecol. Risk Assess. 6:945-963 http://dx.doi.org/10.1080/10807030091124383

Goyer R (1991).Toxic effects of metals. In: Amdur MO, Doull JD Klaassen CD, editors. Casarett and Doull's toxicology, 4th ed. New York, Pergamon Press. pp. 623-680.

Haijian B, Yanhong W, Zhaobin S, Shuchun Y (2011). Historical trends of heavy metal contamination and their sources in lacustrine sediment from Xijiu Lake, Taihu Lake Catchment, China, J. Environ.Sci. 23(10):1671–1678. http://dx.doi.org/10.1016/S1001-0742(10)60593-1

Hari OS, Nepal MS, Aryo, Sindh N (1994). Combined effect of waste of distillery and sugar mill on seed germination, seeding growth and biomass of okara, J. Environ. Bio. 3(15):171-175.

IARC (1990). Monograph on the evaluation of carcinogenic risks to humans, 49, Lyans, France. 318-411. DHHS (1994). Seventh annual report on carcinogens: Summary 1994. Research Triangle Park, NC, USA: DHHS, National Inst. Environ. Health Sci. pp. 262-269.

Korai AL, Sahato GA, Kazi TG, Lashari KH (2008). Lead concentration in Fresh water, Muscle, and Liver of Catla Catla (Hamilton) from Keenjhar lake. Pak. J. Anal. Environ. Chem. 9(1):11-19.

Lashari KH, Sahato GA, Korai ALS, Habib N, Palh ZA, Urooj N (2012). Heavy metals burden of Keenjhar lake, District Thatta, Sindh, Pakistan. Afr. J. Biotechnol. 11 (59):12305-12313. http://dx.doi.org/10.5897/AJB11.1840

Ley QC, Zavala NAA, Espinosa-Carreon TL, Peckham H, Marquez-Herrera C, Campos-Villegas L (2011). Baseline heavy metals and metalloid values in blood of loggerhead turtles (*Caretta caretta*) from Baja California Sur, Mexico, Marine Poll. Bull. 62(9):1979–1983. http://dx.doi.org/10.1016/j.marpolbul.2011.06.022 PMid:21788056

Malle KG, Zink in der Umwelt (1994). Acta Hydrochim. Hydrobiol. 20:196-204.

Mason CF (1998). Biology of Fresh water Pollution. Longman Scientific and Technical.

Miranzadeh IB, Heidari M, Mesdaghinia AR, Yousesain M (2011). Survey of microbial quality of drinking water in rural areas of kashan Iran. Pak. J. Bio. Sci. 14(1):59- 63. http://dx.doi.org/10.3923/pjbs.2011.59.63 PMid:21913499

Prabu PC (2009). Impact of Heavy Metal Contamination of Akaki River of Ethiopia on Soil and Metal Toxicity on Cultivated Vegetable Crops. Elec. J. Environ. Agric. Food Chem. 8(9):818-827.

Rippey B, Rose N, Yang H, Harrad S, Robson M, Travers S (2008). An

assessment of toxicity in profundal lake sediment due to deposition of heavy metals and persistent organic pollutants from the atmosphere. Environ, Int. 34(3):345–356. http://dx.doi.org/10.1016/j.envint.2007.03.006 PMid:17467798
Sharma RK, Aqrawal M (2005). Biological effects of heavy metals: an overview, J. Env. Bio. 26(2):301–313. PMid:16334259
Shi B, Zhaoyu, Lu Z (2008). Environmental Pollution and Human Health, China Environmental Press.
Taub FB (2004). Fish 430 lectures (Biological Impacts of Pollutants on Aquatic Organisms), University of Washington College of Ocean and Fishery Sciences, Seattle, WA.
UNICEF, WHO (2009). Diarrhea: Why Children are still dying and what can be done. APHA (American Public Health Association), Standard methods for the examination of water and waste water. 1992.
Waheed T, Kausar T (1987). Quality of drinking water in Lahore. Pak. J. Med. Res. 26:162-165.
WHO/UNICEF (2004). Meeting the MDG drinking water and sanitation: a mid-term assessment of progress, Geneva: Switzerland.
World health organization (1996). Guidelines for drinking water quality, Quality health criteria and other supporting information: world health organization, Geneva, Switzerland second edition.
World health organization (2003). Guidelines for drinking water quality: world health organization, Geneva, Switzerland.
World health organization (2004). Guidelines for drinking water quality: world health organization, Geneva, Switzerland. 3rd edition.
Wright DA, Pamela W (2002). Environmental Toxicology. Cambridge University Press, Cambridge, U.K. http://dx.doi.org/10.1017/CBO9780511805998

Phytochemical screening of the leaf extracts of *Hyptis spicigera* plant

Z. Ladan[1], J. O. Amupitan[2], O. A. Oyewale[2], R. G. Ayo[3], E. Temple[2] and E. O. Ladan[4]

[1]National Research Institute for Chemical Technology, Private Mail Bag 1052, Zaria, Nigeria.
[2]Department of Chemistry, Ahmadu Bello University, Zaria, Nigeria.
[3]Division of Agricultural Colleges, Ahmadu Bello University, Zaria, Nigeria.
[4]National Agricultural Extension and Research Liaison Services, Ahmadu Bello University, Zaria, Nigeria.

The present study reports the screening of phytochemical constituents of the leaf extracts of *Hyptis spicigera* using hexane, ethylacetate and methanol and the leaf powder of the plant. Qualitative analysis of phytochemical constituents showed the presence of the following secondary metabolites vitannins, carbohydrates, saponins, flavonoids, steroids, alkaloids, quinones, coumarin, terpenoids, resins and cardiac glycosides. The quantitative analysis of total phenolics, alkaloids, saponins, terpenoids and flavonoids carried out using standard protocols revealed the presence of flavonoids (8.82%), saponins (6.23%), terpenoids (16.10%), alkaloids (7.55%) and phenolics (20.75%) respectively. Phenolics showed the highest content (20.75%) while saponins (6.23%) gave the least content. The high content of phenolics in the plant showed that *H. spicigera* plant may contain antioxidant properties and could be a good source of natural antioxidants. Also, the richness in flavonoids, saponins, alkaloids and terpenoids in this plant can be correlated with its medicinal properties used by traditional herbal healers in Northern Nigeria.

Key words: *Hyptis spicigera*, phytochemical screening, secondary metabolites.

INTRODUCTION

Plants have been the subject of human curiosity and use for thousands of years (Ram et al., 2004) and have played important roles in many countries of the world for centuries by providing food, shelter, clothing, agrochemicals, flavours and fragrances and more importantly, medicines (Gurib-Fakim, 2006).Traditional people have relied on medicinal plants to combat various ailments caused bymicroorganisms such as bacteria, fungi and viruses that infect the body system. Plants have indeed formed the basis of sophisticated traditional medicine systems which will continue to provide mankind with new remedies for all forms of ailments (Gurib-Fakim, 2006).

Bioactive natural products have enormous economic

importance as specialty chemicals as they can be used as drugs, lead compounds, biological or pharmaceutical tools, feed stock products, excipients and nutraceuticals (Pieters and Vlietinck, 2005). In recent times, focus on plant research has increased all over the world and a large body of evidence and knowledge has accumulated in the literature to show immense potential of medicinal plants used in various medical, pharmaceutical, cosmetic, agrochemical applications. An advantage of natural bioactive molecule is that they have a milder side effect on the body in comparison to chemically synthesized drugs (Badisa et al., 2003). With the increasing acceptance of herbal medicines as alternative form of health care delivery, the screening of medicinal plants for bioactive compounds is imperative (Masoko et al., 2005; Cowan, 1999).

Hyptis spicigera belongs to the family Lamiaceae and is commonly known as Black beniseed, or Black sesame. It is an erect aromatic herb, up to 1 m in height, with a terminal inflorescence in which the seeds are packed in quadruplets or more in the flowers. The plant is found around Senegal to western Cameroon, possibly native to Brazil, now widely naturalized in tropical Africa and Asia as well as Nigeria. It grows naturally and commonly as a weed. It prefers roadsides, waste places, cultivated places and often damp places (Burkill, 1995). Generally, the whole plant is used in traditional stores to protect cowpea against damage by *Callosobruchus* species (Lambert et al., 1985). The Bajju and Atyapp people of Kaduna state, northern Nigeria, make use of the inflorescence (where the seeds are packed) to cure headaches by sniffing it and also crushing the leaves and applying to the head to relieve head colds and headaches (Dalziel, 1937).

This paper reports the phyto constituents of the leaf extracts of *H. spicigera* and their potential medicinal applications.

MATERIALS AND METHODS

Collection of the plant

About 500 g of the leaf part of *H. spicigera* was collected in Basawa Village, Zaria, Kaduna state, Nigeria on the 26th November, 2013. It was taxonomically identified and authenticated by Mallam U. S. Gallah of the Herbarium Section, Department of Biological Science, Ahmadu Bello University Zaria, Nigeria, and a sample Voucher No.528 was deposited at the Herbarium section of the Department of Biological sciences.

Extraction and isolation

The plant was dried in the shade for 14 days and pulverized to powder using pestle and mortar in the Laboratory. Approximately 450 g of the powdered plant material was macerated sequentially with hexane (1 L), ethyl acetate (1 L) and methanol (1 L) at room temperature (27°C) and concentrated *in vacuo* to afford the various crude extracts which were stored in the refrigerator (4°C) until needed for further analysis.

Phytochemical screening

Phytochemical analysis of the crude extracts was carried out according to standard methods (Harborne, 1998; Sofowora, 1993; Fanrsworth, 1996; Rangari, 2002).

Salkowski reaction test for phytosterols

To 0.5 ml each of the extracts in a test tube was added 1.0 ml of concentrated H_2SO_4 (conc.) from the sides of the test tube and then 1.0 ml chloroform. Appearance of reddish brown colour in chloroform layer indicates the presence of phytosterols.

Liebermann-Burchard's test for triterpenoids

Extracts were treated with few drops of acetic anhydride, boil and cool. Conc. sulfuric acid was added from the sides of the test tubes which showed a brown ring at the junction of two layers, and formation of deep red color indicated the presence of triterpenoids.

Foam test for saponins

Small amount (0.1 g) of the extracts were taken in test tubes with little quantity (1.0 ml) of water and shaken vigorously. Appearance of foam persisting for 10 min indicated presence of saponins.

Dragendroff's test for alkaloids

About 0.5 g each of hexane, ethyl acetate and methanol extracts were dissolved in 1.0 ml chloroform and evaporated. The residue was acidified by adding few drops of Dragendroff's reagent (Potassium bismuth iodide). Appearance of orange red precipitate indicated presence of alkaloids.

Molisch's test for carbohydrates

About 0.5 g each of the extracts was mixed with Molisch reagent, and then added H_2SO_4 conc. along the sides of the test tube to form layers. Appearance of reddish violet ring the interference indicated the presence of carbohydrates.

Lead acetate test for flavonoids

To 0.1 g each of the extracts were dissolved in ethanol and few drops of 10% lead acetate solution were added. Appearance of yellow precipitate indicated presence of flavonoids.

Legal's test for lactones

To 0.1 g each of the extracts1.0 ml sodium nitroprusside and1.0 ml pyridine were added in test-tubes. The mixtures were treated with 0.01 moldm^{-3} NaOH. Appearance of deep red colour indicated the presence of lactones.

Ferric chloride test for phenolic compounds and tannins

About 2.0 ml of each extract was measured in a test tube and 0.01 mol dm^{-3}Ferric chloride solution was added drop by drop. Appearance of bluish black precipitate indicated presence of phenolic compounds and tannins.

Ninhydrin test for proteins

Few drops of ninhydrin were added to the extracts. Appearance of blue colour indicated presence of amino acid. Proteins may rarely give positive result with this test.

Keller-Killiani test for glycosides

About 1 ml of glacial acetic acid, few drops of 0.01 mol dm^{-3}Ferric chloride solution and H_2SO_4 (Conc) slowly through the sides of the test tube) were added to the extracts. Appearance of reddish brown ring at the junction of the liquids indicated the presence of de-oxysugars.

Quantitative determination of phytochemical constituents

Determination of total phenolic compound (TPC)

Total phenolic content of the hexane, ethylacetate and methanolic extracts was determined by standard method (Makkar et al., 1993) with little modifications, using tannic acid as a standard phenolic compound. The extracts were diluted with distilled water to a known concentration in order to obtain the readings within the standard curve range of 0.0 to 600 µg of tannic acid/ml. 250 µl of diluted extract or tannic acid solution was mixed with 1 ml of distilled water in a test tube followed by the addition of 250 µl of Folin-Ciocalteu reagent. The samples were mixed well and then allowed for 5 min at room temperature in order to allow complete reaction with the Folin-Ciocalteu reagent. Thereafter, 2.5 ml of 7% sodium carbonate aqueous solution was added and the final volume was made up to 6.0 ml with distilled water. The absorbance of the resulting blue colour solution was measured at 760 nm using spectrophotometer after incubating the samples for 90 min. All the experiment was conducted in three replicates.

Determination of alkaloids

About 5.0 g of the dried powdered plant was weighed into a 250 ml beaker and 200 ml of 10% acetic acid in ethanol was added. The mixture was covered and allowed to stand for 4 h. This was filtered and the extract concentrated on a water bath to one-quarter of the original volume. Concentrated ammonium hydroxide solution was added dropwise to the extract until the precipitation was complete. The solution was allowed to settle and the precipitate was collected and washed with dilute ammonium hydroxide and finally filtered, dried, weighed and the percentage alkaloid was calculated (Harborne, 1998).

Determination of total terpenoids

About 2 g of the plantleaf powder was weighed and soaked in 50 ml of 95% ethanol for 24 h. The extract was filtered and the filtrate extracted with petroleum ether (60 to 80°C) and concentrated to dryness. The dried ether extract was treated as total terpenoids (Ferguson, 1956).

Determination of saponins

About 15 g of each sample was placed into a conical flask and 100 ml of 20% aqueous ethanol was added. The samples were heated over a hot water bath for 4 h with continuous stirring at 55°C. The mixture was filtered and the residue re-extracted with another 200 ml and 20% ethanol. The combined extracts were reduced to 40 ml over water bath at 90°C. The concentrate was transferred into a 250 ml separating funnel and 20 ml of diethyl ether was added and shaken vigorously. The aqueous layer was recovered while the ether layer was discarded. This purification process was repeated and 60 ml of n-butanol was added. The combined n-butanol extracts were washed twice with 10 ml of 5% aqueous sodium chloride. The remaining solution was heated in a water bath. After evaporation the samples were dried in the oven to a constant weight and the percentage saponin was calculated (Obdoni and Ochuko, 2001).

Determination of flavonoids

About 5.0 g of the plant sample was weighed and extracted repeatedly with 100 ml of 80% aqueous methanol at room temperature. The whole solution was filtered through whattman filter paper No 41. The filtrate was evaporated into dryness over a water bath and weighed to a constant weight. The percentage flavonoids was then calculated (Soni and Sosa, 2013).

RESULTS AND DISCUSSION

The extracts of the leaves of *H. spicigera* were screened for the presence ofthe following secondary metabolites: alkaloids, glycosides, flavonoids, carbohydrates tannins, steroids, terpenoids and resins, coumarins, saponins and quinines. The results of the phytochemical screening showed the presence of all the secondary metabolites analyzed in ethylacetate and methanol extracts while hexane extract showed only the presence of alkaloids, glycosides, carbohydrate and resins. Other secondary metabolites such as flavonoids, tannins, steroids, terpenoids, coumarins, saponins and quinones were absent in the hexane extract.

The phytochemical content was found to be similar to that obtained by other authors (Onayade et al., 1991) with different extracts revealing the different partitioning abilities of the different solvents used. The presence of these phytochemicals in all the extracts is quite instructive as this lends credence of the use of the plant for medicinal purposes. A lot of plants contain non-toxic glycosides that can be hydrolyzed to give phenolic compounds that are toxic to microbial pathogens (Abaoba and Efuwape, 2001). The saponin content in the ethylacetate and methanol extracts were found to be present in these extracts, respectively. Saponins possess the property of precipitating and coagulating red blood cells (Sodipo et al., 1991). It also foamed in aqueous solution and has hemolytic effect and can also bind on cholesterol sites. These properties make saponins present in the plant to exhibit medicinal properties (Sodipo et al., 1991) and this therefore supports the findings in this present study that extracts of the plants may be useful in chemotherapy of mycotic infections which the antimicrobial studies revealed (Ladan et al., 2009). Alkaloids were found present in hexane, ethylacetate and methanol extracts and this can be corroborated with literature reports which indicate that naturally occurring alkaloids and their synthetic derivatives

Table 1. Qualitative phytochemical analysis of the leaf extracts of *Hyptis spicigera.*

Phytochemical constituents	Leaf extracts		
	Hexane	Ethylacetate	Methanol
Tannins	-	+	+
Saponins	-	+	+
Steroids	-	+	+
Flavonoids	-	+	+
Alkaloids	+	+	+
Terpenoids	-	+	+
Glycosides	+	+	+
Coumarins	-	+	+
Carbohydrates	+	+	+
Quinones	-	+	+
Resins	+	+	+

have analgesic, antispasmodic and bactericidal activities (Okwu and Okwu, 2004). They exhibit marked physiological activity when administered to animals. Classes of alkaloids are among the major powerful poisons known and despite being poisonous, some of the alkaloids are known to be useful in correcting renal disorders (Konkwara, 1979). The use of some plants for medicinal purpose, in the traditional treatment of diseases is due to the presence of flavonoids and saponins (Zwadyk, 1992; Othira et al., 2009), hence the use of *H. spicigera* for the treatment of diarrhea, dysentery, colds and several other diseases by local herbalists or traditional healers is not surprising. The presence of flavonoid was evident in methanol and ethylacetate extracts, flavonoid containing plants have been used as diuretic, laxative, emollient and poultice (Baba-Mousa et al., 1999) therefore; the use of *H. spicigera* rich in saponins and other *Hyptis* species in traditional medicine lent credence to the medicinal potentials of the plant. Tannins in some medicinal plants have been found to be responsible for the antiviral and antibacterial activities exhibited by such plants (De-Ruiz et al., 2001; Elegani et al., 2002). Therefore, *H. spicigera* with high tannin content in ethylacetate and methanol extracts could probably be a source of phytochemicals for the treatment of bacterial infections. Phenolic compounds like tannins present in plant cells are inhibitors of many enzymes (proteolytic and hydrolytic) used by plant pathogens. Other compounds such as saponins have antifungal properties (Abaoba and Efuwape, 2001; Mohanta et al., 2007). Therefore, these phytochemicals detected in this study may be responsible for the antimicrobial potency of the leaf extracts of *Hyptis spicigera* and also lend credence to the claims of traditional application of the plant as remedies for various ailments.

Quantitative phytochemical analysis

Results of the quantitative analysis data of the plant material revealed significant levels of phytochemical constituents present in the leaf as evident in the qualitative analysis data (Table 1). Phenolic content (20.75%) is the highest followed by terpenoids content (16.10%) while flavonoid (8.82%), saponins (6.23%) and alkaloids (7.55%) followed respectively. Subhashini et al. (2013) and Soni and Sosa (2013) have reported various phyto constituents in the leaves of *Ecbolium viride* (Forks) Merrill plant and the methanolic and ethyl acetate extracts of the leaves of *Anogeissus leiocarpus* and found the following values: terpenoids (0.3034 w/w), saponins (0.1100 w/w), alkaloids (0.1340 w/w), flavonoids (0.0884 w/w) and phenols (0.03045) for the *E. viride* (Forks) while alkaloids (152.0 ± 0.1 mg/g), phenolics 1294.81± 3.0 mg/g), flavonoids (330.7 ± 3.0 mg/g) in the methanol extract and alkaloids (80.20 ± 0.0 mg/g), phenolics (616.5 ± 4.4 mg/g), flavonoids (202.5 ± 4.0 mg/g) in the ethyl acetate extract of the *A. leiocarpus* plant have been reported. The quantitative values of these metabolites reported in the leaf part of *H. spicigera*are higher (Table 2) than those reported for *E. viride* (Forks) Merrill and *A. leiocarpus* plants. Phenolic compounds are one of the most important constituents of plant secondary metabolites with marked physiological properties.

The phyto constituents found in the plant may be responsible for its biological properties such as anti-oxidative, anti-inflammation, anti-carcinogenic, anti-hypertensive, anti-diabetic, anti-cancer, cardiovascular protection and improvement of endothelial function (Han et al., 2007). Several studies have described the anti-oxidant properties of different parts of various medicinal plants which are rich in phenolic compounds (Brown and Evans, 1998; Krings and Berger, 2001; Malencic et al., 2007). Natural anti-oxidants mainly come from plants in the form of phenolic compounds, such as flavonoids, phenolic acids, tocopherolsetc (Ali et al., 2008) and used for the treatment of degenerative diseases. The anti-oxidative properties of flavonoids are due to several

Table 2. Quantitative analysis of the leaf part of *Hyptisspicigera*.

Phytochemical constituents	Yield (g)	Yield (%)
Total Flavonoids	0.44	8.82
Terpenoids	0.32	16.10
Total Saponins	0.94	6.23
Total Alkaloids	0.38	7.55
Total Phenolics	0.62	20.75

different mechanisms, such as scavenging of free radicals, chelation of metal ions, such as iron and copper and inhibition of enzymes responsible for free radical generation (Akinmoladun et al., 2007; Benavente-Garcia et al., 1997). This plant (*H. spicigera*) will provide the natural anti-oxidant needed to enhance good living by scavenging free radicals that cause ill health in humans.

Conclusion

Phytochemical screening of the leaf part of *H. spicigera* revealed the presence of tannins, carbohydrates, saponins, flavonoids, steroids, alkaloids, quinones, coumarin, terpenoids, resins and cardiac glycosides which are important secondary metabolites. The richness of the plant in phenolic contents and other secondary metabolites affirmed its medicinal efficacy and potentials. The finding from this study therefore suggests that the leaf could be a potential source of natural anti-oxidant that could have great importance as therapeutic agents in preventing or slowing ageing associated with oxidative stress and related degenerative diseases. It is recommended that further investigation on the isolation and characterization of the bioactive constituents of the leaf leading to structural elucidation is necessary.

Conflict of Interests

The author(s) have not declared any conflict of interests.

ACKNOWLEDGEMENT

The authors are grateful to the Management of National Research Institute for Chemical Technology (NARICT), Zaria, Nigeria for providing necessary chemicals and equipment for the conduct of this research work.

REFERENCES

Abaoba OO, Efuwape BM (2001). Antibacterial properties of some Nigerian species. Biol. Res. Comm. 13:183-188.

Akinmoladun AC, Ibukun EO, Afor E, Akinrinlola BL, Onibon TR, Akinboboye AO (2007). Chemical constituents and antioxidant activity of Alstoniaboonei. Afr. J. Biotechnol. 6(10):1197-1201.

Ali SS, Kasoju N, Luthra A, Singh A, Sharanabasava H, Sahuand A (2008). Indian medicinal herbs as source of antioxidants. Food Res Int. 41:1-15. http://dx.doi.org/10.1016/j.foodres.2007.10.001

Baba-Mousa F, Akpagana K, Bouchet P (1999). Antifungal activities of seven West African Combretaceace used in traditional Medcine. J. Ethnopharm. 66:335-338. http://dx.doi.org/10.1016/S0378-8741(98)00184-6

Badisa RB, Tzakou O, Couladis M, Pilarinou E (2003). Phytotherapy Research "Cytotoxic Activities of some Greek labiatae Herbs" 17(5).Wiley and sons Ltd. U.S.A. pp. 472-475.

Benavente-Garcia O, Castillo J, Marin FR, Ortuno A, Del-Rio JA (1997). Uses and properties of Citrus flavonoids. J. Agric. Food Chem. 45(12):4505-4515. http://dx.doi.org/10.1021/jf970373s

Brown JE, Rice-Evans CA (1998). Luteolin rich artichoke extract protects low density lipoprotein from oxidation *in vitro*. Fr. Rad. Res. 29:247-255. http://dx.doi.org/10.1080/10715769800300281

Burkill HM (1995). The useful plants of West Tropical Africa Royal Botanic Gardens Kew 4(2):88-144.

Cowan MM (1999). Plant products as antimicrobial agents. Clin. Microb. Rev. 12(4):564-582. PMid:10515903, PMCid:PMC88925

Dalziel JM (1937). Useful plants of West Tropical Africa. Crown Agent for Overseas Governments, London. pp. 462-463.

De-Ruiz REL, Fusco RMD, Angela S, Sohar OR (2001). Isolation of flavonoids and anthraquinones of *Amaranthus muricatus* (Moquin) ex Hicken (Amarathaceae). Acta Farm. Bon. 20:9-12.

Elegani AA, El-nima MSEI, Muddathir AK (2002). Antimicrobial activity of some species of the family Combretaceae. Phytoth. Res. 16:551-561.

Fanrsworth NR (1996). The ethno-botanical approach to drug discovery: strengths and Limitations, In: Prance G.T.9Eds), Ethno-botany and the search for new drugs. In: Ciba. pp. 100-150.

Gurib-Fakim A (2006). Medicinal plants: Traditions of yesterday and drugs of tomorrow. Mol. Asp. Medi. 27:1-93. http://dx.doi.org/10.1016/j.mam.2005.07.008, PMid:16105678

Han X, Shen T, Lou H (2007). Dietary polyphenols and their biological significance. Int. J. Mol. Sci.950-988.

Harborne JB (1998). Phytochemical methods. Chapman and Hall Ltd., London. pp. 100-200.

Konkwara JO (1979). Medicinal plants of East Africa. Literature Burea, Nairobi. pp. 3-15.

Krings U, Berger RG (2001). Antioxidant activity of roasted foods. Food. Chem. 72:223-229. http://dx.doi.org/10.1016/S0308-8146(00)00226-0

Ladan Z, Amupitan JO, Okonkwo EM, Aimola IA, Habila N (2009). Antimicrobial potency of *Hyptis spicigera* leaf extracts against some pathogenic microorganisms. J. Med. P. Res. 3(11):905-908.

Lambert J, Arnason JT, Philoge-Ane BJR (1985). Bruchid control with traditionally used insecticidal plants *Hyptis spicigera* and *Cassia nigricans*. Ins. Sci. Appl. 6:167-170.

Makkar H, Bluemmel M, Borowy N, Becker K (1993). Gravimetric determination of tannins and their correlations with chemical and protein precipitation methods. J. Sci. Food Agric. 61:161-165. http://dx.doi.org/10.1002/jsfa.2740610205

Malencic D, Popovic M, Miladinovic J (2007). Phenolic content and antioxidant properties of soybean (*Glycine max*, L. Merr Seeds. Mol. 12:576-581. http://dx.doi.org/10.3390/12030576

Masoko P, Picard J, Eloff JN (2005). Antifungal activities of six South

African Terminalia species. J. Ethnopharmacol. 99:301-308. http://dx.doi.org/10.1016/j.jep.2005.01.061

Mohanta TK, Patra JK, Rath SK, Pal DK, Thatoi HN (2007). Evaluation of antimicrobial activity and phytochemical screening of oils and nuts of *Semicarpus anacardium* L. Sci. Res.Ess. 2(11):486-490.

Obdoni BO, Ochuko PO (2001). Phytochemical studies and comparative efficacy of the crude extracts of some homostatic plants in Edo and Delta States of Nigeria. Glob. J. Pure Appl. Sci. 8b:203-208.

Okwu DE, Okwu ME (2004). Chemical composition of Spondiasmombinlinn plant parts. J. Sus. Agric. Environ. 6(2):140-147.

Onayade OA, Looma A, Scheffer JJ, Svendsen AB (1991). Composition of the herb essential oil of *Hyptis spicigera* Lam. Flav. Frag. J. 5:101-105. http://dx.doi.org/10.1002/ffj.2730050209

Othira JO, Onek LA, Deng LA, Omolo EO (2009). Insecticidal potency of *Hyptis spicigera* preparations against *Sitophilus zeamais* I and *Tribolium castaneum* (herbst) on stored maize grains. Afr. J. Agric. Res. 4(3):187-192.

Pieters L, Vlietinck AJ (2005). Bio-guided isolation of pharmacologically active plants components, still a valuable strategy for the finding of new lead compounds. J. Ethnopharmacol. 100:57-60. http://dx.doi.org/10.1016/j.jep.2005.05.029

Ram AJ, Bhakshu LM, Raju RRV (2004). *In vitro* antimicrobial activity of certain medicinal plants from Eastern Ghats India used for skin diseases. J. Ethnopharm. 90:353-357.

Rangari VD (2002). Pharmacognosy and phytochemistry. Nasik:Carrier Pub. pp. 100-150.

Sodipo OA, Akanji MA, Kolawole FB, Adetuga OO (1991). Saponins is the active antifungal principle in Garcinia kola, heckle seed. Biol. Sci. Res. Comm. 3:171.

Sofowora A (1993). Medicinal plants and traditional medicine in Africa, New York, John Wiley and Sons, pp. 191-300.

Soni A, Sosa S (2013). Phytochemical analysis and free redical scavenging potential of herbal and medicinal plant extracts. J. Pharm. Phytochem. 2(4):22-29.

Subhashini S, Poonguzhali TV Madha VS (2013). Quantitative phytochemical analysis of *Ecbolium viride* (Forks) Merrill and *Justicia gendarussa* Burm F. Int. J. Cur. Tr. Res. 2(1):34-37.

Zwadyk P (1992). Enteriobactericeae in Zinsser Microbiology, 20th Ed. Gerogthieme Verlag, Stuggart. P. 87.

11

Determination of some heavy metals in selected edible vegetables grown along River Yedzaram in Uba area Adamawa State, Nigeria

Alexander P. and Ubandoma W. H.

Department of Chemistry, Adamawa State University, Mubi, Nigeria.

The levels of some heavy metals were investigated in selected edible potions of the vegetables; *Amarathus caudatus* (Spinach) and *Hibiscus sabdariffa* (Rosella) are grown in Uba area along the Yedzaram River in North Eastern Adamawa State, Nigeria. All samples were randomly collected from two different gardens. The levels of the heavy metals, (Cu, Fe, Cd, Cr and Zn) were analyzed using Atomic Absorption Spectrophotometer (AAS) (BUK 210 model). In all the samples analyzed, Cd and Cr were not detected. The levels of heavy metals in Farm A for *H. sabdariffa* leaves ranges from Cu (30.00 ± 0.15 mg/kg to 31.00 ± 0.18 mg/kg), Fe (37.39 ± 0.02 mg/kg to 48.47 ± 0.10 mg/kg), Zn (13.00 ± 0.01 mg/kg to 25.50 ± 0.48 mg/kg), respectively. In Spinach, the results ranged from Cu (34.33 ± 0.42 mg/kg to 34.50 ± 0.05 mg/kg), Fe (31.72 ± 0.71 mg/kg to 43.33 ± 0.02 mg/kg), Zn (21.17 ± 0.14 mg/kg to 10.83 ± 0.17 mg/kg), respectively. The data were analyzed with t-test and analysis of variance (ANOVA). There were significant differences ($p < 0.05$) between the levels of the heavy metals in the vegetables obtained from Farms A and B. The order of the metal contamination in the vegetables was Fe > Cu > Zn in Farm A and Cu> Fe> Zn in Farm B. The elevated levels of metals in vegetables in the two gardens could be attributed to excessive usage of fertilizers and other agro - chemicals and of course the environmental factors of the areas. The results were however lower than the published threshold values considered toxic for mature plant tissue, except Fe which has higher values. The consumption of these vegetables as food may not pose possible health hazards to human at the time of the study.

Key words: Heavy metals, *Amarathus caudatus*, *Hibiscus sabdariffa*, Uba area, fertilizers.

INTRODUCTION

Vegetable is a plant or part of a plant used as food, typically as accompany to meat or fish, such as cabbage, potato, carrot or beans (Ihekoroye and Ngoddy, 1985). Eating vegetables regularly in diet can have many health benefits by reducing many health related diseases and used to convert the fats and carbohydrates into energy (Milk, 2012). Eating vegetables is one of the most important pathways for the human body to absorbed dietary mineral, necessary for its healthy developmentbut unfortunately harmful element such as heavy metals which may lead to intoxication and with prolong accumulation are being found in these vegetables (Elsevier, 2008).

However, heavy metals concentrations in soil are associated with biological and geochemical cycles and are influenced by anthropogenic activities such as agricultural practices, industrial activities and waste disposal methods (Ndiokwere and Ezeh, 1990; Usman and Ayodele, 2002; Uwah et al., 2009). Contamination and subsequent pollution of the environment by heavy metals have become a global concern due to their distribution and multiple effects on the ecosystem (Nriagu, 1990). Heavy metals are present in agricultural soils at low levels. Due to their cumulative behaviours and toxicity, they have potential, hazardous effect not only on plants but on human health (Das *et al.,* 1997). Distributions of heavy metals in plants depend upon availability and concentration of heavy metals as well as particular plant species (Punz and Seighardt, 1993). Many researchers have shown that some common vegetables are capable of accumulating high levels of metals from the soil (Xiong, 1998; Uwah et al., 2009). Certain species of Brassica (cabbage) are hyper-accumulators of heavy metals in their edible tissues (Xiong, 1998). Many people could be at risk of adverse health effects from consuming common vegetable cultivated in contaminated soil (Nirmal et al., 2007).

The populations mostly affected by heavy metals toxicity are pregnant women or very young children (Boon and Soltanpour, 1992). Neurological disorders, central nervous system (CNS) destruction and cancers of various body organs are some of the report of heavy metals poisoning (Agency for Toxic Substance and the Disease Registry (ATSDR) 1999a, b: 2000). Low birth weight and severe mental retardation of newly born children have been reported in some cases where pregnant women ingest toxic amount of heavy metal through direct or indirect means (Mahaffey et al., 1981).

Heavy metals like Fe, Cu, Zn, and Ni, are important for proper functioning of biological systems and their deficiency or excess could lead to a number of disorders (Ward, 2005; Uwah et al., 2009). Industrial, urban wastes and agricultural application and also mining activities results in an increased concentration of heavy metals in both soil and plant. Heavy metals pollute both soil and plant and so it is necessary to examine the state of the polluted soil and plant and establish what influence heavy metals have on both. Heavy metals have great significance due to their toxicity and accumulative behaviour or and are not biodegradable (Shinggu et al., 2007). Surface soil may act as carriers and possible source of pollution, since the mobility of these metals is such that remain in upper layers without regard to soil type.

Moreover, these metals are not permanently fixed and can be released by changes in climatic or environmental condition such as rainfall (Nriagu and Pacyna, 1990). The main sources of roadside contaminants are the deposition of aerosol particles which are adhesive in nature. But in the urban environment, these particles originate mainly from road traffic, welding, emission from industries, construction activities and flaking of paint (Radojavice and Bashkin, 1999).

The use of polluted water in the immediate surroundings of big cities for growing of vegetables is a common practice in Nigeria. Although this water is considered a rich source of organic matter and plant nutrients, it also contains sufficient amount of soluble salts and metal like Fe, Mn, Cu, Zn, Pb, Ni, Sn, Hg, Cr, As and Al. When such water is used for irrigation of crops for a long period, these heavy metals may accumulate in soil and may be toxic to the plants and also cause deterioration of soil (Kirkham, 1983; Uwah, 2009).

Heavy metals contamination and pollution of environment has become a global concern, due to their distribution and multiple effects on the ecosystem, waste waters are highly use in agricultural irrigation and long-term usage of these waste waters on agricultural lands often results in the build-up of elevated levels of heavy metals in soils (Rattan et al., 2001). Crops usually cultivated on the metals contaminated soils accumulate these metals in excessive quantities are enough to cause clinical problems both to animal and human beings consuming these metals rich plants.

The study is aimed at investigating the levels of some heavy metals such as (Cu, Cd, Fe, Cr and Zn) in edible portions of spinach *(Amarathus caudatus)* and Rosella leaves *(Hibiscus sabdariffa),* cultivated along River Yedzaram in Uba area. Extrapolate the results and ascertained the suitability or otherwise of the vegetables for human consumptions.

This was carried out by analyzing spectrophotometrically the levels of the metals in the vegetable samples.

MATERIALS AND METHODS

Analytical reagent (AnalaR) grade chemicals and distilled water were used throughout the study. All glassware and plastic containers used in this work were washed with detergent solution followed by 20% (v/v) nitric acid and then rinsed with tap water and finally with distilled water.

Study area

Uba region geographically is located in the North-Eastern part of Borno State and Adamawa State, in North-eastern Nigeria. Its geographical coordinates are 10° 27' North and 13° 17' East of the Greenwich meridian. Uba region occupies land area of 2,362 km^2 and a population of 138,091 (Wandeo, 2005).

Table 1. Concentrations of some heavy metals in Spinach and Rosella from Farm A (mg/kg).

Vegetables/Sampling sites	Cd	Cr	Cu	Fe	Zn
Rosella (*Hibiscus sabdariffa*)					
Location 1	ND	ND	31.00 ± 0.18	48.67 ± 0.1	13.00 ± 0.01
Location 2	ND	ND	30.00 ± 15.00	37.39 ± 0.02	25.50 ± 0.48
Spinach (*Amaranthus caudatus*)					
Location 1	ND	ND	34.33 ± 0.42	43.33 ± 0.02	10.83 ± 0.17
Location 2	ND	ND	34.50 ± 0.05	31.72 ± 0.71	21.17 ± 0.14
WHO/FAO			20 - 100	10 -18	3 - 20
NAFDAC			0 - 40	10 - 20	0 - 50

All values represent mean ± standard deviation of triplicate determination. **ND** = Not detected, **WHO** (1996) = World Health organization, **NAFDAC** = National Agency for Food and Drug Administration and control.

Sampling and sample treatment

The samples analyzed include *H. Sabdariffa* (Spinach) and *A. caudatus* (Rosella) leaves. Samples were collected from May to September, 2012 from two different Farms (A and B) along the River Yedzaram Uba area. Edible portions of the fresh samples of *A. caudatus* (Rosella) and *H. sabdariffa* (Spinach) were randomly collected (handpicked) from two different vegetable Farms (A and B), which supply most of the vegetables consumed in Uba. The samples were wrapped in big brown envelopes and labeled. Only fresh vegetables in good conditions were collected in order to produce good quality dried product (Audu and Lawal, 2005). A total of 10 samples each of *A. caudatus* and *H. sabdariffa* from each of the vegetable farms along River Yadzaram in Uba were collected. Samples from each of the two farms were pooled together to obtained two homogenous samples.

In the laboratory, vegetable samples were washed with tape water and thereafter with distilled water and the water was allowed to drip out and were then sliced into smaller portion and then dried in an oven at 80°C for hours (AOAC, 2000). At the end of the drying, the oven turned off and left overnight to enable the sample cool to room temperature. Each sample was grounded into a fine powder, sieved and finally stored in a 250 cm^3 screw capped plastic jar appropriately labeled (AOAC, 2000).

Digestion procedure

1.0 g of each powdered leaves samples were weighed out into Kjeldahl digestion flask mixed with 10 cm^3 of concentrated sulphuric acid, concentrated perchloric acid and concentrated nitric acid in the ratio 1: 2: 20 by volume respectively and left to stand overnight. Thereafter, the flask was heated at 70°C for 40 min and then, the heat was increased to 120°C. The mixture turned black after a while (Jeffery et al., 1989). The digestion was completed when the solution became clear and white fumes appeared. The digest was diluted with 20 cm^3 of distilled water and boiled for 15 min. This was then allowed to cool, transferred into 100 cm^3 volumetric flasks and diluted to the mark with distilled water. The sample solution was then filtered through a filter paper into a screw capped polyethylene bottle.

Determination of heavy metals

Levels of Cd, Cu, Fe, Zn and Cr in the vegetable samples were determined using Buck 210 model Atomic absorption spectrophotometer (AAS) equipped with an air-acetylene burner and hollow cathode lamps. Working standards were also prepared by further dilution of 1000 ppm stock solution of each of the metals and a calibration curve was constructed by plotting absorbance versus concentration. By interpolation, the concentrations of the metals in sample digests were determined. The mean values of six determinations per sample were recorded.

Statistical analysis

All analysis was performed in triplicates. Results were expressed by mean of ± SD. Statistical significance was established using one way analysis of variance (ANOVA). Means were separated according to Duncan's multiple range analysis ($p < 0.05$) using software SPSS 16.0.

RESULTS AND DISCUSSION

The levels of heavy metals (Cu, Cd, Cr, Fe and Zn) in Rosella (*H. sabdariffa*) and Spinach (*A. caudatus*) are as shown in Tables 1 and 2 of Farm A and B. In Rosella (*H. sabdariffa*), obtained from Farm A, the metal levels were: Cu, 30.00 ± 0.15 mg/kg to 31.00 ± 0.18 mg/kg; Fe, 37.39 ± 0.02 mg/kg to 48.67 ± 0.10 mg/kg and Zn, 13.00 ± 0.01 mg/kg to 25.50 ± 0.48 mg/kg. In those obtained from Farm B, the metal levels were: Cu, 31.33 ± 0.25 mg/kg to 33.83 ± 0.03 mg/kg; Fe, 25.06 ± 0.22 mg/kg to 28.47 ± 0.09 mg/kg and Zn, 7.33 ± 0.02 mg/kg. In Spinach (*A. caudatus*) obtained from Farm A, the metal levels were: Cu, 34.33 ± 0.42 mg/kg to 34.50 ± 0.05 mg/kg; Fe, 31.72 ± 0.71 mg/kg to 43.33 ± 0.02 mg/kg and Zn, 10.83 ± 0.17 mg/kg to 21.17 ± 0.14mg/kg. In those obtained from Farm B, the metal levels were: Cu, 35.03 ± 0.50 mg/kg to 38.00 ± 0.10 mg/kg; Fe, 28.47 ± 0.09 mg/kg and Zn, 10.50 ± 0.09 mg/kg to 25.00 ± 0.44 mg/kg. In both Farms A and B chromium and cadmium were not detected. The analysis revealed that Spinach contained higher concentration of copper than Rosella. Although, the maximum values recorded in both the vegetables are within the National Agency for Food and Drug Administration and control's (NAFDAC) maximum tolerable Cu concentration of 40 mg/kg in fresh vegetables. On the other hand, the results

Table 2. Concentration of some heavy metals in Spinach and Rosella from Farm B (mg/kg)

Vegetables/Sampling sites	Cd	Cr	Cu	Fe	Zn
Rosella (*Hibiscus Sabdariffa*)					
Location 1	ND	ND	31.33 ± 0.25	28.47 ± 0.09	7.33 ± 0.02
Location 2	ND	ND	33.83 ± 0.03	25.06 ± 0.22	7.33 ± 0.02
Spinach (*Amaranthus caudatus*)					
Location 1	ND	ND	38.00 ± 0.10	28.47 ± 0.09	25.00 ± 0.44
Location 2	ND	ND	35.03 ± 0.50	28.47 ± 0.09	10.50 ± 0.09
WHO/FAO			20 - 100	10 - 18	3 - 20
NAFDAC			0 - 40	10 - 20	0 - 50

All values represent mean ± standard deviation of triplicate determination. **ND** = Not detected, **WHO** (1996) = World Health organization, **NAFDAC** = National Agency for Food and Drug Administration and control.

were also lower than the published threshold values for mature plant tissue, except Fe with higher level.

The published threshold values are: As, 5 to 10 mg/kg; Fe, 10 to 20 mg/kg; Cu, 20 to 100 mg/kg; Pb, 30 to 300 mg/kg and Zn, 100 to 400 mg/kg (Kabata-Pendias and Pendias, 1984). The critical values or values regarded as excessive are: Zn, >50 – 100 µg/g; Mn, >1000 – 4000 µg/g; Fe, >200 - 500 µg/g; Cu, >7 – 20 µg/g; Pb, >4 – 30 µg/g and Cd, >1 – 3 µg/g; depending on the plants (vegetables) in question (EC–UN/ECE, 1995). The order of the metals contamination in the vegetables was Fe > Cu > Zn in farm A and Cu > Fe > Zn in Farm B.

Statistical test of significance using the Student t-test and ANOVA, showed significant differences ($p < 0.05$) between the levels of the heavy metals in vegetables obtained from the sample sites in Farm A and those from Farm B, with exception of Cu which showed no significant differences ($p > 0.05$). The elevated level of Fe in vegetables in the two gardens could be attributed to excessive usage of fertilizers and other agro-chemicals, as well as the use of waste water in irrigating the soil and of course, the environmental factors in the areas (Uwah et al., 2011). Similarly, the elevated levels of the metals in the vegetables obtained in Farms A and B could be due to possible pollution as a result of the vast agricultural activities going on in the area, and downstream deposition of fertilizers and other agro- chemicals as the Yedzaram River flows into the area. The consumption of these vegetables as food may not pose possible health hazards to human at the time of the study.

Conclusion

Considering the health risk's encountered in diets as a result of high levels of heavy metals in vegetables, the maximum allowable levels of these metals in vegetables should not exceed levels that reflect good agricultural practices. Farmers should be educated on the problems associated with excessive usage of fertilizers and other chemicals, as well as irrigating the crops with waste and all sorts of polluted water and the needs to grow crops with safe levels of heavy metals.

The vegetables contained variable levels of heavy metals (Cd, Cr, Cu, Fe, and Zn), with the exception of those of Fe, the metals levels were lower than the published threshold values considered toxic for mature plant tissue. Similarly, the levels of some of the metals were lower that the established critical limits causing toxicity in plants.

Agronomic practices such as application of fertilizers and use of waste water can affect bioavailability and crop accumulations of heavy metals.

Consumption of these vegetables as food may not constitute possible health hazards to humans at the time of the study.

The results obtained in this study would go a long way in providing a baseline data for the assessment of the distribution of these metals in Spinach (*A. caudatus*) and rosella (*H. sabdariffa*) grown in Uba area in Adamawa State. Further, studies will be carried out on the concentration of the heavy metals in soil of the studied areas.

Conflict of Interests

The author(s) have not declared any conflict of interests.

REFERENCES

Analysis of Association of Official Analytical Chemists (AOAC). International (2000). 17th Edition, Vol I and II, Gaitherburg, Maryland, USA.Chapter 32 pp 2, 3, 17 and 20 .

ATSDR (Agency for Toxic Substance and Disease Registry) (1999a). Toxicological profile for cadmium and nickel. Agency for toxic substance and disease registry, US Department of Health and Human Service, Public Health Service, 2005-93-0606.

Audu AA, Lawal AO (2005). Variation in metal contents of plants in Vegetable Gardens site in Kano Metropolis. J. Appl. Sci. Environ. Manage. 10(2):105-109.

Boon DY, Soltanpour PN (1992). Lead, Cadmium and Zinc

contamination of Aspen Garden soil and vegetation. J. Environ. Qual. 21:82-86. http://dx.doi.org/10.2134/jeq1992.00472425002100010012x
Das P, Samantarary S, Rout GR (1997). Studies on Cadmium Toxicity in Plant. A review. Environ. Pollut. 98:29-36. http://dx.doi.org/10.1016/S0269-7491(97)00110-3
EC-UN/ECE (Economic Commission – United Nations (1995). Folier Expert Panel. Symposium Paper ICP – Forest, Wien.
Elsevier J (2008). Physical and Chemical Fundamentals of Pollutants, New York. 1-2(7):194-197.
Ihekoroye AI, Ngoddy PO (1985). Integrated Food Science and Technology for the Tropics, Macmillan Press Ltd, London UK. 87:88-92.
Jeffery GH, Bassett J, Mendham J, Denney RC (1989). A Vogel's Textbook of Quantitative Chemical Analysis, 5th Edition, Longman Group Ltd, London, Uk Pp.743-837.
Kabata-Pendias A, Pendias A (1984). Trace Elements in Soils and Plants. CRC Press, Boca Raton, Florida. Pp.321-337.
Kirkham MB (1983). Study on Accumulation of Heavy Metals in soil Receiving Sewage and Effluent Water. Agric. Ecosyst.. Environ. 9:251.
Mahaffey KR, Carpar SG, Caden BC, Fowler BA (1981). Effects on Toxicity and Tissues Metal Concentrations in Rat. J. Lab. Clin. Med. 98(4):463-81.
Milk M (2012). The search for the ideal vegetable www.oole.com). 2003; 186(1- 2):10.
Ndiokwere CC, Ezeh CA (1990). The Occurrence of Heavy Metals in the Vicinity of Industrial Complexes in Nigeria. Environ Int. 16(29):1-295
Nirmal HS, Rita NK (2007). Characterization of Heavy Metals in Vegetable using Inductive Coupled Plasma Analizer (ICPA). J. appl. Sci. Environ. Manage.
Nriagu JO, Pacyna JM (1990). Quantitative Assessment of World Contamination of Air, Water and Soil by Trace Metals. Nature (London) 333:134-139. http://dx.doi.org/10.1038/333134a0
Nriagu JO (1990). Global Metal Pollution Poisoning the Atmosphere. Environ. 7:87-93.
Punz WF, Sieghardt H (1993). The Response of Root of Herbaceous Plant Species to Heavy Metals. Environ. Exp. Bot. 33(1):85-98. http://dx.doi.org/10.1016/0098-8472(93)90058-N
Radojavice M, Bashkin VN (1999). Practical Environmental Analysis. The Royal Society of Chemistry London. P.287.
Rattan RK, Dattan SP, Singh AK (2001). Effect of Long Term Application of Sewage Effluents on available Water Status in Soils under Keshopure Effluent Irrigation Scheme in Delhi. J. Water Manage. 9:21-26.
Shinggu DY, Ogugbuaja VO, Barminas JT, Toma I (2007). Analysis of Street Dust For Heavy Metal pollutants in Mubi, Adamawa State Nigeria. Inter. J. Phy. Sci. 2(1):1-10.
UNEP (1992). United Nation Environmental program Data Report 1992 Basil Blackwell, Oxford.
Usman SOA, Ayodele J T (2002). Bioaccumulation of Four Heavy Metals in Leaves of Calotrophis Procera. J. Chem. Soc. Niger. 27:26-27.(Not cited in the work)
Uwah EL, Ndahi NP, Ogugbuaja VO (2009). Study of the Level of some Agricultural Pollutants in Soils and Water leaf (Talinium triangulare obtained in Maiduguri, Nigeria. J. Appl. Sci. Environ. Sanita 4(2):71-78.
Uwah EL, Ndahi NP, Abdulrahman FI, Ogugbuaja VO (2011). Heavy Metal Levels in Spinach (Amaranthus Caudatus) and Lettuce (Lactuca sativa) grown in Maiduguri, Nigeria. J. Environ. Chem Ecotoxicol. 3(10):264-271
Uwah EL (2009). Concentration Level of some Heavy Metal Pollutants in soil and carrot (*Daucus carota*) obtained in Maiduguri Nigeria. Contin. J. Appl. Sci. Wilodud online J.) 4:76-88
Wandeo Town (2005). Retrieved from http llen.wikipedia.org/w/index.hph?
Ward NI (2005). Environmental Analytical Chemistry in Trace Elements (Eds. Fifields, F.W, F.N and Haines, P.J,) Blackie, Academic and professional, UK, Pp.320-328.
WHO/FAO Dietary recommendations/Nutritional requirement www.WHO.INT/../indexhtml
World Health Organization (WHO) (1996). Trace Elements in Human Nutrition. WHO Technical Report Series No 532. Geneva.
Xiong ZT (1998). Lead Uptake and Effects on Seed Germination and Plant Growth in a Pb Heper accumulator Brassica Pekninensis Repr Bult. Environ. Contaim Toxicol. 60:285-291 fLp2.

Amino acid content, fatty acid content and anti nutritional factor of seeds of new hybrid varieties of *Echinochloa frumentacea* (Sanwa) minor millets

S. Gupta[1], S. K. Shrivastava[1] and M. shrivastava[2]

[1]Department of Applied Chemistry, Jabalpur Engineering College, Jabalpur-482011(M.P.) India.
[2]Department of Chemistry Govt. M. H. College of Home Science and Science for Women Jabalpur (M. P.) India.

Cereals are the staple diet of most of the world's population. The millets are very important staple food in the rural parts of India. Millets can secure India's food and farming in future because it is amazing in their nutrition contents. *Echinochloa frumentacea* (Sanwa) millet is good source of energy and provide protein, fatty acid, minerals, vitamins, dietary fibre and polypheonals. Proteins present in various foods differ in their nutritive value on account of the difference in the amino acid contents. The amino acid content, fatty acid content (TSFA and TUFA) and anti nutritional factor ranged from 0.0008 to 0.522%, 24.2 to 26.0%, 73.5 to 75.4% and 0.301 to 0.302, 0.0202 to 0.0204 g/100 g and 31.95 mg/100 g respectively. No cyanide content and haemagglutinin activity were found. Nutritionally the seeds of *E. frumentacea* variety DFM-1 and HR-374 are rich in aspartic acid (essential amino acid) content and total unsaturated fatty acid content.

Key words: Amino acid, fatty acid content, anti nutritional factor, minor millets variety of *Echinochloa frumentacea*.

INTRODUCTION

Millets are a group of cereal species crops or grain like food that has been used by large group of people in rural, tribal and hilly areas in Asia and Africa (Ravindra et al., 2008; Anonymous, 2006; Rao et al., 2011; Odoemalam and Osu, 2009). Millet is a cereal crop plant belonging to different genera but all within the grass family, Poaceae and subfamily Panicoideae (FAO, 1972; 1991). As the minor millets are consumed by the poor, they guard them against food and nutritional insecurity imposed by various agronomic, socio economic and political factors. Minor millets can thus act as a shield against nutritional deficiency disorders and provide nutritional security. These grains will be used for traditional as well as novel foods (Vanithasri et al., 2012). All of them are small seeded grasses having high capability of resistance to extreme environmental conditions in which major cereals fail to give substantial yields (Amadou et al., 2013; McDonoug et al., 2000; Black et al., 2006; Ahmed et al., 2013). *Echinochloa frumentacea* minor millets are high energy, nutritious foods comparable to other major cereals and some of them are even better with regard to protein and mineral content (Fe, Ca, Mn, Mg). Fat is one of the major nutrients which provide energy, promote body growth, maintain and repair body tissue, promote reproduction and lactation and regulate body process. Fats are carriers of fat soluble vitamins. Dietary fat must also provide essential fatty acids (EFA) which are the functional components of membrane lipids and have other important metabolic function. Fats are made up of fatty acids which include saturated fatty acids like palmitic and stearic, monounsaturated fatty acids (MUFA) like oleic and polyunsaturated fatty acid (PUFA) likes linoleic acid and linolenic acid (Singhai and Shrivastava, 2002; Nagraj, 1995). Lipids are relatively minor constituents in cereal grains; however, they contribute significantly to diet as a source of invisible fat and essential fatty acid (Achaya, 1986, 1987). Nevertheless *E. frumentacea* represents a good source of essential amino acid and essential fatty acid (linoleic acid and linolenic acid). However, it must be pointed out that, *E. frumentacea* also contains some anti-nutritional factors which inhibits proteolytic and amylolytic enzymes, limits mineral, protein and starch digestibility and makes poor human bioavailability of proteins.

This study was therefore conducted to assess the levels of amino acid content, fatty acid content and antinutrinational factors in the seeds of BMVL-29 and BMVL-172 variety of *E. frumentacea* for awareness and exploitation.

MATERIALS AND METHODS

In the present study two new hybrid, authentic, healthy and matured seeds of minor millets viz., *E. frumentacea* (variety BMVL-29 and BMVL-172), under investigation were procured from Agriculture Research Station of Jawaharlal Nehru krishi Vishwavidyalaya, Dindori (M.P.) and were studied for their amino acid, fatty acid and antinutritional factors.

Amino acid analysis

The amino acid composition of seeds of hybrid variety BMVL-29 and BMVL-172 of *E. frumentacea* was analyzed by using liquid chromatography mass spectroscopy (LC-MS).

Solvent extraction and sample preparation

Solvent extraction was done by Soxhlet apparatus and stock solution was prepared by dissolving 10 mg of each amino acid in 100 ml of diluents (acetonitrile/formaic acid) and it was properly shaken. Working standard solution of 1 mg/L was prepared by this stock solution.

LC-MS analysis

LC-MS analysis of sample was done by using C18 column (Brava Amino 5 µ, 4.6 × 250 mm). Column temperature was maintained at 40°C. 10 µl of sample was injected for 10 min, 0.1% Formic acid in water and 0.1% Formic acid in acetonitrile (95+5) were used as mobile phase and its flow rate was 0.8 ml/min. Ionization of sample component were performed on electron spin resonance (ESR) mode (70 eV).

Fatty acid analysis

The hybrid variety (BMVL-29 and BMVL-172), of *E. frumentacea* seeds were studied for their fatty acid composition by gas chromatography. Powdered sample of experimental seeds were subjected to solvent extraction in Soxhlet apparatus for 20 h, using petroleum ether (40 to 60°C) as solvent. Lipids were then estimated gravimetrically by the method of Colowick and Kaplan (1957). Methyl esters of the lipids were prepared by the method of Chowdhary et al. (1984) and analysed by gas liquid chromatogram (GLC). Gas chromatograms were recorded using flame ionization detector (FID) with split ratio 1:50.

Antinutritional factors

The seeds of *E. frumentacea* variety BMVL-29 and BMVL-172 were studied for their tannin content, oxalate content, trypsin inhibitor activity, cyanide content and haemagglutinin activity. Cyanide and tannin contents of seeds were determined by the method of AOAC (1970). The total oxalate content in the form of oxalic acid was determined by using the method of Talpatra et al. (1948). Trypsin inhibitor activity was determined according to the method as described by Kakade et al. (1969) with certain modifications by Gupta and Deodhar (1975). Haemagglutinin activity was determined by the method as given by Liener (1955).

RESULTS AND DISCUSSION

The results of amino acid composition of seed protein of *E. frumentacea* variety BMVL-29 and BMVL-172 are given in the Table 1. The seeds of *E. frumentacea* variety BMVL-29 and BMVL-172 were found to have highest amount of Aspartic acid (0.522%), whereas Lysine content was reported 0.047 and 0.046% in *E. frumentacea* variety BMVL-29 and BMVL-172 respectively. In both the variety of *E. frumentacea* other amino acid in the decreasing order were glutamic, methionine, L-omithine HCl, alanine, arginine HCl, DL-Tryptophan, serine, glycine, proline = valine, threonine, tyrosine, phenylalanine, leucine = L-Hydroxyproline>isoleucine.

From the perusal of the data it appears that both the varieties of *E. frumentacea* minor millets seeds are lacking in Cystine, Histidine, 2-Aminobutaric and L-cysteine amino acids. It has been found that the amount of aspartic acid was maximum while the quantity of isoleucine was minimum but methionine levels of these variety of minor millets was more than the amount present in cereal grains. Methionine is of special importance to animals as a therapeutic and nutritional factor. It protects animals against liver injuries by chloroform, industrial halogenated fumes, and protein deficient diets and prevents the great loss of body nitrogen in the case of fractures, burns and surgical operations (Crocker and Barton, 1952). However, the amino acid composition of seed protein of both the variety (BMVL-29 and BMVL-172) of *E. frumentacea* under study was found to be in general accordance with reported values (FAO, 1970; Glew et al., 2008; Hui, 1996).

Table 1. Amino acid composition of *E. frumentacea* variety BMVL-29 and BMVL-172.

S/No	Amino acid	Amino acid analysis (content in %)	
		Echinochloa frumentacea BMVL-29	*Echinochloa frumentacea* BMVL-172
1	Alanine	0.015	0.016
2	Arginine HCl	0.014	0.013
3	Aspartic	0.522	0.522
4	Cystine	ND	ND
5	Glutamic	0.027	0.034
6	Glycine	0.006	0.006
7	Histidine	ND	ND
8	Isoleucine	0.0009	0.0008
9	Leucine	0.001	0.002
10	Lysine	0.047	0.046
11	Methionine	0.027	0.025
12	Phenylalanine	0.002	0.002
13	Proliney	0.004	0.004
14	Serine	0.011	0.011
15	Threonine	0.003	0.002
16	Tyrosine	0.002	0.003
17	Valine	0.004	0.004
18	2-Aminobutaric	ND	ND
19	L-Omithine HCl	0.023	0.023
20	L-Cysteine HCl	ND	ND
21	DL-Tryptophan	0.012	0.012
22	LHydroxyproline	0.001	0.001

Table 2. Saturated fatty acid composition of *E. frumentacea* variety BMVL-29 and BMVL-172.

Name of variety	Saturated fatty acid %							
	Caprylic acid	Lauric acid	Myristic acid	Palmitic acid	Stearic acid	Arachidic acid	Bahenic acid	Total saturated fatty acid (%)
E. frumentacea BMVL-29	2.1	-	0.1	15.8	5.0	0.9	0.3	24.2
E. frumentacea BMVL-172	1.1	0.1	0.1	17.1	6.1	1.1	0.4	26.0

Tables 2 and 3 showed the variation of fatty acid content of hybrid *E. frumentacea* variety BMVL-29 and BMVL-172. The saturated fatty acid, Caprylic acid was found to be highest (2.1%) in variety *E. frumentacea* BMVL-29 and lowest (1.1%) in variety *E. frumentacea* BMVL-172. The Palmitic acid content was reported higher (17.1%) in variety *Echinochloa frumentacea* BMVL-172 and lower in the variety *E. frumentacea* BMVL-29. The percentage of Stearic acid was found maximum (6.1%) in the variety *E. frumentacea* BMVL -172 and minimum (5.0%) in the variety *E. frumentacea* BMVL-29. The Arachidic acid was found to be maximum (1.1%) in the variety *E. frumentaca* BMVL-172, while minimum in the variety *E. frumentacea* BMVL-29. The variety *E. frumentacea* BMVL-172 has maximum (0.4%) Bahenic acid content while minimum (0.3%) in the variety of *E. frumentacea* BMVL-29. The total saturated fatty acid (TSFA) content was to be greater (26.0%) in the variety of *E. frumentacea* BMVL-172 than the variety *E. frumentacea* BMVL-29(24.2%).

The unsaturated fatty acid, *E. frumentacea* variety BMVL-172 contain maximum amount (29.5%) of Oleic acid (MUFA). Whereas the Linoleic acid was found to be highest (46.9%) in the variety *E. frumentacea* BMVL -29. The Linolenic acid and Ecosenoic acid content was found to be greater (1.0 and 0.5% respectively) in the variety *E. frumentacea* BMVL-29 than the variety *E. frumentacea* BMVL-172 (0.7 and 0.4% respectively). The variety *E.*

Table 3. Unsaturated fatty acid composition of *E. frumentacea* variety BMVL-29 and BMVL-172.

Name of variety	Unsaturated fatty acid (%)				
	Oleic acid	Linoleic acid	Linolenic acid	Ecosenoic acid	Total unsaturated fatty acid
E. frumentacea BMVL-29	27.0	46.9	1.0	0.5	75.4
E. frumentacea BMVL-172	29.5	42.9	0.7	0.4	73.5

Table 4. Anti nutrients of *E. frumentacea* variety BMVL-29 and BMVL-172.

Name of variety	Tannin content g/100 g	Oxalate content g/100 g	Trypsin inhibitor activity TIU/mg protein	Cyanide content	Haemagglutinin activity		
					Human blood	Goat blood	Hen blood
Echinochloa frumentacea BMVL-29	0.301	0.0202	31.95	ND	ND	ND	ND
Echinochloa frumentacea BMVL-172	0.302	0.0204	ND	ND	ND	ND	ND

*ND- not detected, * The values given in the tables are the mean of the triplicate values obtained.

frumentacea BMVL-29 contain minimum value of mono unsaturated fatty acid (MUFA) and maximum value of polyunsaturated fatty acid (PUFA). Total unsaturated fatty acid (TUFA) content was found to be highest (75.4%) in the variety *E. frumentacea* BMVL-29 and lowest (73.5%) in the variety *E. frumentacea* BMVL-172.

E. frumentacea variety BMVL -29 was found to be superior to the variety *E. frumentacea* BMVL-172 under investigation. It contains highest content of linoleic acid (46.9%). The amount of linoleic acid in millet oil is higher in comparison with most other types of vegetable oils (Ravindra et al., 2008). The linoleic acid is one of the most important polyunsaturated fatty acid in human food, because of its prevention of distinct heart vascular disease (Boelhouwer, 1983). This acid is most important essential fatty acid required for growth, physiological function and maintenance, which cannot be synthesized by the human body and one, has to depend on dietary source for their adequate supply. The body metabolizes linoleic and linolenic acid into arachiodonic acid and docosahexaenoic acid (DHA) respectively which are essential to the normal development of central nervous system (Brich et al., 2007; Jacobson et al., 2008). Various developmental problems including attention-deficit/hyperactivity disorder (ADHD) in children have been linked to biological deficiencies in polyunsaturated fatty acids. Additionally, there is evidence that symptoms may be reduced with PUFA supplementation (Sinn and Bryan, 2007).

The result of anti nutritional factors of the *E. frumentacea* varieties are shown in Table 4. The tannin content of different varieties of *E. frumentacea* ranged from 0.301 to 0.302 g/100 g. These values are lower than the earlier findings of Pasala and Bjorn (1989) and are well below the fatal dose (Sarjekar and Shrivastava, 1994). The total oxalate content (in terms of oxalic acid) was found to be maximum (0.0204 g/100 g) in the variety of *E. frumentacea* BMVL-172 while it was minimum (0.0202 g/100 g) in the variety of *E. frumentacea* BMVL-29.

The Trypsin inhibitor activity was found to be maximum (31.95 mg/100 g) in the variety *E. frumentacea* BMVL-29. However, no trypsin inhibitor activity was reported in the variety of *E. frumentacea* BMVL-172. No Cyanide content and haemagglutinin activity were found in the varieties of *E. frumentacea* under study. The value of anti nutritional factors reported in study was lies within the leather dose. These anti nutritional factors may be reduced by simple soaking, heating and germination or fermentation. It is now established that phytates, polyphenols and tannins can contribute to antioxidant activity of the millet foods, which is an important factor in health, aging and metabolic diseases (Bravo, 1998).

Conclusion

Cereals and millets constitute a major component of diet consumed in developing countries like India. *E. frumentacea* (Sanwa) millets are the staple food for millions of poor people in the world. The seed of *E. frumentacea* (variety BMVL-29 and BMVL-172) millets contain significant quantities of essential amino acids particularly the sulphur containing amino acid (methionine), essential fatty acids (PUFA and MUFA) and leather amount of antinutritional factor. It will be a useful and economical source of protein provided that some legume or milk is consumed along with these minor millet, that is, suitable for good nutritional supplementation. Variations in the various constituents of the *E. frumentacea* millets seeds have been attributed to variety, conditions, fertilizer treatments and climatic conditions. Most of the anti nutritional factors are heat-labile and since only humans consume millets after cooking, it would not constitute any major health hazard. It can be concluded that nutritional benefit of. *E. frumentacea* minor millets can be enhanced when all type of processing treatments employed at domestic levels were effective in reducing the biological active factors and therefore could be used to enhance better quality in food materials. Compared to rice and wheat, minor millets contain little high amount of anti nutritional factors. But these anti nutritional factors are plant based phyto chemicals that possess therapeutic qualities and hence are recommended by doctors for various diseases. Diabetics need to control their blood sugar, hypertension as well as cholesterol level and that is why doctors recommend minor millets for these problems.

Conflict of Interest

The authors have not declared any conflict of interest.

REFERENCES

AOAC (1970). Washington DC, Official Method of Analysis: 240-438.

Achaya KT (1986). Indivisible fats revisited. Proc. Nutr. Soc. India 32:1-17.

Achaya KT (1987). Fat status of Indians A review. J. Sci. Ind. Res. India. 46:112-126.

Anonymous (2006). QRT report of millets, Director of millets development, Rajasthan.

Amadou I, Gounga ME, Guo-Wei L (2013). Millets Nutritional composition,some health benefits and processing-A (Review). Emir. J. Food Agric. 25(7):501-508.

Ahmed SM, Saleh QZ, Jing C, Qun S (2013).Millet Grains: Nutritional Quality, Processing, and Potential Health Benefit (Comprehensive Reviews). FSFS 12(3):281-295.

Bravo L (1998). Polyphenols chemistry, dietary sources, metabolism and nutritional significance. Nutr. Rev. 56:317–333. http://dx.doi.org/10.1111/j.1753-4887.1998.tb01670.x

Black M, Bewley JD, Halmer P (2006). Millets -in: The encyclopedia of seeds: science technology and uses. CABI, Wallingford, UK, P. 41.

Brich EE, Garfield S, Casteneda Y, Hughbanks-Wheaton D, Uauy R, Hoffman D (2007). Visuaacuity and Cognitive outcome at 4 years of age in a double-blind, randomized trial of long-chain polyunsaturated fatty acid supplemented infant formula. Early Human Dev. 83:279-284. http://dx.doi.org/10.1016/j.earlhumdev.2006.11.003

Boelhouwer C (1983). Trades in chemistry and technology of lipids. J. Am. Oil Chem. Soc. 60(2):457-462. http://dx.doi.org/10.1007/BF02543540

Colowick SP, Kaplan NO (1957). Methods in Enzymology, III Academic Press Inc., New York: P. 85.

Chowdhury AR, Banerji R, Misra G, Nigam SK (1984). Studies on leguminous seeds. J. Am. Oil Chemists. Soc. 61(6):1023-1024. http://dx.doi.org/10.1007/BF02636208

Crocker W, Barton LV (1952). Physiology of seeds. Chemical Composition of seed. 1: 35.

FAO (1970). Sorghum & millets in human nutrition –FAO corporate document repository.

FAO (1972). Sorghum and Millets in Human Nutrition. Food and Nutrition Series: P. 27.

FAO (1991). Amino Acid Scoring Pattern. In: Protein quality evaluation, FAO/WHO Food and Nutrition Paper, Italy: 12-24.

Gupta AK, Deodhar AD (1975). Variation in trypsin inhibitor activity in soybean (Glycine max). Ind. J. Nutr. Dietet. 12(3):81-84.

Glew RS, Chuang LT, Robert JL, Glew RH (2008). Amino acid,Fatty acid and mineral content of black Finger Millet cultivated on the Jos plateau of Nigeria. Food. 2(2):115-118.

Hui YH (1996). Bailey's Industrial Oil and Fat Products. Edition 5:20-62.

Jacobson JL, Jacobson SW, Muckle G, Kaplan-Estrin M, Ayotte P, Dewailly E (2008). Benificial effects of a Polysaturated fatty acid on infant development evidence from the Inuit of arctic Quebec. J. Pediatrics 152:356-364. http://dx.doi.org/10.1016/j.jpeds.2007.07.008

Kakade ML, Simsons NR, Liener IE (1969). An evaluation of natural and synthetic substances for measuring the antitryptic activity of soybean samples. Cereal Chem. 46:518-526.

Liener IE (1955). The photometric determination of haemagglutinating activity of soyin and crude soybean extracts. Arch. Biochem. Biophys. 54:223. http://dx.doi.org/10.1016/0003-9861(55)90025-4

McDonough, CM, Rooney, LW, Serna-Saldivar SO (2000).The millets - in: Kulp, K. & Ponte, J.G. Handbook of Cereal science and technology. Marcel Dekker, New York: pp. 177–181.

Nagraj G (1995). Quality and Utility of Oil Seeds, Directorate of Oil Seeds Research (ICRA) Hyderabad, India.

Odoemalam SA, Osu CI (2009). Evalution of the phytochemical content of some edible grains marketed in Nigeria. E. J. chem. 6:1193-1199.

Pasala G, Bjorn OE (1989). Nutrient composition and protein quality of minor millets. Plant Foods Human Nutr. 39:201-208. http://dx.doi.org/10.1007/BF01091900

Ravindra U, Vijaykumari SS, Raghnprasad KP, Kavaloor R (2008). Newcastle disease virus as an oncolytic agent. Tropical Agric. Res. 20:115-122.

Rao BR, Nagasampige MH, Ravikiran M (2011). Evaluation of nutraceutical properties of selected small millets. J. Pharm. Bioallied. Sci. 3:277-279. http://dx.doi.org/10.4103/0975-7406.80775

Sarjekar P, Shrivastava SK (1994). Anti nutrtional factors in some cultivated and wild leguminous seeds. Ann. Plant Physiol. 8(2):198-200.

Singhai B, Shrivastava SK (2002). Studies on fatty acid composition of seeds of improved varieties of Cicer arietinum. Asian J. Chem. 14:1080-1082.

Sinn N, Bryan J (2007). Effect of Supplementation with Polyunsaturated Fatty Acids Associated with Child ADHD. J. Dev. Behavioral Pediatrics 28:82–91. http://dx.doi.org/10.1097/01.DBP.0000267558.88457.a5

Talpatra SK, Roy SC, Sen KC (1948). A new method estimation of oxalic acid in biological materials and the oxalic acid content of Indian feeding stuffs. Indian J. Vet. Sci. Anim. Husb.18:99-108.

Vanithasri J, Kanchana S, Hemalatha G, Vanniarajan C, Sahul Hameed M (2012). Role of millets and its importance in new mellinium. Int. J. Food Sci. Technol. 2(1):35-47.

Kinetic and equilibrium studies of the heavy metal remediation potential of *Helix pomentia*

Adeoye, Moriam Dasola[1], Azeez, Lukman Adeyemi[1], Lawal, AbdulAzeez Tunbosun[2], Olayiwola, Olajumoke Abidemi[3] and Shitta, Olanrewaju Razaq[1]

[1]Industrial and Environmental Chemistry Unit, Chemical Sciences Department, Fountain University, Osogbo Osun State, Nigeria.
[2]Industrial and Environmental Chemistry Unit, Chemical Sciences Department, Al-Hikmah University, Ilorin Kwara state, Nigeria.
[3]Chemical Sciences Department. Oduduwa University, Ipetumodu, Ile-Ife, Osun State, Nigeria.

Heavy metals are one of the major pollutants in water and are very toxic. However, chitosan – a waste biomass of agricultural products possess the ability to chelate heavy metals in water. In this study, the application of snail (*Helix pomatia*) shell based chitosan for remediating heavy metals contaminated Osun River (OSR), Odo-ebu stream (OBS) and Aso well water (AWW), all in Osogbo community, Nigeria; and the binding efficiency of chitosan for heavy metals in these water sources were evaluated. The result shows the prevalence of Iron (Fe), Cadmium (Cd), Zinc (Zn), Copper (Cu), Chromium (Cr), and Lead (Pb) ions in the three water samples with OBS and OSR having higher metal concentrations in the order: Pb > Cd > Fe > Zn > Cr > Cu compared with AWW which follows the order: Fe > Zn > Pb > Cr > Cd > Cu. The concentration range of heavy metals in OBS and OSR were (0.0011 -2.831) mg/L and (0.0012 - 1.687) mg/L respectively, while for AWW it was (0.0004-0.0237) mg/L and below the Standard Organization of Nigeria (SON) 2007 and World Health Organization (WHO) 2004; maximum permissible level of metal ions in drinking water. The prepared chitosan is a selective candidate for remediating four heavy metals ions from these water samples, with the maximum sorption capacities for the metal ions in the sequence: Pb > Cd > Fe > Cr. The sorption behaviour followed a pseudo-second-order model, while the equilibrium data correlated well with the Langmuir isotherm models with the R_L values in the range $0 < R_L < 1$.

Key words: Heavy metals, *Helix pomatia*, chitosan, dumpsite, equilibrium studies.

INTRODUCTION

Heavy metals are of great concern due to their contrasting toxicities, long-term mobilities, bio-availabilities and their non-degradable nature in the environment. They enter the environment through a

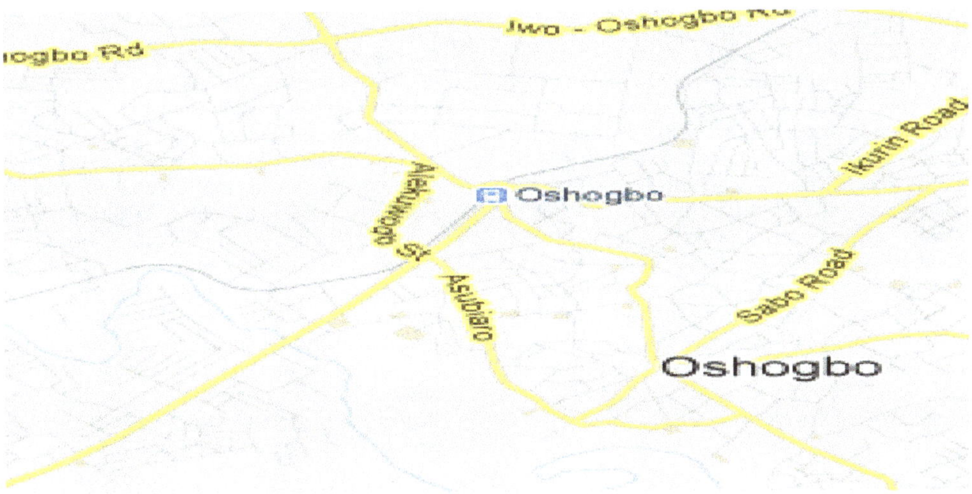

Figure 1. Map of Osogbo showing sampling area.

variety of anthropogenic and industrial activities, poor waste disposal mechanism; and pose serious threats to plants and human health (Kanamadi et al., 2003; Jude and Augustin, 2007). Heavy metals such as copper, zinc, nickel, chromium, cadmium and iron present in most Nigerian rivers have been found in concentrations well above acceptable and permissible levels (Eniola et al., 2010). Although, degradation of water quality has been claimed to be most severe in the four states that contain 80% of the nations industries that is, Lagos, Rivers, Kano and Kaduna state but this has extended to other regions. Thus, the importance of environmental quality of Osogbo (an urban settlement) in Osun state (south-western Nigeria) has attracted a great deal of interest due to the increasing population of the inhabitant, increasing industrialization, poor land use system, agricultural activities, industrialization and anthropogenic impact which has made the water and land being polluted (Olajire and Imeokparia, 2001). This is of great concern due to the great threat on health and economic development (Mustapha, 2008).

Bioremediation, the use of waste-derived materials or microbes to detoxify and degrade environmental contaminants provide safe and economic alternatives for the removal of toxic metal ions from wastewater or any aquatic environment and soil (Volesky, 2001). It also converts wastes into useful materials for soil and water decontamination. Chitosan, a poly-b (1-4)-2-amino-2-deoxy-D-glucosed derived from chitin - a natural biopolymer found in the outer shell of crustaceans such as crabs, shrimps and prawns has been found to have sorption capacity for several metal ions (Adewuyi et al., 2009; Yildiz et al., 2010). The metal sorption capacity of chitosan varies with crystallinity, affinity for water, deacetylation degree and amino group content. Kinetic studies have demonstrated that the rate of metallic ion sorption onto chitosan differs depending on the raw material (snail, shrimp, crab or lobster shells), preparation method, chemical modification, and chitosan particle shape (Martha, 2008).

This study presents the kinetic and uptake performance of the binding potential and selectivity of prepared biopolymers- chitosan for heavy metal ions (Cr, Pb, Cd, Fe) in the Osun River, Odo Ebu and Aso Hotel well water which serve as the major sources of water for the Isale-Osun community. The importance of this study lies on the direct evaluation of the metal sorption capacity of chitosan from real contaminated water samples from Isale-Osun community.

MATERIALS AND METHODS

Chitosan preparation

Helix pomatia (snail) shell procurement, deproteination, demineralization and N-deacetylation of the chitin were carried out using the methods of Adewuyi et al. (2009) and Taboada et al. (2003). All the chemicals used were of analytical grade products of British Drug House (BDH), Poole, England. De-ionized-distilled water was also used throughout the whole processes. The resulting chitosan was collected, washed and oven dried at 60°C for 4 h. The degree of deacetylation of the prepared chitosan was carried out using literature method of Tolaimate et al. (2002). Ultra Violet-Visible spectra analysis (using JENWAY spectrophotometer) of the prepared chitosan sample, covering the wavelength range of 190 to 800 nm with quartz cells (with a thickness of 0.2 cm) before and after treatment with the different water samples to determine its wavelength of adsorption.

Sampling and metal analysis

Surfaces water samples from Aso Hotel well water, Ebu stream (Odo Ebu) and Osun River water, all in the vicinity of dumpsite located at Isale Osun axis of Asubiaro area (Lat 07,44° N Long 04.74°E) in Osogbo, Osun state were collected (Figure 1). The pH and temperature of these water samples was determined using

Jenway 3505 pH- portable meter. The water samples were digested using the method of Sallau et al. (2011). 50 cm^3 of each sample was treated with 5 cm^3 of conc. HNO_3 and heated on a hot plate with gradual addition of conc. HNO_3 as necessary until the solution boils. It was then evaporated to about 20 cm^3; 5 cm^3 of conc. HNO_3 was finally added, covered and allowed to cool and then filtered. The pH of these solutions was initially fixed at a pH higher than that for the chitosan solubilization threshold (pH around 6.2). The filtrate was poured into a 50cm^3 standard volumetric flask and made up to the mark with distilled water. Metal analyses of the digested samples were determined with Solaar AAS series 711047v1.22 atomic absorption spectrometer. Detection limits were estimated from digested blank (deionized water) which was run during the analysis. Triplicate digestions and analyses were run and average values were reported. Same procedure was repeated after the samples had been treated with chitosan.

Equilibrium studies

Equilibrium studies were carried out using the method of Adeogun et al. (2010). Adsorption was performed in a set of 100 ml flasks wherein the water samples (100 ml of each sample) were placed. Equal mass of 0.2 g of the prepared chitosan of particle size 40 mesh was added to the samples and kept on an isothermal shaker (orbital shaker) at 25 ± 1°C for 48 h for equilibrium to be reached between the solid-solution mixture. Similar procedure was followed for another set of 100 mls flasks containing the water samples without adsorbate (to be used as a control). The pH was adjusted to 4.8 - 6.0 by adding few drops of diluted HCl or NaOH (0.1 $moldm^{-3}$). The flasks were then removed from the shaker and the final concentrations of heavy metals in the solutions were determined by Atomic Absorption Spectroscopy (AAS). Each experiment was duplicated under identical conditions. The amount of adsorption at equilibrium, $q_e^{/}$ (mg/g), was calculated by:

$$q_e^{/} = \frac{(C_0 - C_e)V}{W} \tag{1}$$

Where: C_o and C_e (mg/ml) are the concentrations of metal ion at initial and equilibrium stage respectively, V is the volume of the solution (ml), and W is the mass of dry adsorbent used (g).

Adsorption kinetics

The kinetic experiments were identical to those of equilibrium tests except that the aqueous samples were taken at different time intervals from the solid-solution mixture, and the concentrations of heavy metals in the solutions were similarly measured (Adeogun et al., 2010). The amount of adsorption q_t (mg/g), at time t, was calculated by:

$$q_t = \frac{(C_0 - C_t)V}{W} \tag{2}$$

C_o and C_t (mg/ml) are the concentrations of the metal ions in the samples at initial and any time t, respectively, V is the volume of the solution (ml) and W is the mass of dry adsorbent used (g). An ideal sorbent for metal decontamination should not only have a large sorbate capacity but also a fast sorption rate (Crini and Badot, 2008). According to Sud et al. (2008), predicting the rate at which sorption takes place and the binding mechanism are essential to determine the efficiency of a sorption process. These may be controlled by: (i) solute transfer from the bulk solution to the boundary film that surrounds the sorbent's surface, (ii) solute transport from the boundary film to the sorbent's surface, (iii) solute transfer from the sorbent's surface to the active intraparticle sites and (iv) interaction(s) between solute and binding sites of the sorbent (intra-particle diffusion). The pseudo- first order and pseudo-second order kinetic and intra-particle diffusion (chemical binding reaction) kinetic models were employed to determine the rate constant and the controlling mechanism of the sorption process. These were done to further confirm the best fit kinetic model (s) for the sorption process. Although, several metal sorption studies have shown that most sorption kinetics usually followed the pseudo-second-order kinetic model. The pseudo-first-order and intra-particle diffusion equations were generally acclaimed to be applicable over the initial stage of the adsorption process (Mohan et al., 2006; Pan et al., 2009). The linear forms of pseudo- first order pseudo- second order the intra-particle diffusion equations are as expressed in Equations 3, 4 and 5 respectively (Lagergren, 1898 Kamari, 2011; Ho and McKay, 2000).

$$\ln(q_e^{/} - q_t) = \ln(q_e) - k_1 t \tag{3}$$

$$\frac{t}{q_t} = \frac{1}{K_2 q_e^2} + \left(\frac{1}{q_e}\right)t \tag{4}$$

$$q_t = K_{id} t^{0.5} \tag{5}$$

Where: $q_e^{/}$ and q_t (mg/g) are the amount of metal ions sorbed (mg/g) at equilibrium and at time t (min) respectively while k_1 (g/mg/min), k_2 (g/mg/min) and k_{id} (mg/g/$min^{0.5}$) are the rate constants of the pseudo-first –order, pseudo-second-order and intra-particle diffusion equations for the bio-sorption processes respectively (Kamari et al., 2011).

Adsorption isotherm

Adsorption isotherm study was carried out using the Langmuir adsorption isotherm (Langmuir, 1916). Langmuir isotherm assumes monolayer adsorption onto a surface containing a finite number of adsorption sites of uniform strategies of adsorption with no transmigration of adsorbate in the plane of surface (Fytianos et al., 2003). The energy term in Langmuir equation varies as a function of the surface coverage (Fytianos et al., 2003). The applicability of the isotherm equation was judged by the correlation coefficients, R^2. The linear form of Langmuir's isotherm model is given by the equation:

$$\frac{C_e}{q_e^{/}} = \frac{C_e}{q_{max}} + \frac{1}{q_{max} \cdot b} \tag{6}$$

C_e is the equilibrium concentration of the adsorbate (heavy metal ion) (mg/L); q_e, the amount of adsorbate adsorbed per unit mass of adsorbent (mg/g); q_{max} and *b* are Langmuir constants which are related to monolayer adsorption capacity and affinity of adsorbent towards adsorbate respectively. The important characteristic of the Langmuir isotherm is expressed in terms of a dimensionless equilibrium constant separation parameter (R_L) (Saifuddin and Kumaran, 2005), which is defined as:

$$R_L = \frac{1}{(1 + b \cdot C_0)} \tag{7}$$

Where: b = the Langmuir constant, C_o = the highest metal ion concentration (mg/L).

Table 1. Weight of bio-sorbent obtained during production.

Stages of production	Dry weight (g)
Ground Shell	200.00
Demineralized Shell	56.23
Deproteinized Shell (Chitin)	50.26
Deacetyled Shell (Chitosan)	40.50

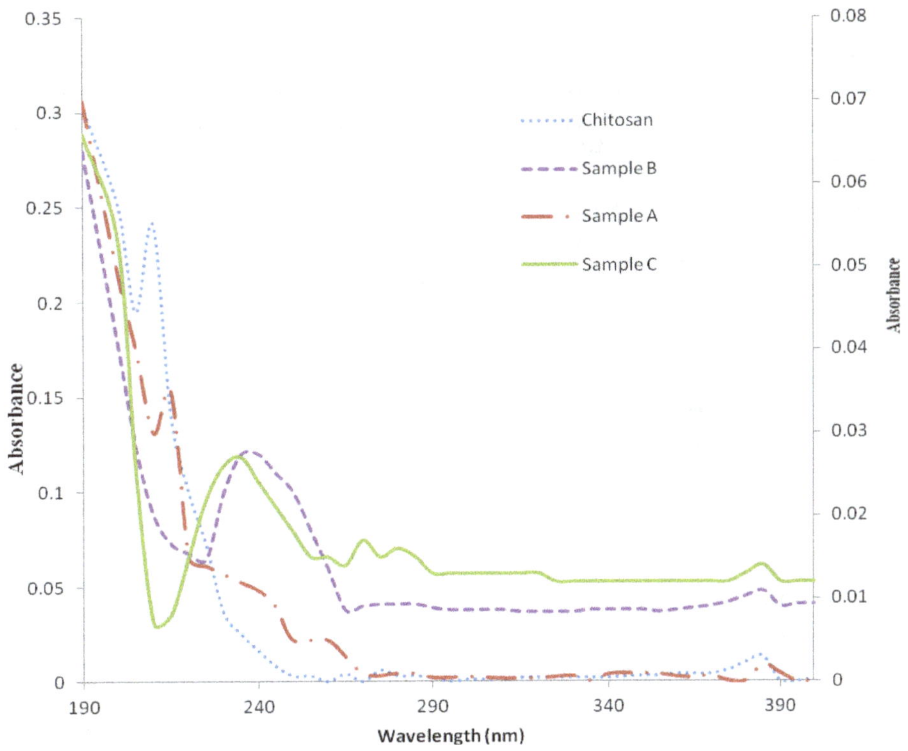

Figure 2. Electronic absorption spectra showing variation of the absorbance maximum as a function of metal ion concentration in solution. Sample A: Aso well water; Sample B: Ebu Stream water; Sample C: Osun river water. The curves of Samples A and C are plotted against the secondary axis.

The value of R_L indicates the shape or type of the isotherm to be either unfavourable ($R_L > 1$), linear ($R_L = 1$), favourable ($0 < R_L < 1$) or irreversible ($R_L = 0$) (Saifuddin and Kumaran, 2005).

RESULTS AND DISCUSSION

Chitosan yield and characterisation

Yield of chitosan

The yield after each process is as presented in Table 1. The snail shell yielded approximately 20% of chitosan. This implies that snail shells (*Helix pomatia*) can be considered as a better source of chitosan compared to freshwater crab (*Potamon potamios*) whose yield is about 4.65% (Bolat et al., 2010), however, the yield can be improved if loss of mass due to washing can be reduced to the minimum possible. A contributing factor to this significant amount of yield is the non-edibility of the snail shell itself.

Characterisation of the prepared chitosan

The degree of deacetylation (%DD) of this snail based as calculated using potentiometric titration following Tolaimate et al. (2002) was found to be 65%. The electronic absorption spectra of pure and used chitosan sample depicting the effect of chelation on the spectra properties of chitosan is as shown in Figure 2. The

Table 2. The pH of the water samples and the initial concentration of heavy metal ions in the samples.

Sample	Concentration (mg/L)							
	pH	T(^{0}C)	Fe	Cd	Zn	Cu	Cr	Pb
A	6.94	27.7	0.0237	0.0007	0.0077	0.0004	0.0016	0.0019
B	8.90	29.4	0.0702	0.288	0.0275	0.0011	0.0036	2.831
C	8.33	28.7	0.0710	0.151	0.0139	0.0012	0.0040	1.687

Sample A: Aso well water; Sample B: Ebu Stream water; Sample C: Osun river water.

Table 3. Conc. of heavy metals ion in samples after treatment with chitosan.

Sample	Concentration (mg/L)					
	Fe	Cd	Zn	Cu	Cr	Pb
A	nd*	nd*	nd*	nd*	0.0001	0.0002
B	0.0163	0.0008	nd*	nd*	0.0001	0.0133
C	0.0153	0.0001	nd*	nd*	0.0004	0.0135

Sample A: Aso well water; Sample B: Ebu Stream water; Sample C: Osun river water, nd*: not detected.

Table 4. Maximum permissible level of metal ions in drinking water.

Metals	SON (mg/L)	WHO (mg/L)
Pb	0.01	0.01
Cd	0.003	0.003
Cr	0.05	0.05
Cu	2	1
Zn	-	3
Fe	-	0.3

(NIS, 2007; WHO, 2004).

prepared chitosan absorbs at a wavelength of about 210 nm which is close to the characteristic bands (208 nm) reported for highly deacetylated chitosan (Tolaimate et al., 2002). There is shift in the wavelength (λ_{max}) of absorption of the treated chitosan with each water sample (215, 235 and 236 nm for Aso well, Ebu stream and Osun river water treated chitosan respectively). This bathochromic shift indicated the binding of metal ions with chitosan which is one of its potential applications.

Evaluation of heavy metals in the sample

The concentration of heavy metals in the prepared chitosan-treated and untreated water samples as reported in Tables 2 and 3 indicated the prevalence of six heavy metals (Fe, Cd, Zn, Cu, Cr, and Pb) in all the three samples with Odo-ebu stream (OBS) and Osun River (OSR), having higher metal concentrations compared with AWW water. The pH and temperatures are also functions of these metal concentration accumulation. The level of contamination by heavy metal in Ebu stream and Osun River follow the order: Pb > Cd > Fe > Zn > Cr > Cu while for Aso well water, it was Fe > Zn > Pb > Cr > Cd > Cu. OSR and OBS samples whose sources are not too far from dumpsite contain high concentrations of Pb ions, followed by Cd, thus making both samples the most contaminated of the three water samples. Other metal ions concentration in OBS and OSR samples are also very high when compared with AWW sample. Although, sample from AWW can be said to be unpolluted since the concentrations of most of the analyzed heavy metals are below the allowable standards of SON and WHO limits for drinking water (Table 4), but it may not be declared as being totally safe for drinking due to the risk of bioaccumulation in the human body (Majolagbe et al., 2013; Sekabira, 2010). The high concentrations of these metals can be attributed to leaching from dumpsite which is about 50 m to Ebun stream, and 100 m to Osun River. The high Pb concentration could be coming from exhaust from heavy vehicles (trucks) that ply the routes and burning of refuse at the dumpsite. The activity of

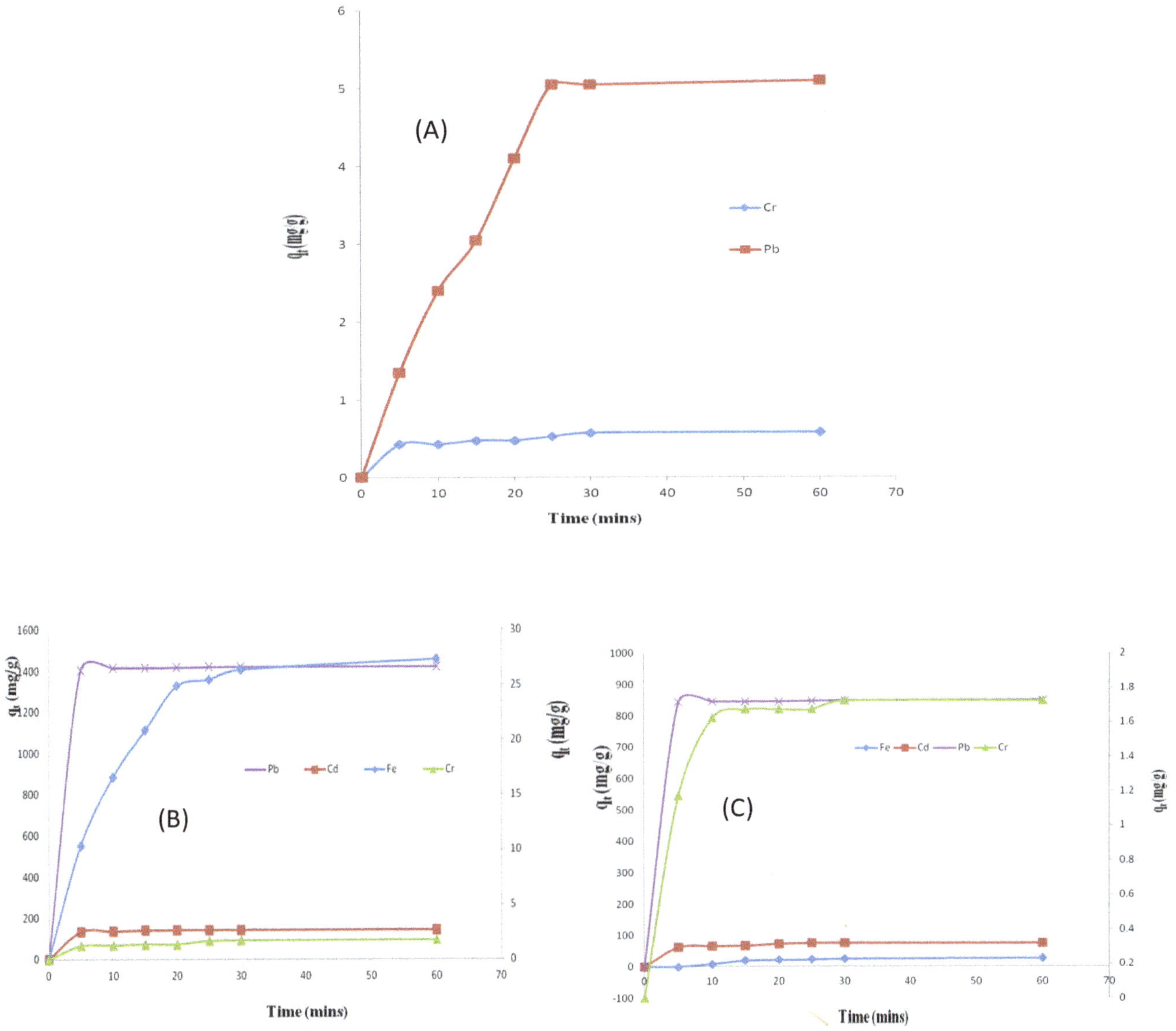

Figure 3. (a) Variation in Adsorption Capacity of chitosan with time for Sample A (Aso well water), **(b)** Variation in Adsorption Capacity of chitosan with time for Sample B (Ebu stream), **(C)** Variation in adsorption capacity of chitosan with time for sample C (Osun river water). The absorption of Cd and Fe were not considered in sample A as the initial concentration of these metals in the sample was significantly low (Table 2). In Figure 3b, Fe and Cr were plotted via the secondary axis ditto Cr in Figure 3c.

inhabitant e.g. washing of clothes and vehicles along the bank of Osun River could also be responsible for high concentration of the other metals. Some of these heavy metals have been reported to cause vomiting, dizziness, mortality, morbidity, pulmonary disorder and haematological disorder and cancer (Majolagbe et al., 2013).

Effect of contact time and adsorption capacity of chitosan

The time profiles of metal ions sorption by chitosan carried out on the three water samples at 25°C are presented in Figure 3. The amount of metal ions sorbed increased with contact time before plateauing, beyond which no more ions were removed from the samples (the concentrations of heavy metal ions left in the sample after treatment with chitosan are presented in Table 3). At this point, equilibrium has been reached between the amount of the heavy metal ions desorbed from the chitosan and heavy metal ions sorbed onto it. The selectivity of chitosan for the heavy metals ions (Pb, Cd, Cr, Fe) vary while Cu and Zn were below the detection limit. This can be explained on the basis of these metal ions ionic radii, hydration energies and their concentration in solution

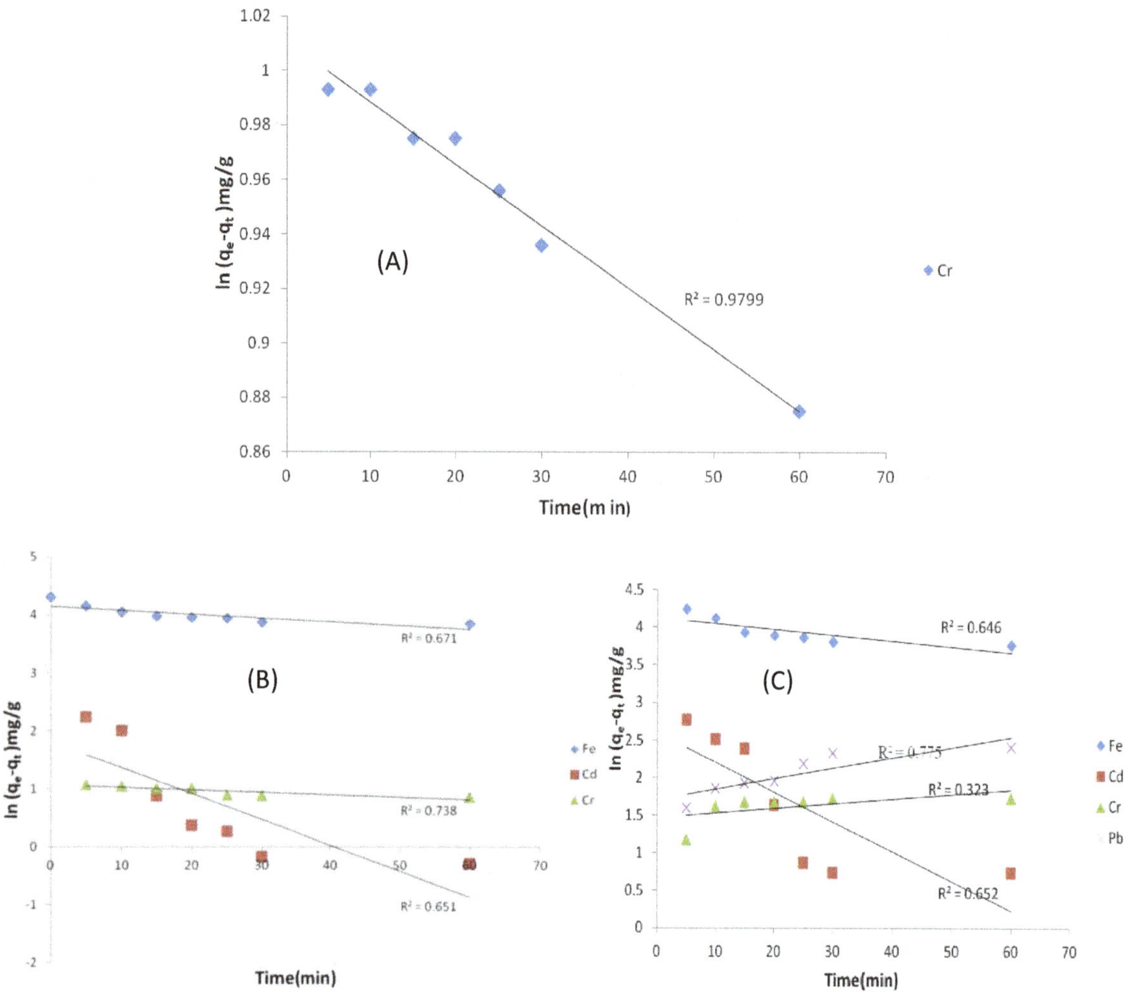

Figure 4. **(a)** Pseudo-first order adsorption kinetic plot for Sample A (Aso well), (b) Pseudo-first order adsorption kinetic plot for Sample B (Ebu stream), **(c)** Pseudo-first order adsorption kinetic plot for Sample C.

(Rhazi et al., 2002). The time required to attain this state of equilibrium is termed the equilibrium time, and the amount of heavy metal ions adsorbed at the equilibrium time reflects the maximum adsorption capacity of the adsorbent under those operating conditions. It was found that 96.04% of Pb, 99.83% of Cd, 94.47% of Cr and 77.6% of Fe were bound to chitosan within 15 min. The initial high amount of metal ions sorbed indicates instantaneous sorption, which can be attributed to the availability of binding sites on the sorbents. However, as these sites progressively react, the sorption of metal ions slowed before attaining equilibrium. The ability of chitosan to bind a large amount of metal ions within 15 min suggests that they are effective sorbents. Furthermore, the rapid kinetics has significant practical importance as it will facilitate the application to smaller reactor volumes ensuring efficiency and economy (Kamari et al., 2011).

The maximum adsorption capacities of each heavy metal based on the samples are presented in the Table 1. This further shows the selectivity of chitosan for adsorbing particular heavy metals ions when more than one heavy metal ions is present in solution. Chitosan showed the highest maximum adsorption capacity for Pb amongst the four heavy metal ions detected after metals sorption in OBS and OSR water samples. The selectivity of the snail shell based chitosan in these samples for the metal ions follow the sequence: Pb > Cd > Fe > Cr.

Adsorption kinetics

The kinetic parameters and their values obtained for the heavy metals sorption processes of the prepared chitosan on the three water samples are presented in Figures 4(a – c), 5(a - c) and Table 5. The experimental equilibrium sorption capacities (q_e experimental) determined from the contact time study were in good agreement with the theoretical equilibrium sorption

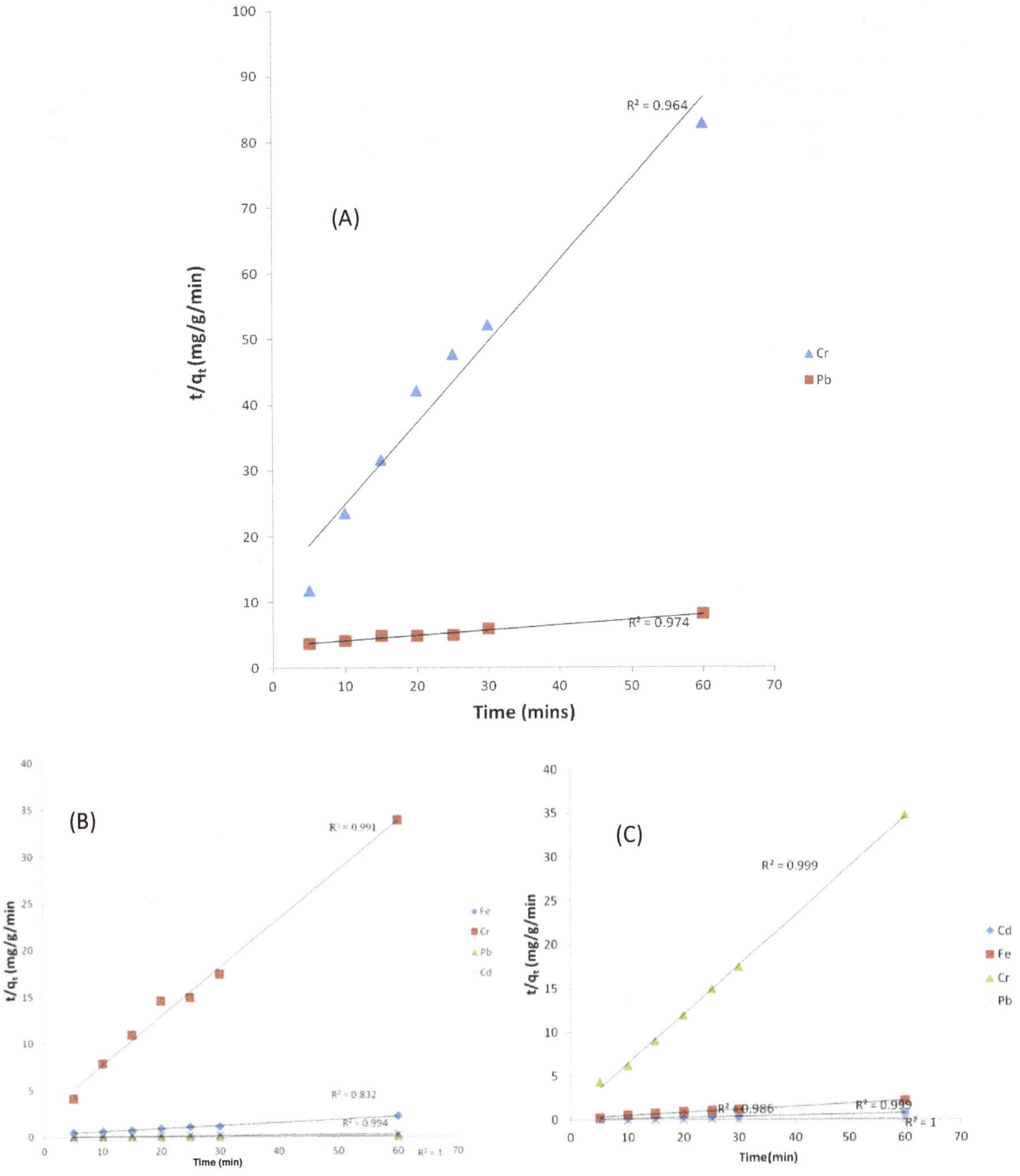

Figure 5. (**a**) Pseudo-second order adsorption kinetics plot for Cr and Pb in sample A, (b) Pseudo-second-order adsorption kinetic plot for Sample B, (C) Pseudo-second order adsorption kinetic plot for Sample C.

capacities (q_e theoretical) calculated from the pseudo-second-order kinetic model than for the pseudo-first-order. Moreover, experimental sorption data correlated well to the pseudo-second-order kinetic model (R^2 values being greater than 0.95 in most cases except for Cd metal sorption and range from 0.984 to 0.999). Lagergren's pseudo-first order and Intra-particle diffusion equations did not fit well for those metals they can be plotted, the R^2 values are relatively small and in some cases, having negative values. This had earlier been reported by Kamari et al. (2011) and Mohan et al. (2006) who concluded that pseudo-first-order equations did not fit well to the whole range of contact time for heavy metal sorption processes of chitosan but were only applicable over the initial stages of the processes of adsorption The results suggest that the binding of metal ions studied onto

Table 5. Comparison of the Intra-particle diffusion, pseudo first- and pseudo second-order adsorption rate constants, calculated and experimental q_e, values for the heavy metals ions in the water samples.

Metal	$q_{e\ exp.}$ (mg/g)	Intra-particle diffusion		Pseudo-first order model			Pseudo-second order model		
		k_{id} (mg/g/min $^{0.5}$)	R^2	$q_{e\ cal.}$ (mg/g)	K_1 (10^{-03}) (g/mg/min)	R^2	$q_{e\ cal.}$ (mg/g)	K_2 (10^{-03}) (g/mg/min)	R^2
				Sample A					
Pb	5.05	0.805	0.67	-	-	-	12.59	1.87	0.97
Cr	0.58	0.106	0.61	2.75	2.03	0.98	0.81	123	0.96
				Sample B					
Pb	1.42×10^3	274.8	0.092	12.86	-14.00	0.37	1.0×10^3	25.03	0.99
Cr	1.73	0.365	-4.81	2.90	4.00	0.74	1.92	1110	0.99
Cd	1.44×10^2	27.61	0.17	6.11	44.00	0.65	333.3	0.173	0.83
Fe	2.64×10^1	4.554	0.52	63.50	6.03	0.67	32.26	3.44	0.99
				Sample C					
Pb	846.33	164.2	0.089	5.55	-13.22	0.775	733	0.15	0.99
Cr	1.73	0.320	0.43	4.34	-6.01	0.323	1.78	395	1.00
Cd	75.63	13.96	0.40	13.49	39.04	0.652	83.30	7.23	0.99
Fe	25.68	4.089	0.53	62.18	7.05	0.646	30.30	5.45	0.99

Table 6. Langmuir Isotherm constant for heavy metal analysed in the samples.

Metal	b(X 10^3)	q_{max}	R_L(X10^{-7})
	Sample B		
Pb	7.00	1428.57	1.00
Cd	18.5	135.14	4.00
Cr	26.7	1.25	30.0
	Sample C		
Pb	40	833.33	0.30
Cd	3.18	62.89	50.00
Cr	4.50	1.11	2,000.00

chitosans was best described by the pseudo-second order kinetic model and that the chemical binding reaction was the rate-limiting step as discussed by Ho and McKay (2000).

Adsorption isotherm

The adsorption isotherm indicates how the adsorbed molecules distribute themselves between the liquid phase and the solid phase when the adsorption process reaches an equilibrium state. The analysis of equilibrium adsorption data by fitting them to an isotherm models is an important step to show that the Langmuir model can be used for design purpose (Haghseresht and Lu, 1998). When C_e/q_e was plotted against C_e, a straight line with slope, $1/q_{max}$ was obtained, and are presented in Figure 6(a – c). The Langmuir constants 'b' and 'q_{max}' were calculated from this isotherm and their values are presented in Tables 6. The large values of b signify that the chitosan has a very high affinity for the heavy metals coupled with a high binding energy. The conformation of the experimental data, q_{max}, to that obtained using Langmuir isotherm model indicated the homogeneous nature of snail shell based chitosan surface, i.e., each metal ion molecule/chitosan biosorption has equal adsorption activation energy. The linear plot obtained further buttress the monolayer form of adsorption. Also from Table 4(a and b), values of R_L were found to be less than 1 but greater than zero in all cases. This confirmed that the biosorbent prepared from the snail shell is favourable for biosorption of Pb, Cd, and Cr ions

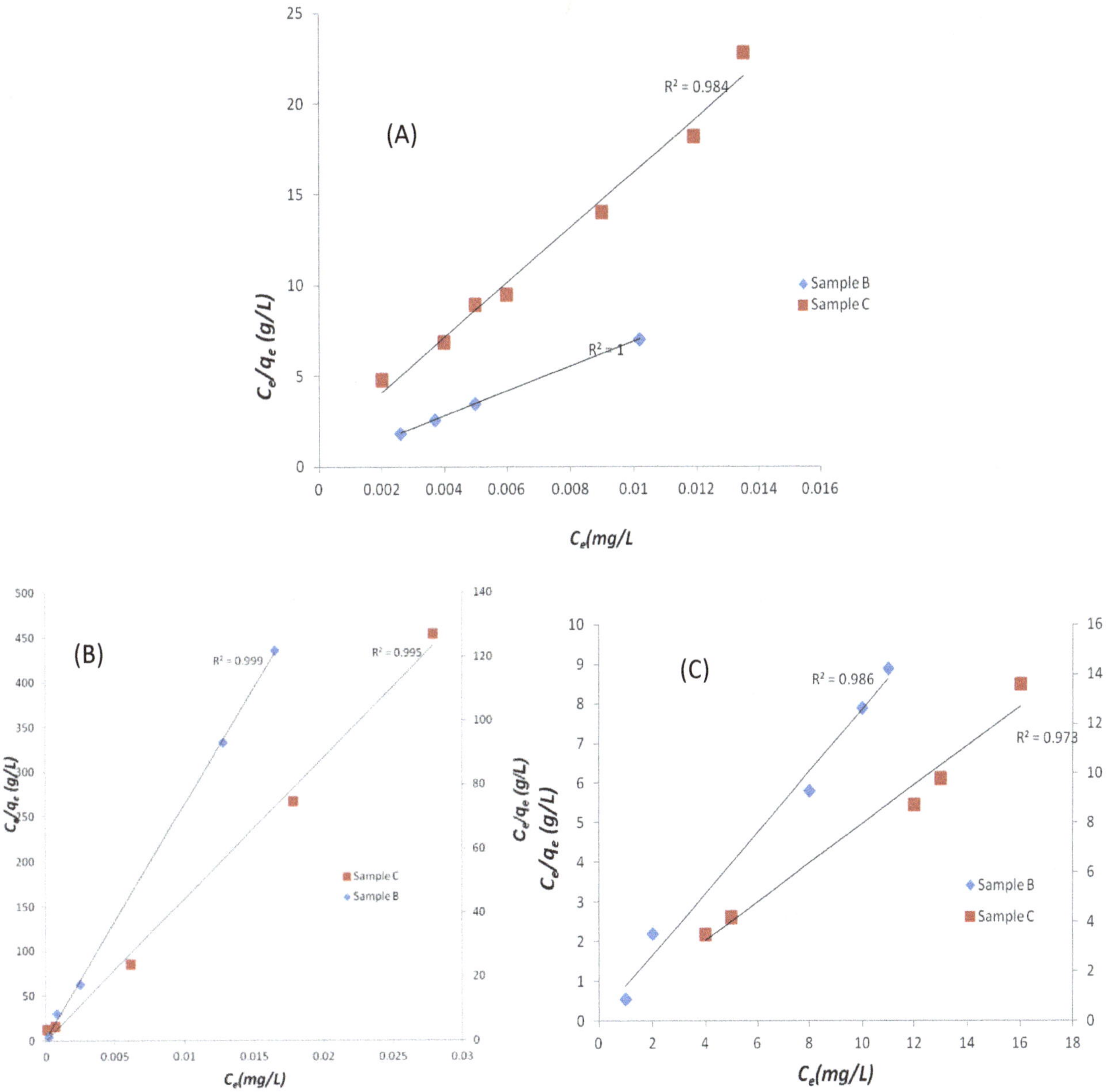

Figure 6 (a) Langmuir adsorption isotherm for the biosorption of lead ions, (b) Langmuir adsorption isotherm for the biosorption of Cadmium ions. sample B is plotted against the secondary axis, (c) Langmuir adsorption isotherm for the biosorption of chromium ions at 25 °C. Sample C is plotted against the secondary.

under the conditions used in this study.

Conclusion

The findings from this study have shown that the Ebu stream and Osun river water samples were highly polluted with heavy metals and chitosan has great affinity for these metal ions in the analyzed samples in the following order: Pb > Cd > Fe > Cr. The adsorption kinetics follows pseudo-second order and the experimental equilibrium sorption capacities determined from the contact time study were in good agreement with the theoretical equilibrium sorption capacities calculated from the kinetic model. The Langmuir model fitted the isotherm equilibrium data with $R^2 \geq 0.985$. The values of maximum sorption capacities also correlated well with those determined theoretically and experimentally. R_L values obtained also confirmed that the prepared biosorbent is favorable for biosorption of heavy metals. Similar results were reported by Kamari et al. (2011) for crab shell based chitosan and Adeogun et al. (2010)

using adsorbent obtained from plumb shell.

Conflict of Interest

The authors have not declared any conflict of interest.

REFERENCES

Adeogun AI, Bello OS, Adeoye MD (2010). Biosorption of lead ions on biosorbent prepared from plumb shells (*Spondias mombin*): Kinetics and equilibrium studies. Pak. J. Sci. Ind. Res. 53(5):246-251.

Adewuyi S, Adeoye MD, Sanyaolu NO (2009). Equillibruim removal efficiency of Nickel metal ion by chitosan. Proc. of the 33rd ann. Inter. Conf. Chem. Soc. Nig. (CSN), Abeokuta, Nigeria. pp. 403-405.

Bolat YŞ, Bilgin A, Günlü L, Izci SB, Koca SÇ, Koca HU (2010). Chitin Chitosan yield of freshwater crab (*Potamon potamios*, Olivier 1804) shell. Pak. Vet. J. 30(4):227-231.

Crini G, Badot PM (2008). Application of chitosan, a natural polysaccharide, for dye removal from aqueous solutions by adsorption process using batch studies: a review of recent literature. Prog. Poly. Sci. 33:399-447. http://dx.doi.org/10.1016/j.progpolymsci.2007.11.001

Eniola EB, Chukwu LO, Olaide BS (2010). Hydro-chemistry, macro-invertebrate fauna and fish production of acdja fishing sites in a Tropical Lagoonal Ecosystem. J. Am. Sci. 6(1):22-28.

Fytianos K, Voudrias E, Kokkalis, E (2003). Sorption-desorption behaviour of 2,4-dichlorophenol by marine sediments. Chemosphere 40:3-6. http://dx.doi.org/10.1016/S0045-6535(99)00214-3

Haghseresht F, Lu GQ (1998). Adsorption characteristics of phenolic compounds onto coal-reject-derived adsorbents. Ener. Fuels. 12:1100-1107. http://dx.doi.org/10.1021/ef9801165

Ho YS, McKay G (2000). The kinetics of sorption of divalent metal ions onto sphagnum moss peat. Water Res. 34:735-742. http://dx.doi.org/10.1016/S0043-1354(99)00232-8

Jude CI, Augustin AA (2007). Equilibrium sorption isotherm studies of Cd(II), Pb(II) and Zn(II) ions detoxification from waste water using unmodified and EDTA-modified maize husk. Elect. J. Biotech. 10(4): ISSN: 0717-3458. DOI: 10.2225/vol10-issue4-fulltext-15. http://dx.doi.org/10.2225/vol10-issue4-fulltext-15

Kamari A, Pulford ID, Hargreaves JSJ (2011). Binding of heavy metal contaminants onto chitosans: An evaluation for remediation of metal contaminated soil and water. J. Env. Mgt. 92:2675-2682. http://dx.doi.org/10.1016/j.jenvman.2011.06.005

Kanamadi RD, Ahalya N, Ramachandra TV (2003). Review Paper: Biosorption of Heavy Metals. Res. J. Chem. Environ. 7(4):71-79.

Lagergren S (1898). Zur theorie der sogenannten adsorption gelöster stoffe. Kungliga Svenska Vetenskapsakademiens. *Handlingar* 24:1-39.

Langmuir I (1916). The constitution and fundamental properties of solids and liquids. J. Am. Chem. Soc. 38:2221-2295. http://dx.doi.org/10.1021/ja02268a002

Majolagbe TA, Azeez L, Salawu HO, Adeoye MD, Lawal AT, Agbaogun BKO, Tijani KO (2013). Assessment of heavy metal pollution in soils and wells around municipal dumpsites in Osogbo metropolis. Environ. Sci.-An Ind. J. 8(2):56-61.

Martha B (2008). Adsorption of Metallic Ions onto Chitosan: Equilibrium and Kinetic Studies Royal Institute of Technology. Department of Chemical Engineering and Technology. Division of Transport Phenomena, Stockholm, Sweden. Licentiate Thesis pp. 3-4.

Mohan D, Pittman Jr CU, Steele PH (2006). Single, binary and multi-component adsorption of copper and cadmium from aqueous solutions on Kraft lignin-a biosorbent. J. Colloid. Interf. Sci. 297:489-504. http://dx.doi.org/10.1016/j.jcis.2005.11.023

Mustapha MK (2008). Assessment of the water Quality of Oyun Reservoir, Offa, Nigeria, Using Selected Physico-Chemical Parameters. Tur. J. Fish. Aq. Sci. 8:309-319.

Olajire AA, Imeokparia FE (2001). Water quality assessment of osun river: studies on inorganic nutrients. Environ. Monit. Assess. 69:17-28. http://dx.doi.org/10.1023/A:1010796410829

Pan X, Wang J, Zhang D (2009). Sorption of cobalt to bone char kinetics, competitive sorption and mechanism. Desal. 249:609-614 http://dx.doi.org/10.1016/j.desal.2009.01.027

Rhazi M, Desbrieres J, Tolaimate A, Rinaudo M, Vottero P, Alagui A El Meray M (2002). Influence of the nature of the metal ions on the complexation with chitosan. Application to the treatment of liquid waste. Euro. Poly. J. 38:1523-1530 *www.elsevier.com/locate/europolj*.

Saifuddin MN, Kumaran P (2005). Removal of heavy metal from industrial wastewater using chitosan coated oil palm shell charcoal Elec. J. Biotech. 8(1):22-25 ISSN: 0717-3458 DOI: 10.2225.

Sallau AA, Ekanem EO, Abubakar AB (2011). The 34th Annual International Conference of the Chemical Society of Nigeria (CSN) at Unilorin, EN001- EN007 Sattelite map of osogbo. <a href=http://www.maplandia.com/nigeria/osun/osogbo/.

Sekabira K , Origa HO, Basamba TA , Mutumba, G, Kakudidi E (2010). Int. J. Environ. Sci. Tech. 7(3):435-446 http://dx.doi.org/10.1007/BF03326153

Standard Organization of Nigeria (2007). Nigerian Standard for Drinking Water Quality. pp.1-30.

Sud D, Mahajan G, Kaur MP (2008). Agricultural waste material as potential adsorbent for sequestering heavy metal ions from aqueous solutions - A review. Bioresour. Technol. 99:6017-6027 http://dx.doi.org/10.1016/j.biortech.2007.11.064

Taboada S, Carbrera G, Cardenas G (2003). Retention capacity of chitosan for copper and Mercury ions. J. Chilian Chem. Soc. 48(1):10.

Tolaimate A, Rhazi M, Desbrieres J, Rinaudo M, Vottero P, Alagui A, El-Meray M (2002). Influence of the nature of the metal ions on the complexation with chitosan. Application to the treatment of liquid waste. Eur. Polymer J. 38:1523-1530. http://dx.doi.org/10.1016/S0014-3057(02)00026-5

Volesky B (2001). Detoxification of metal-bearing effluents: biosorption for the next century. Hydromet. 59:203-216. http://dx.doi.org/10.1016/S0304-386X(00)00160-2

World Health Organization (WHO) (2004). The Guidelines for Drinking Water Quality, Third edition. *www.who.int/water_sanitation_health/dwq/gdwq3rev*

Yildiz B, Şengül B, Ali G, Levent I, Seval BK, Soner Ç, Habil UK (2010). Chitin-Chitosan Yield of Freshwater Crab (*Potamon potamios*, Olivier 1804) Shell. *Pak. Vet. J.* ISSN: 0253-8318 (PRINT), 2074-7764 (ONLINE). Accessible at: www.pvj.com.pk. 675-2682

14

Photolysis of 4-oxazoline-2-one derivative in diverse solvent environments

Singh I. S.

Department of Physical Sciences, School of Mathematics and Natural Sciences, The Copperbelt University, P. O. Box 21692, Kitwe, Zambia. E-mail: indra@cbu.ac.zm.

The photolysis of 4,5-diphenyl-3-(4-methylphenyl)-4-oxazolin-2-one in benzene or cyclohexane in the presence of singlet oxygen using UV light affords benzil, N-4-methylphenyl benzamide and N,N-dibenzoyl-4-methyl aniline. When the same reaction is carried out in protic solvents, the formation of benzil, N-4-methylphenyl benzamide, N,N-dibenzoyl-4-methyl aniline and benzoic acid is observed. The characterization uses spectral data and comparison (undepressed mixed melting point and identical IR spectra) with authentic samples. This study derives an interesting route for photooxidative fragmentation of oxazoline-2-one ring systems. The proposed mechanism involves the dioxetane derivative which after fragmentation yields radicals to afford aforementioned products.

Key words: Irradiation, column, chromatographic separation, evaporation, solvents.

INTRODUCTION

The oxazoline ring system has assumed marked significance due to their wide application (Frump, 1971). The synthetic utility of 2-oxazolines has been reviewed for aromatic systems in which they go for direct o-lithiation of benzene molecule (Reuman and Meyers, 1985). The reaction of dianions from ketomonoimines with ethyl chloroformate gives substituted 4-oxazoline-2-ones (Mehrotra et al., 1985). The thermal rearrangement of isoxazoline-3-one to 4-oxazoline-2-one has also been observed (Gagneux and Goschke, 1966). Because of the wide application of such compounds, it would be interesting and extremely significant to study the photochemistry of such systems. The photosensitized oxidation of 3,4,5-triphenyl-4-oxazoline-2-one affording a mixture of benzoic acid, benzanilide and *ortho-* and *para-*benzamide benzophenone has been reported (Tsugea et al., 1976). A similar photodecarbonylation of a cyclic carbonate of 1,2-diketone derivative has also been reported (Stahlke et al., 1972). The reactions of singlet oxygen with organic substances are very sensitive to the experimental environment (Chan et al., 1990). As such, the present study attempts to explore the effect of several protic and aprotic solvent systems.

This study uses singlet oxygen and aims to explore the effects of protic and aprotic solvents on photooxidative cleavage of 4-oxazolin-2-one ring system. In order to provide further insight in the mechanistic details of the fragmentation pattern of substituted 4-oxazoline ring systems, different protic and aprotic solvent combinations were used. In this study, photooxidative cleavage of 4,5-diphenyl-3-(4-methylphenyl)-4-oxazolin-2-one in protic and aprotic solvents has been discussed. The reaction route explored in this study demonstrates an interesting photochemistry.

MATERIALS AND METHODS

Instrumentation

Chemicals and reagents were purchased from Aldrich Chemical Co. and used without further purification. Melting points were determined in capillaries on melting point apparatus and are uncorrected. The NMR spectra in $CDCl_3$ were recorded on Varian A-60D and VN 1009 (S-60T) spectrophotometer with TMS as an internal standard. The IR spectra were measured on a Perkin Elmer 720 spectrophotometer and UV spectra on a Beckman DB-G spectrophotometer. Photochemical experiments were carried out using Hanovia type A, 450 W medium pressure mercury immersion lamp as light source.

Preparation of 2-Imino-1,2-diphenylethanone

The starting 2-imino-1,2-diphenylethanone was obtained by heating

Table 1. Results of chromatographic separation of irradiated sample in benzene.

Fraction (50 ml)	Eluent	Product type	Yield
1 - 10	n-hexane	No product	No yield
11 - 18	n-hexane	Solid	0.18 g
19 - 30	n-hexane-benzene (1:1)	Solid	0.11 g
31 - 36	Benzene	Viscous liquid	No yield

Table 2. Results of chromatographic separation of irradiated sample in methanol-benzene mixture.

Fraction (50 ml)	Eluent	Product type	Yield
1 - 10	Pet. Ether (petroleum ether)	No product	No yield
11 - 15	Pet. ether	Solid	0.08 g
16 - 20	Pet. ether-benzene(4:1)	Solid	0.06 g
21 - 30	Pet.ether benzene(1:1)	Solid	0.16 g
31 - 40	Benzene	Viscous liquid	No yield

an equimolar mixture of benzil and 4-methylphenylamine at 150℃ for 2 to 3 h following the reported method and purified by fractional crystallization from ethanol (75 to 78%) (Padwa, 1965).

Preparation of 4,5-diphenyl-3-(4-methylphenyl)-4-oxazoline-2-one (1)

The starting material (**1**) was prepared by the reaction of dianions from a suitable 2-imino-1,2-diphenylethanone with ethyl chloroformate and purified by using a reported method by Mehrotra et al. (1985) . UV (Ethanol): 294(ε1.89 × 10^4) nm; v_{max}(Nujol): 1600 cm^{-1} (C = C) and 1750 cm^{-1}(C = O); δ_H ($CDCl_3$): 7.42 (m, 14H, aromatic protons); 2.33 (s, 3H, methyl); v_{max}(Nujol): 1600 cm^{-1} (C = C) and 1750 cm^{-1} (C = O).

General procedure for irradiation

A dilute solution containing 0.5 g of oxazoline (**1**) in 300 mL of benzene (thiophene free) or cyclohexane or a mixture of benzene and protic (methanol, ethanol or isopropanol) solvent was irradiated with UV light from a 450W Hanovia lamp for 9.5 h while oxygen was continuously bubbled into the solution to ensure proper mixing as well. The course of the reaction was monitored by UV and TLC until the spot corresponding to the starting material disappeared.

Irradiation of 4,5-diphenyl-3-(4-methylphenyl)-4-oxazoline-2-one (1)in benzene in presence of oxygen

A dilute solution containing 0.5 g of (**1**) in 300 mL of benzene was irradiated for 9.5 h according to the general procedure as described earlier. Three such runs were combined. The solvent was removed under reduced pressure and the residue was triturated with ethanol to give 0.28 g (19.5%) of solid, mp 142 to 44℃, iden tified as N, N-dibenzoyl-4-methylaniline; δ_H ($CDCl_3$): 7.4 (m, 14H, arom); 2.3 (s, 3H, CH_3); v_{max} (Nulol): 1700 cm^{-1}(CO). The mother liquor was concentrated and chromatographed on a column of silica gel (30 g, 10°, 2.2 cm) and the results are given in Table 1.

Evaporation of solvent from the fractions 11 to 18 under reduced pressure afforded 0.18 g (19%) of a solid, mp 95℃, i dentified as benzil. The IR spectrum was superimposable to that of an authentic sample. Evaporation of solvent from the fractions 19 to 30 under reduced pressure gave 0.11g (11.3%) of a solid identified as N-4-methylphenyl benzamide; mp 157℃; δ_H ($CDCl_3$): 7.50 (m, 9H, arom); 3.4 (bs, H, NH, D_2O exchangeable) and 2.3 (s, 3H, CH_3); v_{max} (Nujol): 3300 cm^{-1} (NH), 1620 cm^{-1} (CO). Evaporation of the solvent from the fractions 31 to 36 under reduced pressure gave a viscous liquid which, however, could not be identified.

Irradiation of 4,5-diphenyl-3-(4-methylphenyl)-4-oxazoline-2-one (1) in cyclohexane in presence of oxygen

A dilute solution containing 0.50 g of (**1**) in 300 mL of cyclohexane was irradiated for 9.5 h following the general procedure as described earlier. Three such runs were combined. The solvent was removed under reduced pressure and the residue was triturated with ethanol to give 0.29 g (20%) of solid, identified as N,N-dibenzoyl-4-methylaniline as described earlier. The mother liquor was concentrated and chromatographed on a column of silica gel (30 g, 12 × 2.2 cm) as described earlier. Two products were obtained which were identified as benzil 0.19 g (19.7%) and N-4-methylphenyl benzamide, 0.12 g (12.4%) as described earlier when benzene was used as the solvent.

Irradiation of 4,5-diphenyl-3-(4-methylphenyl)-4-oxazoline-2-one (1) in (1:1) methanol-benzene mixture in the presence of oxygen

A dilute solution containing 0.50 g of (**1**) in 300 mL of methanol-benzene mixture was irradiated for 9.5 h according to the general procedure. Three such runs were combined. The solvent was removed under reduced pressure. The residue was triturated with ethanol to give 0.2 g (13.8%) of a solid, mp 142 to 44℃, identified as N,N-dibenzoyl-4-methylaniline on the basis of spectral data as described earlier. The mother liquor was concentrated and chromatographed on a column of silica gel (30 g, 10 × 2.2 cm) and the results are given in Table 2.

Evaporation of the solvent from the fractions 11 to 15 afforded a solid 0.08 g (8.3%), mp and mixed mp 95℃. The IR s pectrum was superimposable to that of an authentic sample of benzil.

Evaporation of the solvent from the fractions 16 to 20 under reduced pressure gave a solid 0.06 g (10.7%), mp and mixed mp 121℃. The IR spectrum was superimposable to that of an authentic sample of benzoic acid. Evaporation of the solvent from fractions 21

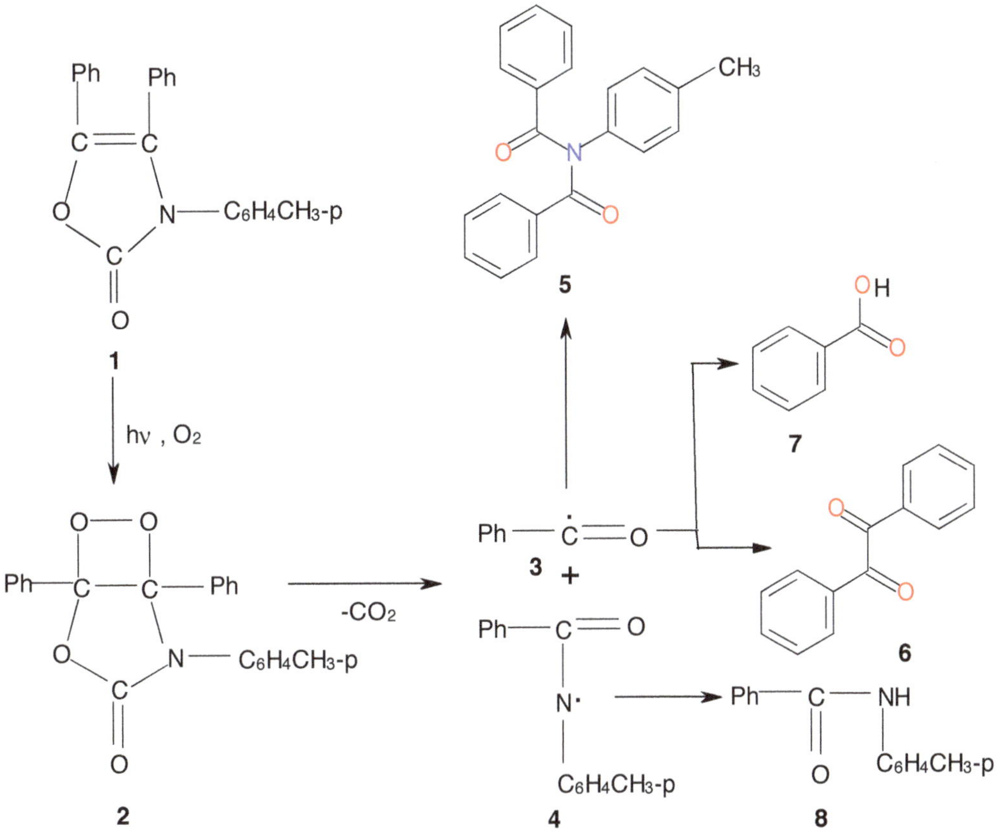

Scheme 1. The reaction route for the formation of products.

to 30 gave 0.16 g (16.5%) of a solid, mp 157℃, identified as N-4-methylphenyl benzamide on the basis of spectral data as described earlier. Evaporation of the solvent from the fractions 31 to 40 under reduced pressure gave a viscous liquid which, however, could not be identified.

Irradiation of 4,5-diphenyl-3-(4-methylphenyl)-4-oxazoline-2-one (1) in ethanol-benzene(1:1) mixture in the presence of oxygen

A similar irradiation of a solution containing 0.50 g of (1) in 300 mL of ethanol benzene mixture for 9.5 h according to the general procedure and usual work up of three such runs, as described earlier, gave four products which were identified as benzil 0.1 g (10.4%), benzoic acid (0.06 g, 10.7%), N,N-dibenzoyl-4-methylaniline 0.19 g (13.1%) and N-4-methylphenyl benzamide 0.28 g (28.9%).

Irradiation of 4,5-diphenyl-3-(4-methylphenyl)-4-oxazoline-2-one (1) in 2-propanol in the presence of oxygen

A similar irradiation of a solution containing 0.50 g of (**1**) in 300 mL of 2-propanol according to the general procedure for 9.5 h and the usual work up of three such runs, as described earlier, gave four solid products which were identified as benzil 0.08 g (8.3%), benzoic acid 0.1 g (17.8%), N,N-dibenzoyl-4-methylaniline 0.15 g (10.3%) and N-4-methylphenyl benzamide 0.38 g (39.3%).

RESULTS AND DISCUSSION

The starting material, 4,5-diphenyl-3-(4-methylphenyl)-4-oxazoline-2-one(1) was prepared using previously reported method by Mehrotra et al. (1985). Irradiation of a dilute solution of (1) in benzene in the presence of oxygen with UV light afforded benzil (6), N-4-methylphenyl benzamide (8) and N, N-dibenzoyl-4-methylaniline (5). Benzil was characterized by comparison (undepressed mixed melting point and superimposable identical IR spectrum) with authentic sample which was procured from Aldrich Chemical Company. N-4-methylphenyl benzamide was characterized by spectral data and comparison (undepressed mixed melting point) with authentic samples. The authentic sample of N-4-methylphenyl benzamide was prepared from the reaction of benzoyl chloride and p- toluidine in the presence of potassium hydroxide (Furniss et al., 1989). N, N-dibenzoyl-4-methylaniline has been characterized on the basis of spectral data. The reaction route for the formation of products by photooxidative fragmentation of (**1**) can be depicted in Scheme 1.

The addition of singlet oxygen to (**1**) gives the dioxetane derivative (2). The dioxetane derivative (**2**) after fragmentation and simultaneous loss of carbon dioxide could give rise to benzoyl radical (**3**) and N-4-methylphenyl benzamide radical (4). The benzoyl radical (3) dimerizes to yield benzil (6), and the amide radical (4) abstracts hydrogen to give N-4-methylphenylbenzamide (8). The radical (3) and (4) combine to result in N, N-dibenzoyl-4-methylaniline (5).

Irradiation of a dilute solution of (1) in protic solvents in the presence of singlet oxygen with UV light gave four products which were identified as benzil, benzoic acid (7), N-4-methylphenyl benzamide and N, N-dibenzoyl-4-methylaniline. The product benzil and benzoic acid were characterized by comparison (undepressed mixed melting point and identical IR spectra) with authentic samples. As reported earlier, the products (5) and (8) have been characterized on the basis of its spectral data. The reaction route for the formation of products by the fragmentation of (1) in protic solvents seems to be similar to Scheme 1. The additional product, benzoic acid may arise from the benzoyl radical (3).

The oxygenation of similar systems gave a mixture of benzoic acid, benzamide and ortho- and para-benzamide benzophenone (Tsugea et al., 1976). It has been observed that the reaction route involves a dioxetane derivative which is proposed on the basis of IR spectrum of photolysis mixture during the reaction. The IR spectrum exhibits a weak band at 880 cm^{-1} which gives a strong support for the presence of dioxetane which, however, could not be isolated (Bellamy, 1958). A similar photodecarbonylation of a cyclic carbonate to 1,2-diketone derivative has been reported (Stahlke et al., 1972). In this study, N-4-methylphenyl benzamide (8) is obtained by hydrogen abstraction by radical (4). It was also found that the yield of (8) increases substantially in protic solvents as compared to that in aprotic solvents which offers a strong support for the proposed mechanism in Scheme 1.

Conclusion

The photolysis of substituted 4-oxazoline-2-ones in aprotic solvents in oxygen atmosphere yields a 1,2-diketone, N-substituted benzamide and N, N-dibenzoyl derivative. The similar photolysis in protic solvents affords benzoic acid in addition to the aforementioned three products. The reaction route involves a dioxetane intermediate which subsequently cleaves into radicals to lead to the final products.

ACKNOWLEDGEMENTS

This work was sponsored by the Copperbelt University (CBU). The author is thankful to the Department of Chemical Engineering and Physical Sciences Department at the CBU for providing numerous facilities. The use of lab facilities at the Chemistry Department at B.H.U., India, and useful discussion with Prof. K.N Mehrotra is greatly acknowledged.

REFERENCES

Bellamy LJ (1958). The infrared spectra of complex molecules. 2nd Ed John Wiley and sons: Inc, New York, p. 313.

Chan YY, Li X, Zhu C, Liu X, Zhang Y, Leungg H (1990). Sensitized photooxygenation. 3.Mechanistic studies on the singlet oxygenation of 5,6-disubstituted-3,4-dihydro-2H-pyran. J. Org. Chem. 55: 5497.

Frump JA (1971). Oxazolines. Their preparation, reactions and applications. Chem. Rev., 71: 483.

Furniss BS, Hannaford AJ, Peter W, Smith G, Austin RT (1989). Vogel's Textbook of Practical Organic Chemistry. 5th Ed. Longman Scientific and Technical, Essex, England, p. 916.

Gagneux AR, Goschke R (1966). The thermal rearrangement of n-arylisoxazoline-3-ones. Tetrahedron Lett., p. 5451.

Mehrotra KN, Singh IS, Roy J (1985). Cyclization of Dianions from 2-Imino-1,2-diphenylethanone to Oxazolines, Oxazocines, and Oxazonines. Bull. Chem. Soc. Jpn., 58: 2399.

Padwa A (1965). Epoxidation studies III. The peracid oxidation of substituted benzoylimines. J. Am. Chem. Soc., 87: 4365.

Reuman M, Meyers AI (1985). The synthetic utility of oxazolines in aromatic substitution. Tetrahedron, 41: 837.

Stahlke KR, Heins HG, Hartman W (1972). Zur photochemie arylsubstituierter vinylencarbonate. Liebigs Ann. Chem., 116: 764.

Tsugea O, Oe K, Ueyama Y (1976). The photochemical reaction of 3,4,5-triphenyl-4-oxazolin-2-one. Chem. Letts., 1976: 425.

15

Adsorption of heavy metal ions onto chitosan grafted cocoa husk char

Aderonke A. Okoya[1], Abimbola B. Akinyele[1], Ifeanyi E. Ofoezie[1], Omotayo S. Amuda[2], Oluwagbemiga S. Alayande[3] and Oladotun W. Makinde[3]

[1]Institute of Ecology and Environmental Studies, Obafemi Awolowo University, Ile – Ife, Nigeria.
[2]Ladoke Akintola University of Technology, Ogbomoso, Nigeria.
[3]Center for Energy and Research Development, Obafemi Awolowo University, Ile – Ife, Nigeria.

The use of chitosan-grafted cocoa husk char (CCH) and unmodified cocoa husk char (CH) as adsorbents for the removal of Cr^{6+} and Pb^{2+} were studied. The adsorption efficiencies were compared with that of the commercial activated carbon (GAC F-300). Energy Dispersive X-ray spectroscopy (EDS) was used to determine the elemental compositions of the adsorbents while Scanning Electron Microscope (SEM) was used for the surface texture and morphological characteristics of the cocoa husk. The study optimised metal ion concentration, adsorbent dosage, agitation time and adsorbent particle size for the removal of the metals from simulated wastewater using Atomic Absorption spectroscopy (AAS) and removal efficiencies were calculated. Langmuir and Freundlich models were employed to study the adsorption pattern. The equilibrium experimental data results of Pb^{2+} fitted well to Langmuir and Freundlich the two adsorption isotherms. The coefficient of correlation (R^2) for both models falls within $0<R^2<1$. The experimental data obtained from the sorption of Cr^{6+} by the adsorbents, except CH, fitted Langmuir model than Freundlich model. The Langmuir isotherm fits the experimental data than Freundlich isotherm for Cr^{6+} sorption by the CCH modified cocoa husk and the activated carbon except the unmodified cocoa husk. The study concluded that cocoa husk, an agricultural waste material, have potential as an adsorbent to remove toxic heavy metal like chromium and lead from industrial waste water. Modification of this waste with chitosan also enhances its performance in the removal of the metals.

Key words: Adsorption, cocoa husk, chitosan, heavy metal, atomic absorption spectroscopy, isotherm.

INTRODUCTION

Heavy metals contamination has been a serious environmental concern due to its toxicity, persistency and accumulation. This contamination has increased over decades due to increase in population which has increased the consumption rate and has led to rapid industrialization. The metals through industrial processes are released into the air, soil, and water (Deniz et al., 2012). It has been consistently desired that their concentrations be lowered in industrial and municipal effluents before ultimate disposal into the environment

(Muthusamy et al., 2012).

Researchers have developed various methods such as chemical precipitation, coagulation, adsorption, ion exchange and membrane filtration for the removal of heavy metals from waters and wastewaters (Deniz et al., 2012; Onundi et al., 2010; Georg Steinhauser, 2008). The limitations associated with these methods are high operational cost, sludge production and they differ in their effectiveness, especially when metals are present in solution at very low concentrations. Of all the treatment methods mentioned above, adsorption had been reported as an efficient method for the removal of heavy metals from aqueous solution because of their effectiveness even at low concentration (Liang et al., 2009; Amuda and Edewor, 2013).

The commercial activated carbon most widely used as an adsorbent has become economically less feasible owing to its relatively high cost. Recently, different low cost biosorbent sourced from natural materials and biological wastes from industrial processes has been studied for the development of cheap and effective metal sorbents (Onwu and Ogah, 2010; Sethu et al., 2010). Also, the development of surface modified activated carbon has produced a variety of activated carbon with far surpassing adsorption capacity (Amuda et al., 2007; Lwin et al., 2008). Surface modification of natural adsorbent derived from agricultural waste improves their metal uptake performance which would add to economic value (Amuda and Edewor, 2013; Lwin et al., 2008). Biochar is a charcoal made from carbon rich material (biomass). Latest developments in renewable bioenergy technologies make it possible to convert waste biomass, for instance agricultural waste, into value-added biochar and at the same time produce bioenergy (Dong et al., 2011). In any agricultural country such as Nigeria, accumulation of waste material is always one of the by-products of farming activities. The uses of such agricultural by-products have been in their conversion into usable forms. It is the policy of the Nigerian government to encourage and promote the utilization of the country's raw material and waste such as cocoa husk, for other industrial processes. Caustic soda is a widely used industrial alkali, and it serves at least 61 major industries in the country (Essien et al., 1986; Okoya and Ogunkoya, 2009). It has numerous applications in rayon and other textile manufacture, pulp and paper production, reclaiming of rubber, food production, detergents and soap production. Despite all these reports, cocoa husk, the major raw material in the production of caustic alkali, still constitute an environmental problem in the light of the way it is being disposed in Nigeria, just as in many other tropical countries in which the disposal problem of agricultural wastes is widely recognized, efforts to find solutions are not equal for different sectors (Ulloa et al., 2003). Mosquitoes breed inside them wherever they are heaped up due to under utilization for the production of caustic alkali (Okoya and Ogunkoya, 2009), local black soap, medicinal and other purposes by the indigenous people. Hence there is need to exploit the possibility of utilizing this underutilized material as an adsorbent for the removal of metals from waste water.

Biochar shows a great affinity for heavy metals and their sorption capacity is comparable with other biosorbents. For instance, the ranges of Pd sorption capacity of different biochar are from 2.4 (rice husk) to 20.5 mg/g (sugarcane bagasse) (Mohan et al., 2007; Liu and Zhang, 2009; Ding et al., 2010; Cao et al., 2009). Even though little Cr removal by biochar information is available, the effectiveness of its removal by various biomaterials has been investigated (Park et al., 2008; Abdullah and Prasad, 2009). The production of biochar of different waste biomass has help reduce the cost involved in waste disposal and provides potentially inexpensive alternatives for commercial activated carbon (Amuda and Edewor 2013; Lwin et al., 2008).

In recent years, development of surface modified activated carbon has generated a diversity of activated carbon with far superior adsorption capacity. Among these low cost adsorbents, chitosan has the highest sorption capacity for several metal ions (Amuda et al., 2007; Amuda et al., 2009). Chitin (2-acetamido-2deoxy- -d-glucose-(N-accetylglucan) is the main structural component of molluscs, insects, crustaceans, fungi, algae and marine invertebrates like crabs and shrimps. Chitosan (2-acetamido-2-deoxy-B-d-glucose-(N-acetyl glucosamine) is a partially deacetylated polymer of chitin and is usually prepared by deacetylation of chitin with a strong alkaline solution as shown in Scheme 1 (Amuda et al., 2009).

Yang and Zall (1984) reported that chitosan can chelate five to six times greater concentrations of metals than chitin. They further reported this property to be related to the free amino groups exposed in chitosan because of deacetylation of chitin. Chitosan is slightly soluble at low pHs and posesses problems for developing commercial applications. Also, the active binding sites of chitosan are not readily available for sorption. The sites are reported to be soft and have a tendency to agglomerate or form gel in aqueous solutions. Transport of metal contaminants to the binding sites plays a very important role in process design. Therefore, it is necessary to provide physical support and increase the accessibility of the metal binding sites for process applications.

However, minimal work has been done on modifying cocoa husk biochar, comparing the removal efficiency of chitosan modified biochar with commercial activated carbon and comparing their effectiveness in removing different toxic metal. The purpose of this study is to evaluate the efficiency of cocoa husk char as adsorbent for Lead and Chromium, to study the effect of surface modification of cocoa husk char with chitosan on the removal of the metals from wastewater; and also to compare the efficiency of CCH and CH with that of GAC F – 300.

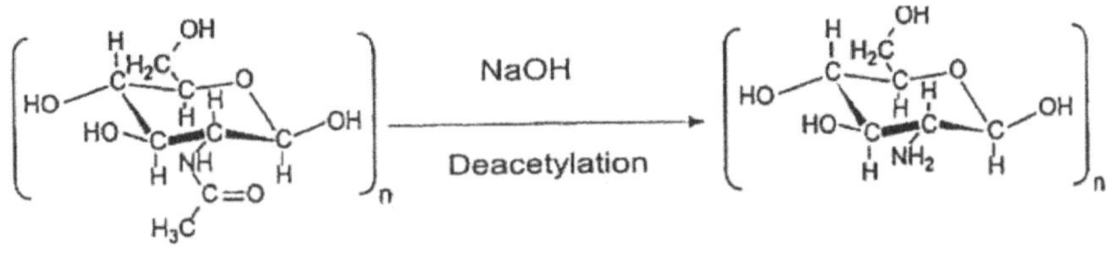

Chitin
2-acetamido-2-deoxy-β-D-glucose-(N-acetylglucan)

Chitosan
2-acetamido-2-deoxy-β-D-glucose-(N-acetylglucosamine)

Scheme 1. Conversion of chitin to chitosan by deacetylation. Source: (Amuda et al., 2007; Souundarrajan et al.,2012)

MATERIALS AND METHODS

The cocoa husks used for this research were collected from Obafemi Awolowo University Teaching and Research Farm while the snail shell used for the extraction of chitosan was collected from a household in Idanre, Ondo State. Activated carbon (GAC F-300) by Boncaki Nig. Ltd was purchased commercially. Synthetic wastewater was employed for the adsorption studies. 500 mg/L each of stock solution of analytical graded $K_2Cr_2O_7$ and $Pb(NO_3)_2$ was prepared by dissolving exact quantity of the salts in deionized water. The range (5 to 25 mg/L) of concentrations of Cr^{+6} and Pb^{+2} was prepared from the stock solution.

Preparation of cocoa husk char

The cocoa husks collected were sun dried, cut, ground and sieved to three particle sizes of 150, 212 and 300 µm. The char of the cocoa husk was prepared using the modified methods of Lwin et al. (2008) and Amuda and Edewor (2013). About 15 g of each of the raw material was placed in a crucible, and put into a furnace. The furnace temperature was increased at a rate of 10°C/min from room temperature to 500°C in a nitrogen environment to ensure nitrogen free environment and the temperature was held at this temperature for 2 h after which they were washed with distilled water to remove any leachable impurities due to free acid and adherent powder.

Adsorbent characterization

The elemental composition of the prepared husk was determined using Energy Dispersive X-ray Micro Analysis (EDX) while Scanning Electron Microscope (SEM) (High resolution SEM EVO MA 10 carl Zeiss) was used to study the surface morphological characteristics.

Activation of cocoa husk char

The char samples were treated with 2% H_2SO_4 (v/v) in an incubator for 24 h at a temperature of 110°C. The treated char sample was then soaked in deionized water until the pH was stabilized. It was then soaked in 2% $NaHCO_3$ to remove the residual acid. The acid-treated cocoa husk carbon was then dried overnight in an oven at 110°C, and stored in desiccators until use (Kadirvelu et al., 2001).

Preparation of chitosan

The snail shells were powdered and sieved with 2 mm sieve. 50 g of the less than 2 mm size powdered snail shell was weighed into a 500 ml beaker and 200 ml of 4% (w/v) KOH was added with constant stirring for 6 h at 80°C and filtered. The residue was washed with distilled water until it is free of base and then dried at 100°C for 2 h. The deproteinized snail shell residue was then poured into a 250 ml conical flask and 100 ml of 3 % (v/v) 1 M HCl was added and placed on a magnetic stirrer for 3 h at 30°C to demineralise it. The content was later filtered and the residue was washed until it is free of acid. The acid free residue was then dried at 90°C for 1 h after which it was decolourized by refluxing in acetone for 3 h at 60°C. The content was filtered and the residue dried. A snow white residue called chitin was obtained.The chitin was poured into a 250 ml conical flask for deacetylation, 50% (w/v) NaOH solution was added, stirred at 30°C for 4 h and filtered. After filtration, the residue, which is chitosan (2-acetamido-2-deoxy-β-D-glucose-N-(acetylglucosamine), was then washed and dried at 90°C for 1 h.

Preparation of chitosan gel

About 5 g of chitosan was slowly added to 100 ml of 10% (w/v) oxalic acid with constant stirring in a beaker. The mixture was also heated to about 50°C to facilitate mixing. A chitosan-oxalic acid mixture was formed (a whitish viscous gel).

Modification of adsorbent

The chitosan gel (100 ml) was diluted with water (~500 ml) and heated to about 50°C. 50 g of cocoa husk char was slowly added to the diluted gel in separate container, and mechanically agitated using a shaker at 200 osc/min for 24 h. The chitosan-coated cocoa husk char were then washed with distilled water and dried after which it was soaked in 0.5 % (w/v) NaOH solution for 3 h. It was then extensively rinsed with deionized water and dried in an oven at 102°C for 2 h, cooled at room temperature and stored in a desiccator.

Batch adsorption studies

Experiments were conducted in 5 different 250 ml Erlenmeyer flasks containing 100 ml varied concentration (5, 10, 15, 20 and 25 mg/L) of Cr^{+6} solutions and CH (150 µm cocoa husk) dosage of 1.5 g was added to the different flasks. Flasks were agitated on a shaker at 400 osc/m constant shaking. The mixture was filtered using Whatman no.1. after 30 min, and the filterate analysed spectrophotometrically using Atomic Absorption Spectroscopy (AAS). The adsorption procedure was repeated using different adsorbents (150, 212 and 300 µm) sizes of unmodified cocoa husk, modified cocoa husk and activated carbon), dosages (0.5, 1.0, 1.5, 2.0 and 2.5 g) and different adsorbent, contact period (30, 60, 90, 120 and 150 min) in turns. Triplicate analyses were carried out for each adsorption. Batch adsorption studies for varied concentration of Pb^{2+} were also carried out. The removal efficiency (E) of adsorbent on Cr^{+6} and Pb^{+2} were calculated as follows:

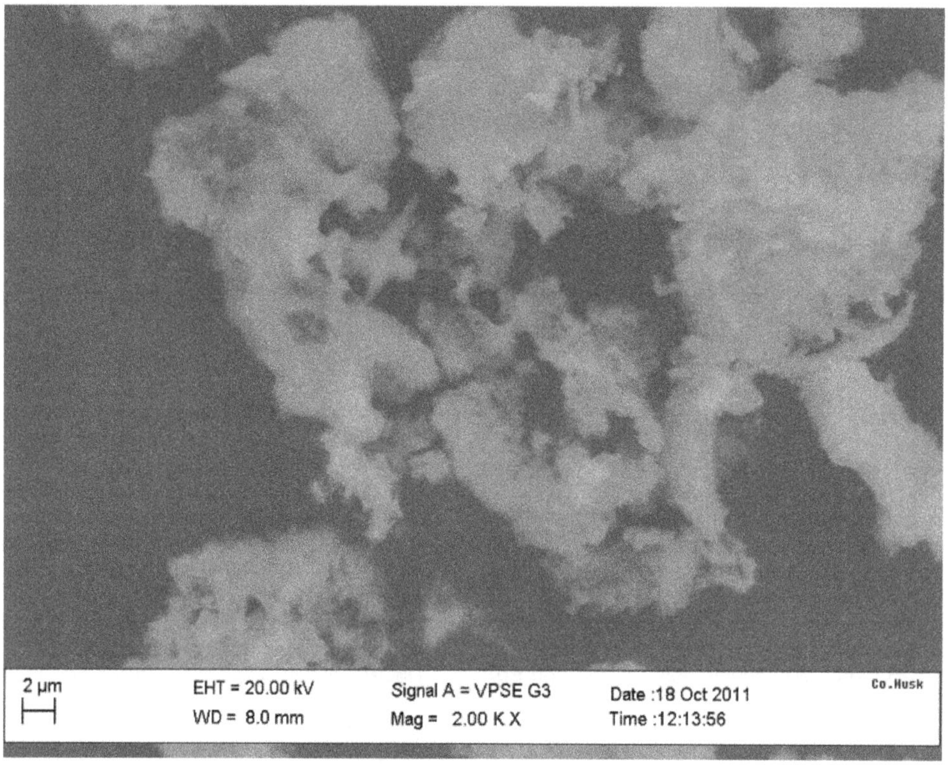

Figure 1. SEM image of cocoa husk char.

$$E\% = ((C_i - C_f)/C_i) \times 100 \quad (1)$$

Where C_i and C_f are the initial and final concentrations of Cr^{+6} and Pb^{+2} (mg/L), respectively, (Amuda et al., 2007).

DATA EVALUATION

The amount of metal bound by the adsorbent was calculated as follows:

$$Q = v\,(C_i - C_f)/m \quad (2)$$

Where Q is the metal uptake (mg metal per g adsorbent), v is the volume of solution containing the metal (ml), C_i is the initial concentration of the metal in the solution (mg/L), C_f is the final (equilibrium) concentration of the metal in the filtrate (mg/L) and m is the amount of the added adsorbent on the dry basis (g) (Jun et al., 2012).

RESULTS AND DISCUSSION

Sample characterisation

Scanning electron micrographs (SEM) clearly revealed the surface texture and morphological characteristics of the adsorbent (cocoa husk char). The SEM image in Figure 1 reveals the porous structure of the cocoa husk char. It also shows the agglomeration of the particles of the cocoa husk char and the distinctive irregular material with large surface area. As shown in Table 1, the Energy Dispersive X- ray spectroscopy (EDX) reveals the presence of elements such as Carbon C, Potassium K, Calcium Ca, Magnesium Mg, Aluminium Al, Silicon Si, Phosphorus P and Chlorine Cl in the adsorbent. The porosity and agglomeration of cocoa husk char particles enhance metal uptake and encourage mass transport of heavy metals through the inside of the sorbent (Grabriela et al., 2006; Edwin, 2008; Zhang et al., 2009; Alayande et al., 2011; Gholami et al., 2006). While the presence of the elements mentioned above had been reported to enhance adsorption metals through ion exchange, chelation, coordination and complexation reactions. Elements such as Al, K, Ca, and Mg could exchange for the heavy metals while elements like P, O and Si could prompt chelation, coordination and complexation (Grabriela et al., 2006; Edwin, 2008; Michaud, 2011) (Figure 2).

Effect of initial concentration

Figures 3 and 4 show the effect of Initial metal ion concentration on the removal efficiency exhibited by the adsorbents with increase in metal ion concentration. From these Figures, commercial activated carbon show an outstanding removal efficiency over modified and unmodified cocoa husk char. In fact it exhibit 100%

Table 1. Elemental composition of cocoa husk char.

Element	Cocoa husk	
	Weight %	Atomic %
C	42.04	54.13
O	37.95	36.69
Mg	2.47	1.57
Al	1.07	0.61
Si	3.36	1.85
P	1.25	0.62
K	4.71	1.86
Ca	6.32	2.44
Fe	0.82	0.23

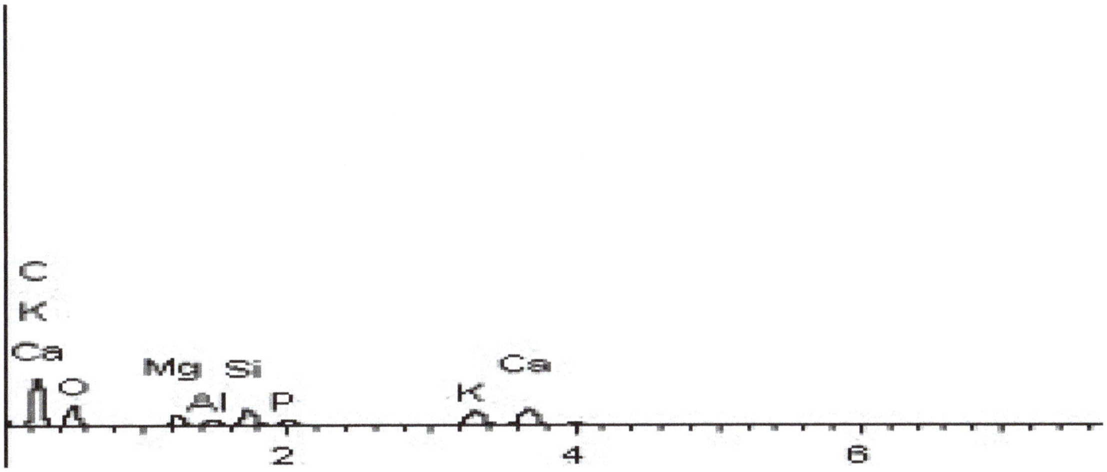

Figure 2. EDX spectra of cocoa husk char.

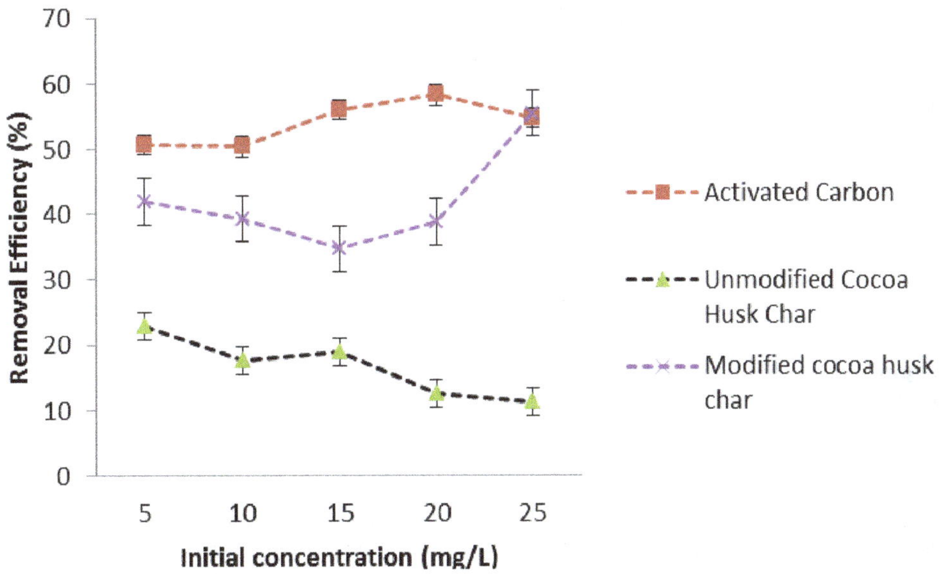

Figure 3. Effect of initial concentration on the removal efficiency of Cr (VI). Agitation time: 60 min; Adsorbent dose: 1.5; Agitation speed 400 osc/min.

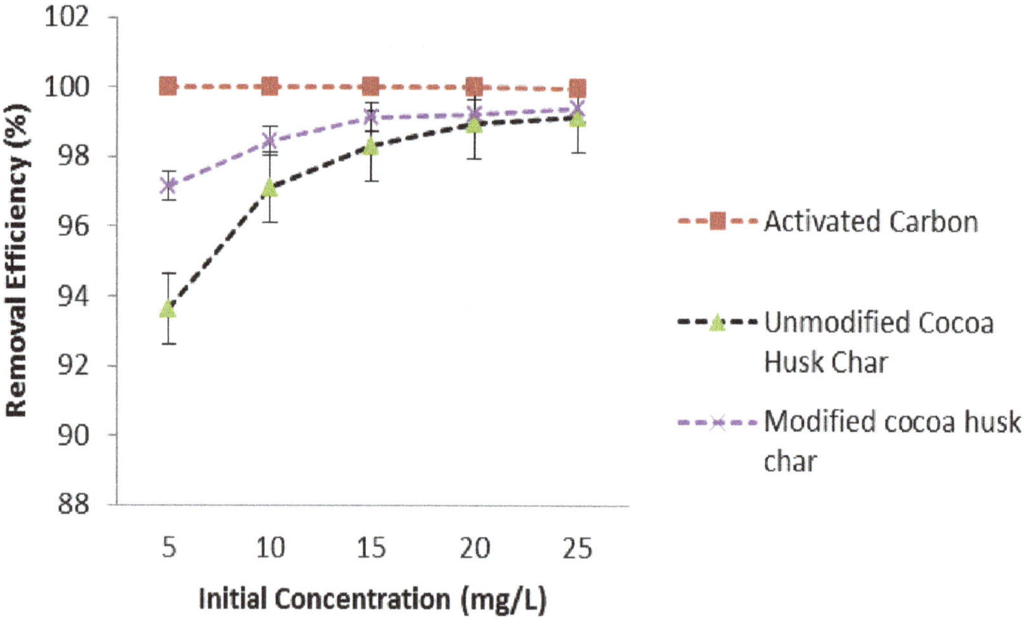

Figure 4. Effect of initial concentration on the removal efficiency of Pb (II). Agitation time: 60 min; Adsorbent dose: 1.5; Agitation speed 400 osc/min.

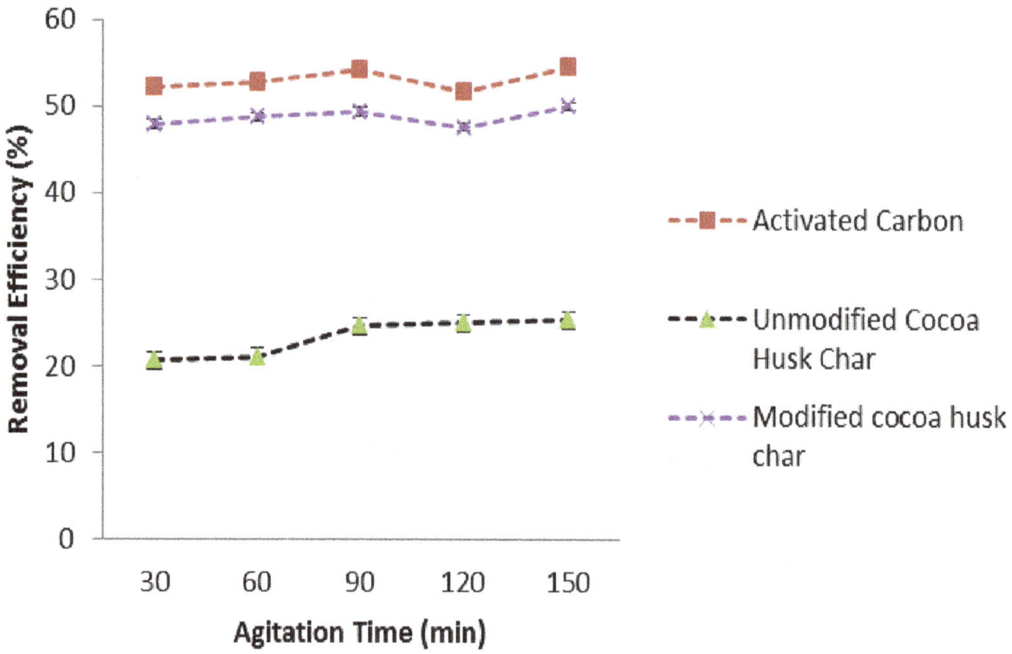

Figure 5. Effect of agitation time on removal efficiency of Pb (II) by the adsorbents. Pb (II) Concentration: 20 mg/L; agitation speed: 400 osc/min; dose of each adsorbent: 1.0 g.

removal efficiency for Pb^{+2} ion (Figure 5). In Figure 4, the removal efficiency of unmodified cocoa husk char decreased as the metal ion concentration increases while the removal efficiency of the modified adsorbent initially decreased as metal ion concentration increased from 5 to 15 mg/L and later increased as metal ion concentration increased. The decrease in removal efficiency may be due to the limited number of active sites in the adsorbent as it attained saturation above certain concentration (Sugashini and Gopalakrishnan, 2012). However, in the

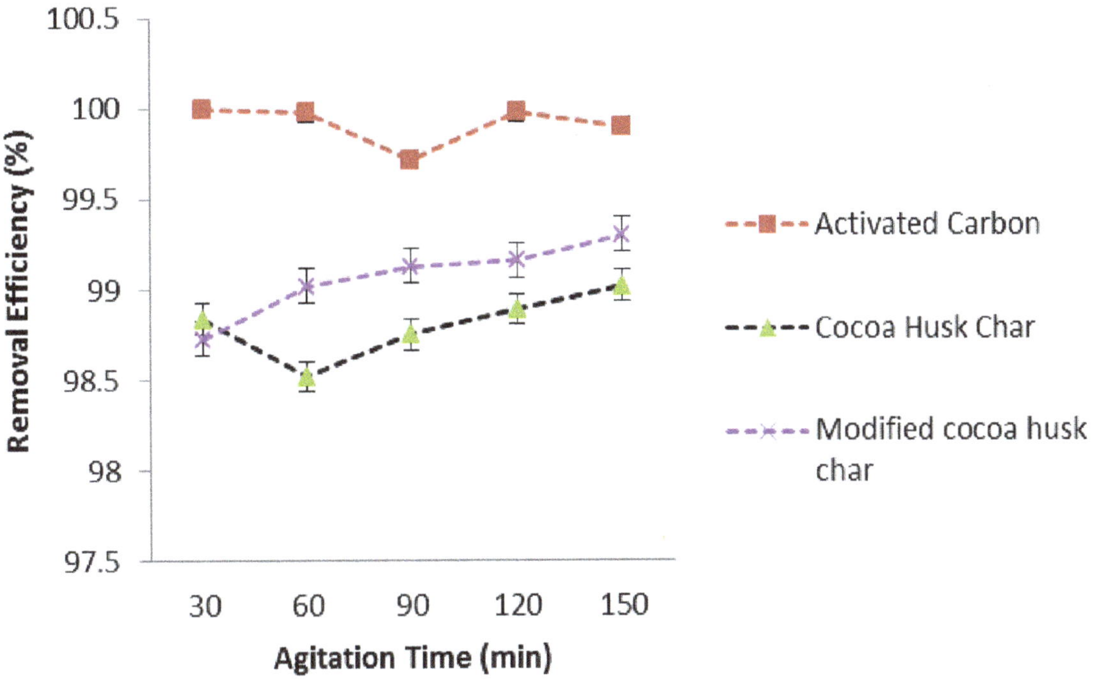

Figure 6. Effect of agitation time on removal efficiency of Cr (VI) by the adsorbents. Cr (VI) Concentration: 20 mg/L; agitation speed: 400 osc/mim; dose of each adsorbent: 1.0 g.

case of Pb^{+2}, the removal efficiency of both modified and unmodified adsorbent increase as initial metal ion concentration increases. The increase in removal efficiency as metal ion concentration increases may be as a result of competition for the available binding site on the adsorbent by the metal ion as the concentration increases (Olu-owolabi et al., 2012; Gholami et al., 2006). Also, the initial metal concentration provides an important driving force to overcome all mass transfer resistances of the metal between aqueous and solid phase (Aksu and Akpinar, 2000). The results show that feasibility and efficiency of adsorption process depends not only on the properties of the adsorbents, but also on the concentration of the metal ion solution. It was observed that the removal efficiency of the chitosan modified cocoa husk char was higher compared to the unmodified adsorbents (Figures 3 and 4). This may be due to the fact that modification increases the available active binding sites for metals thereby enhancing the performance of the adsorbent (Igwe and Abia, 2007; Amuda et al., 2009). Development of surface modified activated carbon has generated a diversity of activated carbon with far superior adsorption capacity. Among these low cost adsorbents, chitosan has the highest sorption capacity for several metal ions (Amuda et al., 2007; Amuda et al., 2009). Yang and Zall (1984) reported that chitosan can chelate five to six times greater concentrations of metals than chitin. They further reported this property to be related to the free amino groups exposed in chitosan because of deacetylation of chitin.

In addition, removal efficiency of Pb^{2+} with the modified cocoa husk char is relatively close to that of activated carbon (Figure 4).

Effect of contact time

As shown in Figure 5, the removal efficiency of Cr^{6+} by unmodified cocoa husk char increases with adsorbent contact time from 30 to 90 min. However, it remained constant within the adsorbent contact time of 120 and 150 min, meanwhile that of modified cocoa husk char and commercial activated carbon increased as contact time increases. In Figure 6, the removal efficiency of Pb^{2+} using modified and unmodified cocoa husk char increased with increase in contact time while for commercial activated carbon, equilibrium was reached within 30 min. These results showed that initially the number of vacant sites was large which later reduced as the vacant active site is been saturated (Sugashini and Gopalakrishnan, 2012). Also, the fast adsorption at the initial stage was probably due to the initial concentration gradient between the adsorbate in solution and the number of available binding sites on the adsorbent surface at the beginning while the decrease in efficiency may be due to limited mass transfer of the adsorbate molecules from the bulk liquid to the external surface of the adsorbents (Onundi et al., 2011). Also greater availability of various functional groups on the surface of the chitosan, which are required for

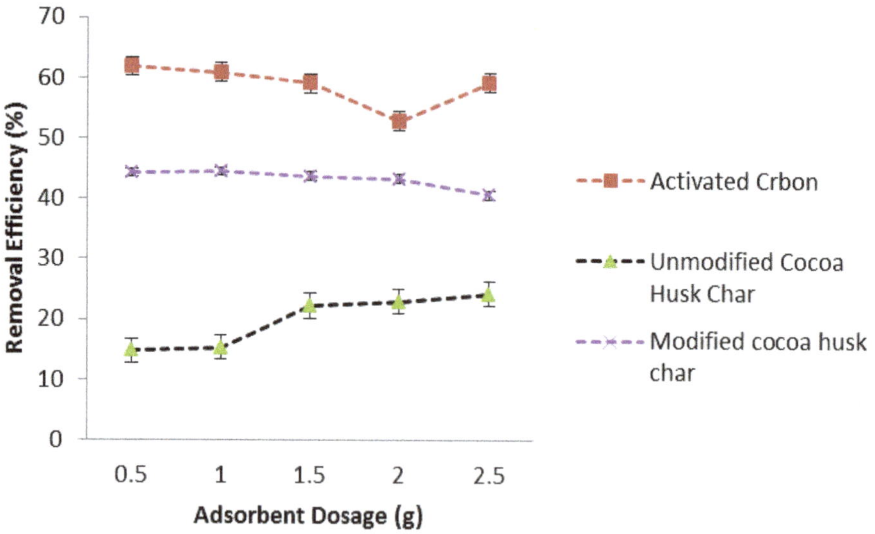

Figure 7. Effect of dose of the adsorbents on the removal efficiency of Cr (VI). Cr (VI) Concentration: 30 mg/L; Agitation time: 60 min; Agitation speed 400 osc/min.

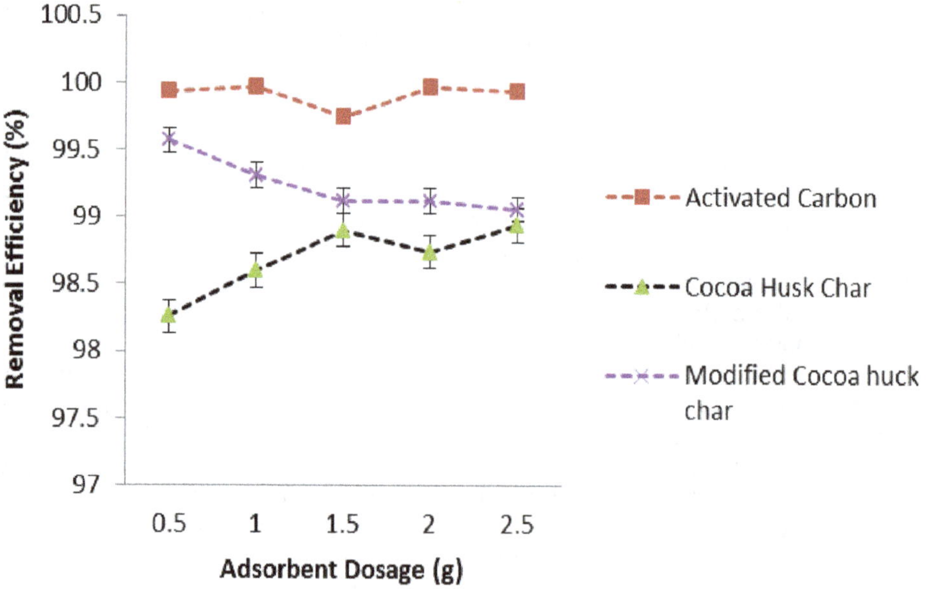

Figure 8. Effect of dose of the adsorbents on the removal efficiency of Pb (II). Pb (II) Concentration: 30 mg/L; Agitation time: 60 min; Agitation speed 400 osc/min.

interaction with anions and cations, could significantly improve the binding capacity of an adsorbent and adsorption proceeds rapidly (Amuda et al., 2009; Nomanbhay and Palanisamy, 2005).

Effect of adsorbent dosage

Effect of adsorbent dosage on adsorption of Cr^{6+} and Pb^{2+} is studied by varying the adsorbent dosage from 0.5 to 2.5 g and other parameters like initial metal ion concentration were kept constant. The Figures 7 and 8 show increase in the removal efficiency of Cr^{6+} and Pb^{2+} as dosage of unmodified cocoa husk char increases. This is because of the availability of more binding sites in the surface of the adsorbent as the dosage increase (Asubiojo and Ajelabi, 2009; Choi et al., 2009; Dhabab, 2011). However, there is a decrease in removal efficiency as commercial activated carbon and modified cocoa husk char increases (Figures 8 and 9). The decrease in the

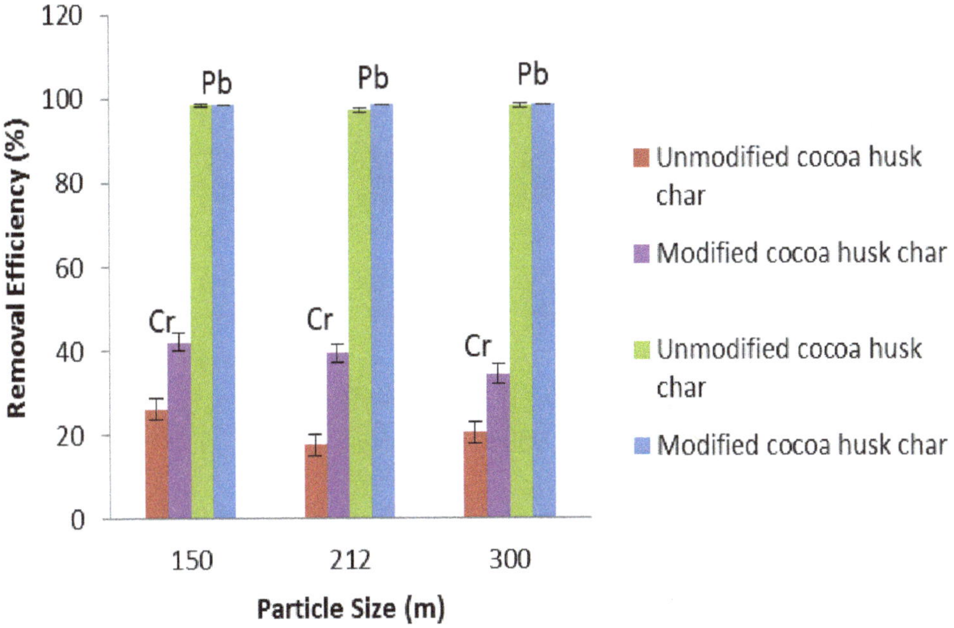

Figure 9. Influence of the adsorbent particle size on the removal efficiency of Cr^{6+} using modified and unmodified cocoa husk char. Metal concentration: 10mg/L; Agitation time: 45 min; agitation speed: 400 osc/min; adsorbent dosage: 1.5 g.

sorption efficiency with the increasing adsorbent dose may be as a consequence of a partial overlapping or aggregation of adsorbent surface area available to ion and an increase in diffusion path length, which occurs at high biomass concentration given rise to a decrease of active sites (Gönen and Serin, 2012; Souundarrajan et al., 2012; Sudha et al., 2012). Also, differential adsorption capacities of the adsorbents depend on the extent of surface modification (Amuda et al., 2009; Nomanbhay and Palanisamy, 2005).

Effect of particle size

The results of the effects of particle size on the removal efficiencies of Cr^{6+} and Pb^{2+} is presented in Figure 10. The particle sizes used were 150, 212 and 300 µm, respectively. Other parameters such as adsorbent dose, initial ion concentration and agitation time were kept constant; temperature and agitation speed were kept at 25°C and 400 osc/min, respectively. The removal efficiency of Cr^{6+} and Pb^{2+} by modified cocoa husk char decreased as particle size increased while for the unmodified cocoa husk char, 300 µm particle size show higher efficiency than 212 µm. This increased efficiency as particle size decreased may be due to the fact that smaller particles give large surface areas. Sometimes, larger particle size can show higher removal efficiency over smaller particle if the particles are spherical in shape. Larger particles with spherical shape, in general, present higher external mass transfer than small particle size which therefore responsible for the higher removal efficiency (Leusch and Volesky, 1995).

Adsorption isotherm

Langmuir and Freundlich models are the most commonly used theoretical model to generate adsorption isotherm. Langmuir model assumption state that uptake of metal ions occurs on a homogenous surface by monolayer adsorption without any interaction between adsorbed ions while the assumption of Freundlich model is based on adsorption on heterogenous surface. The experimental data were analyzed according to the linearized form of the Langmuir and Freundlich model. The linearized equation of Langmuir model is represented by:

$$1/Q = 1/Q_{max} (1/b\ C_f + 1) \quad (3)$$

Where Q_{max} is the maximum metal uptake under the given conditions, b is a constant related to the affinity between the adsorbent and sorbate.

The linearized equation of Freundlich model is represented by:

$$\text{Log } Q = \text{Log } k + 1/n \log C_f \quad (4)$$

Where k and n are Freundlich constant, which correlated to the maximum adsorption capacity and adsorption intensity, respectively.

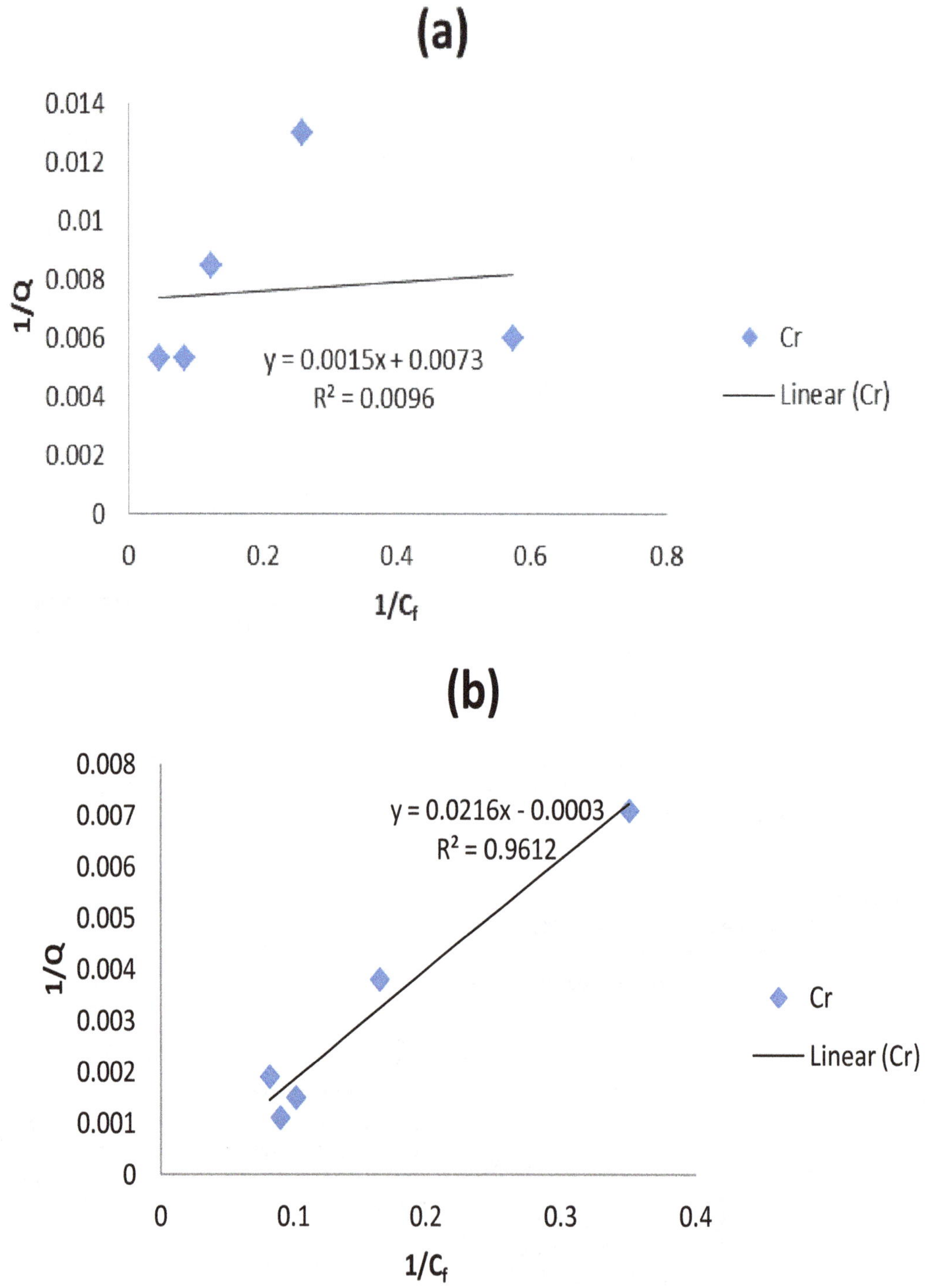

Figure 10. Langmuir adsorption isotherms for the adsorption of Cr^{6+} by (a) unmodified cocoa husk char (b) modified cocoa husk char.

Figures 10 to 13 and Table 2 shows the applicability of Langmuir and Freundlich model to the metals using cocoa husk char adsorbent. The two models fitted well for the adsorption studies of Cr^{6+} and Pb^{2+} using modified adsorbent while Freundlich model fitted well for the adsorption study of the two metals than Langmuir. This is because of the higher correlation coefficient depicted by the two models (Table 2). Q_{max} value which is the adsorption capacity determined from the slope of Langmuir indicate a good adsorbing capacities of the

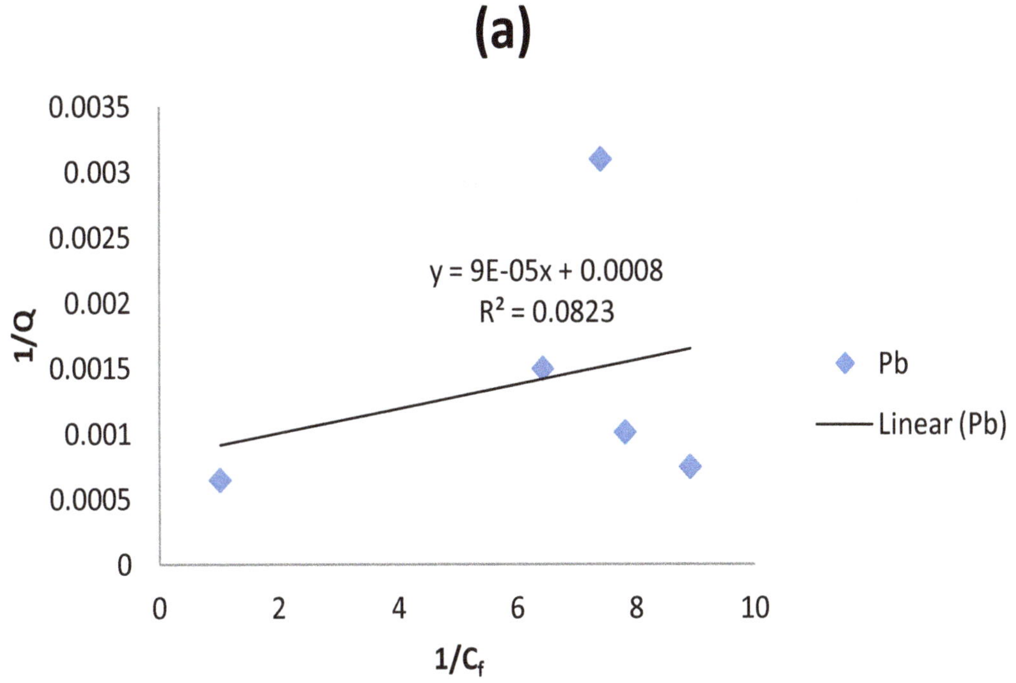

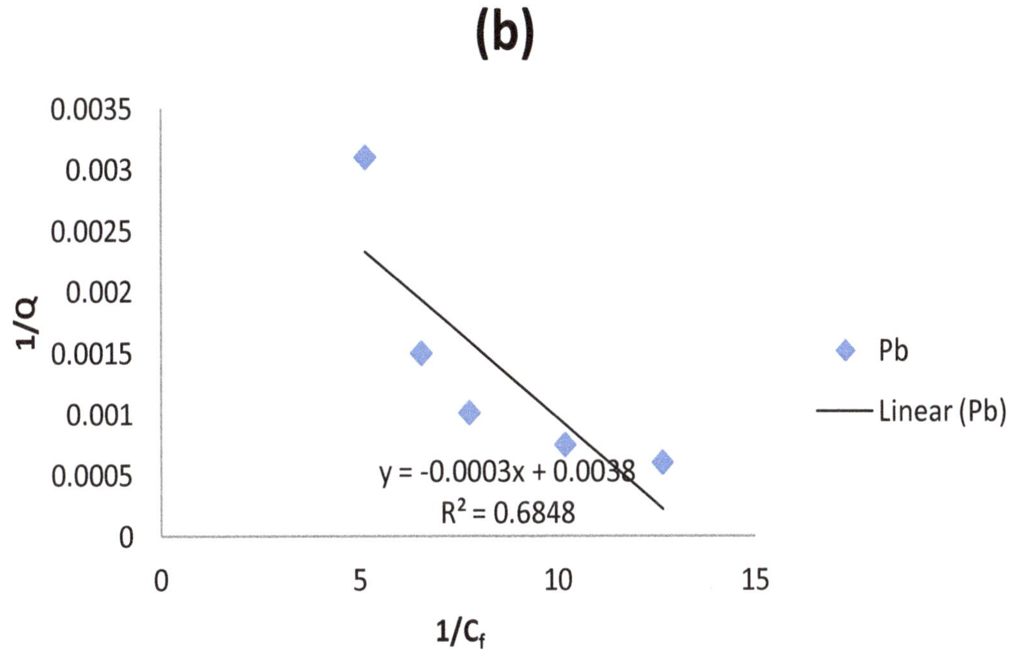

Figure 11. Langmuir adsorption isotherms for the adsorption of Pb^{2+} by (a) unmodified cocoa husk char (b) modified cocoa husk char.

adsorbent for Pb^{2+} than Cr^{6+}. The K values for the sorption of Pb^{2+} by the cocoa husk char adsorbents are higher (1500.7) than that of the sorption of Cr^{6+} (39.138) by the adsorbent (Table 2). The higher the value of K, the greater the adsorption intensity (Vaishnav et al., 2012). The higher K-values exhibit by the sorption of Pb^{2+} suggests that Pb^{2+} has greater sorption tendency towards the adsorbents than Cr^{6+} (Kose et al., 2012).

The fractional value of *1/n* indicates the surface of adsorbent is of heterogeneous type with an exponential distribution of energy sites (Kose et al., 2012). The value of 1/n less than 1 (Table 2) is indicative of favourable

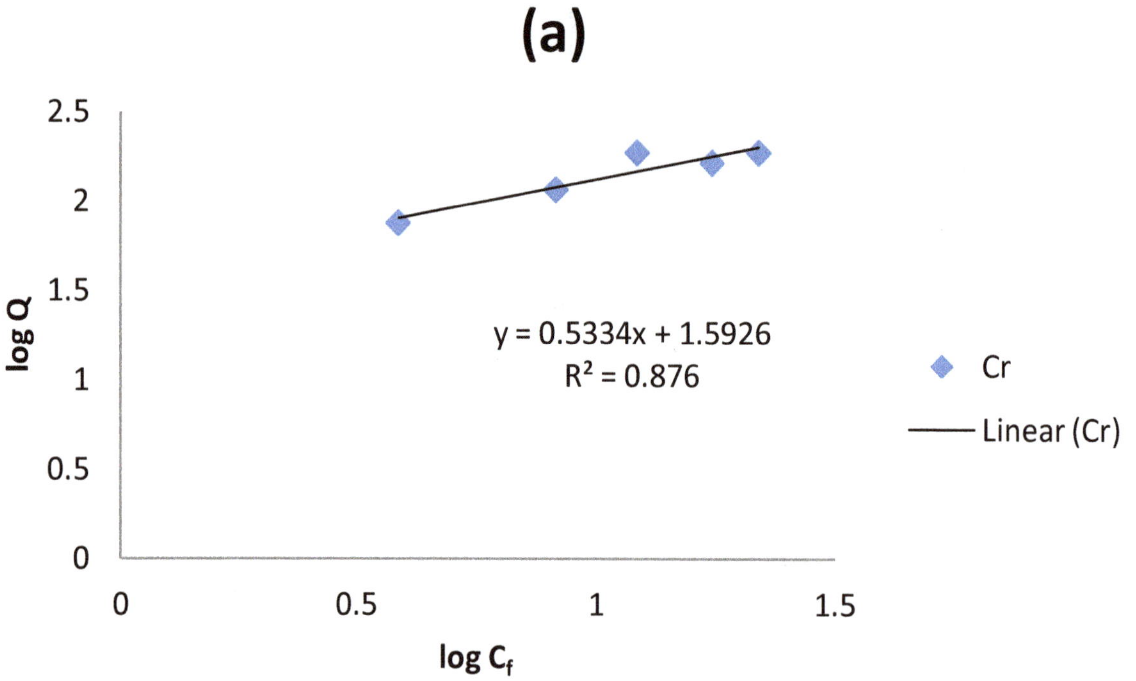

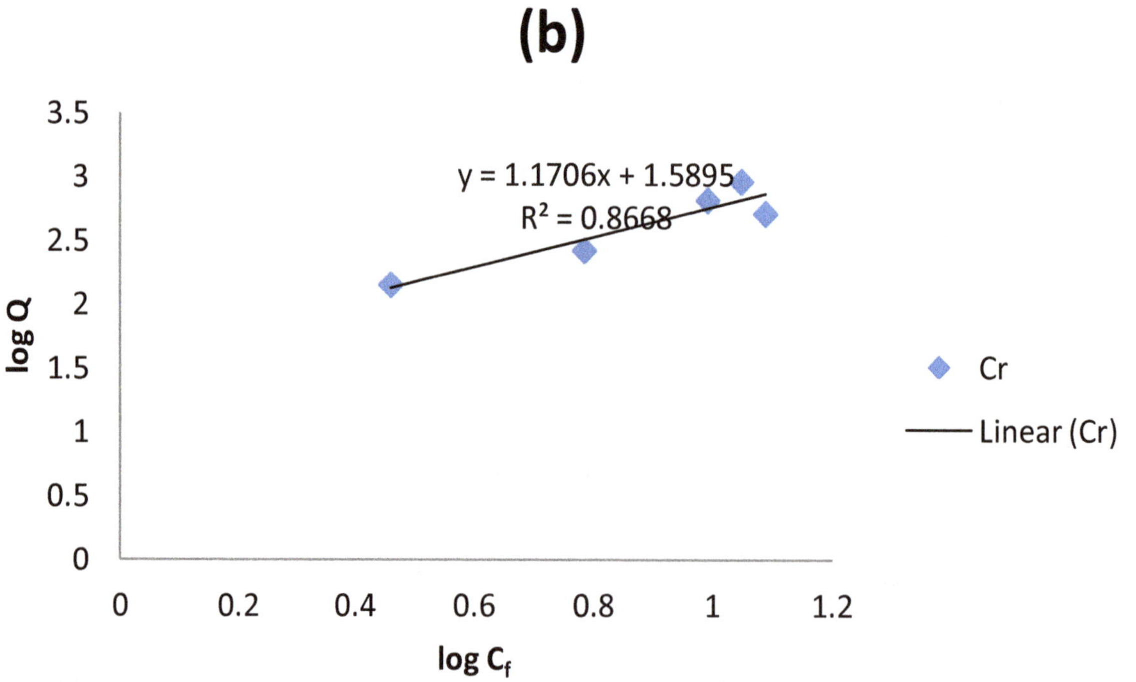

Figure 12. Freundlich adsorption isotherms for the adsorption of Cr^{6+} by (a) unmodified cocoa husk char (b) modified cocoa husk char.

adsorption (Onwu and Ogah, 2010). The fact that the value of 1/n in the adsorption of Pb^{2+} is small (< 1 even some tend towards negative) (-1.7634, 0.3464) compared to 1/n value of the adsorption of Cr^{6+} (1.1706, 0.5334), is indicative of favourable adsorption for Pb^{2+} than Cr^{6+}. A smaller value of 1/n indicates better adsorption mechanism and formation of relatively stronger bond between adsorbate and adsorbent Tables 3 and 4 (Al-Sultani and Al-Seroury, 2012; Alfa et al., 2012; Patil et al., 2006).

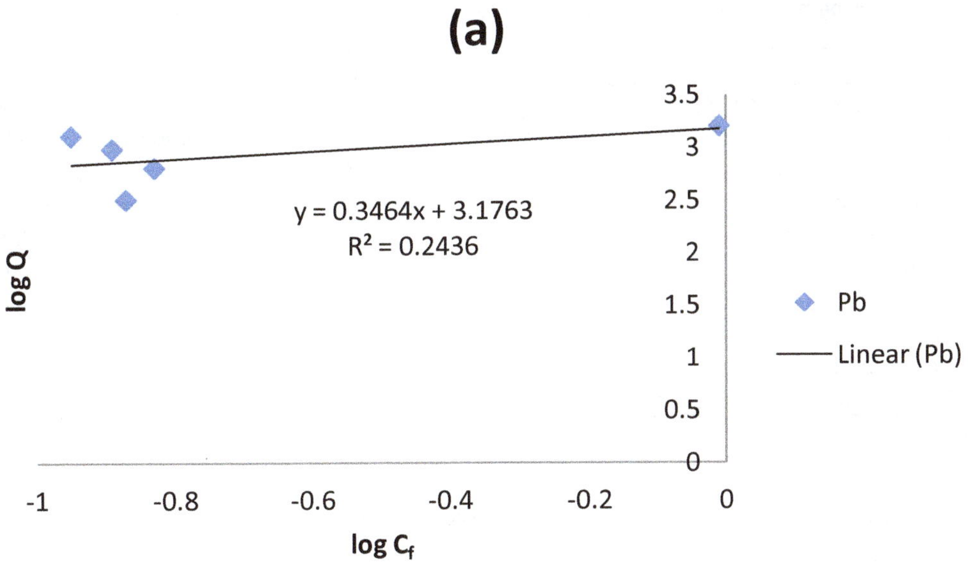

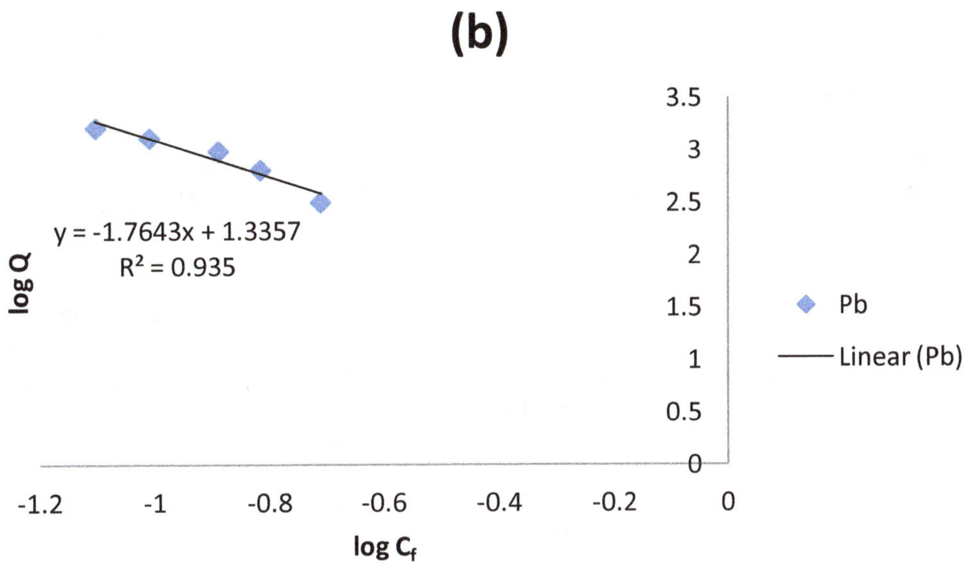

Figure 13. Freundlich adsorption isotherms for the adsorption of Pb^{2+} by (a) unmodified cocoa husk char (b) modified cocoa husk char.

Table 2. Langmuir and Freundlich constants for the adsorption of Cr^{6+} and Pb^{2+} at 25°C using modified and unmodified cocoa husk char.

Element	Langmuir constants			Freundlich constants		
	b (L/mg)	Q_{max} (mg/g)	R^2	K	1/n	R^2
Unmodified cocoa husk char						
Cr	4.867	136.98	0.0096	39.138	0.5334	0.876
Pb	0.1125	1250	0.0823	1500.7	0.3464	0.2436
Modified cocoa husk char						
Cr	-13.903	333	0.9612	38.86	1.1706	0.8668
Pb	-12.667	263.16	0.6848	21.6621	-1.7643	0.935

R^2= correlation coefficient.

Table 3. Comparisons of the Langmuir constants for cr adsorption onto adsorbents.

Adsorbent	Qmax ((mg/g)	References
Caladium bicolor biomass	53.63	Horsfall and Spiff (2005)
Wood ash	73.8	Chirenje et al. (2006)
Oryza Sativa L. husk	8.6	Zulkali et al. (2006)
Sago waste	46.6	Quek et al. (1998)
Sphagnum moss peat	30.7	Ho and McKay (2000)
Groundnut husks	39.3	Okieime et al. (1991)
Corynebacterium glutamicum	2.74	Choi and Yun (2004)
Peat moss	76.7	Akinbayo (2000)
Chrysophyllum albidum shell	103.42	Amuda et al., (2009)
Chitosan modified Cocoa husk carbon	333	This study

Table 4. Comparisons of the Langmuir constants for Pb adsorption onto adsorbents.

Adsorbent	Qmax (mg/g)	References
Chitosan impregnated with Microemulsion	85.59	de Castro Dantas et al., 2001
Chitosan–coated acid treated C. albidum seed shell (CACASC)	84.31	Amuda et al., 2009
Chitosan–coated C. albidum seed shell (CCASC)	76.23	Amuda et al., 2009
Chitosan modified Cocoa husk carbon	263.16	This study

Conclusion

This study shows that the use of chitosan coated cocoa husk char for metal ion removal is technically feasible, eco-friendly and with high efficacy. Besides that, cocoa husk and even the snail shell being agricultural industry waste makes the study to be relevant in the management of cocoa husk and snail shell waste. The preparation of chitosan from snail shell is cost effective. Cocoa husk char could be used for waste water treatment.

Conflict of Interest

The authors have not declared any conflict of interest.

REFERENCES

Abdullah MA, Prasad AGD (2009). Kinetic and equilibrium studies for the biosorption of Cr(VI) from aqueous solutions by potato peel waste. Int. J. Chem. Eng. Res. 1:51–62.

Akinbayo A (2000) Removal of Lead from Aqueous Solution by Adsorption Using Peat Moss. MS Thesis. Department of Environmental System Engineering, University of Regina, Canada.

Aksu Z, Akpinar D (2000) Modelling of simultaneous biosorption of phenol and nickel (II) onto dried aerobic activated sludge. Sep. Puri. Tech. 21(1-2):87-99.

Alayande O, Akinlabi A, Olalekan D, Okesola B (2011). Valuable potentials of cowpea husk waste. Int. J. Chem. Sci. 4(2):358-361.

Alfa YM, Hassan H, Nda-Umar UI (2012). Agricultural Waste Materials as Potential Adsorbent for Removal of Heavy Metals from Aqueous Solutions. Int. J. Chem. Res. 2(2):01-13.

Al-Sultani KF, Al-Seroury FA (2012). Characterization the Removal of Phenol from Aqueous Solution in Fluidized Bed Column by Rice Husk Adsorbent. Rev. J. Recent. Sci. 1:145-151.

Amuda OS, Adelowo FE, Ologunde MO (2009). Kinetics and equilibrium studies of adsorption of chromium(VI) ion from industrial wastewater using Chrysophyllumalbidum (Sapotaceae) seed shells. Colloids and Surfaces B: Biointerfaces 68:184-192.

Amuda OS, Edewor TI (2013). Steam-activated carbon prepared from Chrysophyllum albidum seed shell for the adsorption of cadmium in wastewater: kinetics, equilibrium and thermodynamic studies. Int. J. Environ. Waste Manage. 12(2):213-229.

Amuda OS, Giwa AA, Bello IA (2007). Removal of heavy metal from industrial wastewater using modified activated coconut shell carbon. Biochem. Eng. J. 36:174-181.

Asubiojo OI, Ajelabi, OB (2009). The removal of heavy metals from aqueous solution by natural adsorbent. J. Environ. Chem. Toxi. 91:883-890.

Cao X, Ma L, Gao B, Harris W Harris (2009). Dairy-manure derived biochar effectively sorbs lead and atrazine. Environ. Sci. Tech. 43:3285-3291.

Chirenje T, Ma LQ, Lu, L (2006). Retention of Cd, Cu, Pb and Zn by wood ash, lime and fume dust.Water Air Soil Pollut. 171:301–314.525.

Choi SB, Yun YS (2004). Lead biosorption by waste biomass of Corynebacterium glutamicum generated from lysine fermentation process. Biotechnol. Lett. 26:331–336.

Choi HD, Cho JM, Xang JS, Lee JX (2009). Influence of cationic surfactant on adsorsoption of Cr(VI) onto activated carbon. J. Hazad. Mater. 161:1565-1568.

de Castro Dantas TN, Dantas Neto AA, de A. Moura MCP, Barros Neto EL, de Paiva Telemaco E. (2001). Chromium Adsorption by Chitosan Impregnated with Microemulsion. Langmuir 17:4256–4260.

Deniz B, Merve H, Sermin E (2012). Evaluation of Lead Removal onto Black Cumin by using Multi linear Regression. BALWOIS 2012 - Ohrid, Republic of Macedonia. pp.1-6.
Dhabab JM (2011). Removal of Fe(II), Cu(II), Zn(II), and Pb(II) ions from aqueous solutions by duckweed. J. Oceano. Marine Sci. 2(1):17-22.
Ding W, Gao B, Inyang M, Ma L (2010). Characteristics and mechanisms of lead sorption by biochars derived from sugarcane bagasse at different temperatures, Unpublished manuscript, in 2010.
Dong X, Ma LQ, Li Y (2011). Characteristics and mechanisms of hexavalent chromium removal by biochar from sugar beet tailing. J. Hazad. Mater. 190:909-915.
Edwin VA (2008). Surface Modification of Activated Carbonfor enhancement of Nickel (II) adsorption. Electronic J. Chem. 5(4):814-819.
Essien IO, Udoh AP, Etuk BR (1986). Husk: A source of chemical raw material, Chemical and Industrial News.
Georg-Steinhauser MB (2008). Adsorption of ions onto high silica volcanic glass. Appl. Rad. Iso. 66(1):1-8.
Gholami F, Mahvi AH, Omrani GhA, Nazmara Sh, Ghasri A (2006). Removal of Chromium (VI) from aqueous solution by *ulmus* leaves. Iran. J. Environ. Health Sci. Eng. 3(2):97-102.
Gönen F, Serin DS (2012). Adsorption study on orange peel: Removal of Ni(II) ions from aqueous solution. Afr. J. Biotech. 11(5):1250-1258.
Grabriela HP, Souze de Mesquita LM, Mauricio LT, Gustavo A, Saavedra P (2006). "Biosorption of Cadmium by green coconut shell powder" Sci. Direct Min. Eng. 19:380-387.
Ho YS, Mckay G (2000). The kinetics of sorption of divalent metal ions onto Spagnum moss peat. Water Res. 24:735–742.
Horsfall M, Jr. Spiff AI (2005). Effect of metal ion concentration on the biosorption of Pb2+ and Cd2+ by Caladium bicolor (Wild Cocoyam). African J. Biotechnol. 4:191–196, 535
Igwe JC, Abia AA (2007). Adsorbtion isotherm studies of Cd (II), Pb (II) and Zn (II) ions; Bioremediation from aqueous solution using unmodified and EDTA-modified maize cob. Eclética química año/vol.32 número 001:33-42.
Jun D, FengLian R, ChunYuan T (2012). Adsorption of Cr(VI) and Speciation of Cr(VI) and Cr(III) in Aqueous Solutions Using Chemically Modified Chitosan. Int. J. Environ. Res. Public Health. 9:1757-1770.
Kadirvelu K, Thamaraiselvi C, Namasivayam C (2001). Adsorption of nickel (II) from aqueous solution onto activated carbon prepared from coirpith. Separa. Purifi. Tech. 24:497-505.
Kose TD, Gharde BD, Gholse SB (2012). Studies on albiziaprocera legumes for effective removal of Fe (II) and Mn (II) from aqueous solution. J. Chem. Pharm. Res. 4(4):2021-2028
Leusch A, Volesky B (1995). The influence of film diffusion on cadmium biosorption by marine biomass. J. Biotech. 43:1-10.
Liang S, Guo XY, Feng NC, Tian Q (2009). Application of orange peel xanthate for the adsorption of Pb^{2+} from aqueous solutions. J. Hazard Mater. 170:425-429.
Liu Z, Zhang F (2009). Removal of lead from water using biochars prepared from hydrothermal liquefaction of biomass. J. Hazard Mater. 167:933–939.
Lwin CC, Hnin SM, Oo MM (2008). Preparation of Adsorbent for heavy metal from Industrial Wastewater, GMSARN International Conference on Sustainable Development: Issues and Prosperity. pp.12-14.
Michaud CF (2011). "Ion Exchange Reaction – A Review" Water conditioning and Purification. Spotlight Article. http://www.wcponline.com/pdf/1104Michaud.pdf.
Mohan D, Pittman Jr. CU, Bricka M, Smith F, Yancey B, Mohammad J, Steele PH, Alexandre-Franco MF, Gómez-Serrano V, Gong H (2007). Sorption of arsenic, cadmium, and lead by chars produced from fast pyrolysis of wood and bark during bio-oil production. J. Colloid. Interface Sci. 310:57-73.
Muthusamy P, Murugan S, Manothi S (2012). Removal of Nickel ion from Industrial Waste Water using Maize Cob. ISCA J. Bio. Sci. 1(2):7-11.
Nomanbhay SM, Palanisamy K (2005). Removal of heavy metal from industrial wastewater using chitosan coated oil palm shell charcoal. Electronic J. Biotech. 8(1):43-53.
Okieimen FE, Okundia EU, Ogbeifun DE (1991). Sorption of Cadmium and lead ions on modified groundnut (*Arachis hypogea*) husks.J. Chem. Technol. Biotechnol. 51:97–103.
Okoya AA, Ogunkoya L (2009). Environmental Management and Evaluation of Cocoa – POD husk: In Proceedings of TWOWS African regional Conference, Theme: Gender and the Millennium development goals (MDGS), 16- 20th November
Olu-owolabi BI, Oputu OU, Adebowale KO, Ogunsolu O, Olujimi OO (2012). Biosorption of $Cd^{2+}$and Pb^{2+}ions onto mango stone and cocoa pod waste: Kinetic and equilibrium studies. Sci. Res. Essays. 7(15):1614-1629.
Onundi YB, Mamun AA, Al Khatib MF, Ahmed YM (2010). Adsorption of copper, nickel and lead ions from synthetic semiconductor industrial wastewater by palm shell activated carbon. Int. J. Environ. Sci. Tech. 7(4):751-758.
Onundi YB, Mamun AA, Al Khatib MF, Al Saadi MA, Suleyman AM (2011). Heavy metals removal from synthetic wastewater by a novel nano-size composite adsorbent. Int. J. Environ. Sci. Tech. 8(4):799-806.
Onwu FK, Ogah SPI (2010). Studies on the effect of pH on the sorption of cadmium (II), nickel (II), lead (II) and chromium (VI) fromaqueous solutions by African white star apple (*Chrysophyllum albidium*) shell. Afr. J. Biotech. 9(42):7086-7093.
Park D, Lim SR, Yun YS, Park JM (2008). Development of a new Cr(VI)-biosorbent from agricultural biowaste. Bioresour. Techno. 99:8810-8818.
Patil S, Bhole A, Natrajan G (2006). Scanvenging of Ni (II) metal ions by adsorption on PAC and Babhul Bork. J. Environ. Sci. Eng. 48(3):203-208.
Quek SY, Wase DAJ, Forster CF (1998). The use of sago waste for the sorption of lead and copper.Water SA 24:251–256.
Sethu VS, Goey KS, Iffah FR, Khoo CM, Andresen JM (2010). Adsorption Characteristics of Cu (Ii) Ions in Aqueous Solutions using Mangifera Indica (Mango) Leaf Biosorbents. J. Environ. Res. Develop. 5(2):262-278.
Souundarrajan M, Gomathi T, Sudha PN (2012). Adsorptive removal of chromium (VI) from aqueous solutions and its kinetics study. Archiv. Appl. Sci. Res. 4(1):225-235.
Sudha PN, Ramasubramaniam S, Govindarajan C, Gomathi T (2012). Removal of Chromium (VI) from aqueous solution using chitosan - Starch blend. Der Pharmacia Lettre. 4(1):240-248.
Sugashini S, Gopalakrishnan S (2012). Studies on the Performance of Protonated cross linked Chitosan Beads (PCCB) for Chromium Removal. Res J. Chem. Sci. 2(6):55-59.
Ulloa JB, van Weered JH, Huisman EA, Verreth JAJ (2003). Tropical Agricultural Residues and their Potential uses in Fish Feeds: the Costa Rican situation. Escuela de Ciencias Biologicas, Universidad Nacional, Heredia 86 – 3000, Costa Rica Fish Culture and Fisheries Group, Wageningen University, Wageningen, The Netherlands 2003.
Vaishnav V, Daga K, Chandra S, Lal M (2012). Adsorption Studies of Zn (II) ions from Wastewater using Calotropisprocera as an Adsorbent. Res. J. Recent Sci. 1:160-165.
Zhang J, Huang Z, Ruitao L, Yang Q, Kang F (2009). Effect of growing CNTs onto Bamboo charcoals onadsorption of copper ions in aqueous solution. Langmuir 25(1):269-274.
Zulkali MMD, Ahmad AL, Norulakmal NH (2006). *Oryza sativa* L. husk as heavy metal adsorbent: Optimization with lead as model solution. Biores. Technol. 97:21–25.

16

Phytochemical study on the benzene:acetone extract of the leaves of *Artabotrys odoratissimus*

Faizan Danish Khaleel[1], S. M. Zuber[2], B. K. Mehta[1], Darshina Mehta[1] and S. R. Kolisetty[1]

[1]School of Study in Chemistry and Biochemistry, Vikram University Ujjain (M.P)-456010, India.
[2]Division of Aquatic Sciences, The Himalayan Ecological and Conservation Research Foundation, J&K, 180006, India.

Artabotrys odoratissimus **has been investigated by many workers for its constituents. The compounds so far isolated from the leaves of *A. odoratissimus* are Pentadecyl-7-hydroxy- dodecanote, Pentadecyltritriacontanonte, 4,5-epoxy -26-ol- dopentacontane, β-Sitosterol, Aplysterol, Nonacosanylhexaconsanoate, Pentatetraacont-19-ol, triacont-2-ol, Dotriacont-7-ene, Octacose-7-ene, 1-hydroxy-2,5-dimethyl-9,10-anthraquinone, 1,4,5-trihydroxy-9,10-anthraquinone, 13-hydroxynonacosane, Nonanoicacid, Methylphenyl propanoate, Decanoicacid, Diethylphthalate, Dibutylphthalate, 2-amino-3-ethylbipheny-l, 5-methyl-9-phenylnonane-3-ol, 1-phenylundecane, 2,5-dimethyl-1-phenylheptane1-one, hexadeca-2,7,11-triene, 1-isopropyl-4,6-dimethylnaphthalene, 1-phenyldeca-1-one, 1-phenylundecan-1-one, 5-(2-butylphenyl)pent-3-en-2-ol, 2,5-dimethyltetradecanhydro phenenthrene So far, not many studies have been carried out on this genus but there are some reports on this plant. Previous phytochemical studies have revealed this genus to be rich in secondary metabolites including phenylcoumarins, xanthones and triterpenoids. Our recent study on the benzene:acetone extract of the leaves of *A. odoratissimus*, have led to the isolation of steroidal compound. The structure of the compound has been established by modern spectroscopic techniques such as Infrared Spectrometry (IR), Hydrogen-Nuclear Magnetic Resonance Spectrometry (^{1}H-NMR), Carbon-Nuclear Magnetic Resonance Spectrometry (^{13}C-NMR) and Mass-Spectroscopy and identified as 2β- hydroxy- Stigmasta-4- en-3, 6- dione- Methoxy ester.**

Key words: *Artabotrys odoratissimus*, medicinal plant, new compound, spectral analysis.

INTRODUCTION

Artabotrys odoratissimus, commonly known as Kantili Champa, is an ornamental shrub distributed throughout the country. Leaves are oblong, lanceolate, glabrus, shining acute at the base, petioles are 6 to 10 mm long. The size of the leaves is up to 18 by 3.8 to 5.0 cm. Flowers are acrid, bitter and yellowish white in colour. It is available in Bangladesh and India (Chopra et al., 1956). Ayurvedic and Yunani doctors use the leaves and flowers as a remedy for cholera, vomiting, thirst, headache and volatile oils from the leaves show antifungal and antimicrobial activity. The antifertility activity of *A. odoratissimus*, plant has been reported in albino rats (Chakarabarti et al., 1968).

The fruit extracts showed cardiac stimulatory effects on some animals and cardiac depressant effects on others(Trivedi et al., 1971). Previous Phytochemical studies have revealed this genus to be rich in secondary metabolites including, alkaloids, flavenoids,

Table 1. Column chromatography processing of benzene:acetone extract of *A. odoratissimus* (leaves).

Fraction No.	Eluent	Ratio of eluent (v/v)	Volume collected (ml)	TLC spots	Yield (gram)
1	Hexane	-	2500	3 Spots with streak	11
2	Hexane:benzene	1:1	3000	Dark streak	7
3	Hexane:benzene	2:3	2000	Unresolved streak	8
4	Hexane:benzene	1:4	5000	Single spot	13
5	Benzene:ETOAC	3:1	3000	2 Spots with streak	18
6	Benzene:ETOAC	1:1	3000	Single spot with streak	30
7	Hexane:MeOH	2:1	3000	Single spot with streak	5
8	MeOH	-	3000	Unresolved streak	3.2

Weight of Silica Gel = 900 g; Weight of extract = 100 g.

Table 2. Re-column chromatography processing of fraction No.6 (From Table 1).

Fraction No.	Eluent	Ratio of eluent (v/v)	Volume collected (ml)	TLC spots	Yield (mg)
1	Hexane:benzene	1:1	1000	Unresolved streak	19
2	Benzene:EtOAC	3:2	3000	2 Spots with streak	11.2
3	EtOAC:$CHCl_3$	3:1	3000	Unresolved dark streak	13
4	EtOAC:$CHCl_3$	2:1	5000	Compound S1	60.2

Weight of Silica gel = 200 g; Weight of extract = 30 g.

phenylcoumarins, xanthones and triterpenoids (Connoly et al 1994; Haider et al 1991; Perold et al 1978; Bheemasankara et al 1984; Sharma et al 2002; Singh et al 2009; Chakabati et al 1968). In this paper, we have described the isolation and structural elucidation of the isolated compound from the benzene: acetone extract of the leaves of Artabotrys odoratissimus. The isolated compound is identified by its Spectral data and has not been reported before.

MATERIALS AND METHODS

General

Freshly distilled solvents were used for extraction, isolation and purification. Evaporations were performed under reduced pressure on a Buchii rotary evaporator. Infrared (IR) spectra were recorded (KBr discs) on a Shimadzu UV-168A Spectrophotometer, validation (Vmax in cm^{-1}). Hydrogen-Nuclear Magnetic Resonance Spectrometry (^{1}H-NMR) were recorded on a Bruker R-32 (300 MHz) instrument in $CDCl_3$ and DMSO-d6 with TMS as an internal standard (Chemical Shifts in delta, ppm). All solvents used were of analytical grade. Thin layer chromatography (TLC) was performed using Silica gel GF254.

Plant materials

The leaves of *A. odoratissimus* were collected from the gardens of Ujjain city and university campus and were identified by the authorities of IEMPS, Vikram University Ujjain (M.P).

Extraction of the Compound S1

Dried leaves of the plant (5 kg) were milled into powder and then extracted with hexane (8 L) in a Soxhlet extractor for 36 h. The hexane extract was evaporated in a rotator evaporator and dried by vacuum pump. The hexane extract (200 g) was extracted successively with hexane, benzene, benzene:acetone, benzene:EtOAC and ethanol to yield hexane (17.5 g), benzene (14 gm), benzene:acetone (100 gm), benzene:EtOAC (11 g) and ethanol (2.3 g) soluble fractions, respectively. The benzene:acetone (100 g) soluble fraction on TLC examination showed several spots. To isolate different compounds, the benzene:acetone extract (100 g) was subjected to column chromatography using Silica gel as adsorbent. The column was eluted with different solvents in their increasing order of polarity. Various elutes showed different pattern of spots on TLC examination and therefore, were further separated by rechromatography on Silica gel columns and the results are being reported in Tables 1 and 2.

Thus, the EtOAC:$CHCl_3$ (2:1,v/v) fraction revealed the presence of single Compound S1 with clear spot on TLC plate.

RESULTS AND DISCUSSION

Finally, the Compound S1 was crystallized from benzene:acetone extract to give reddish brown crystals(60.2 mg), m.p. 173 to 179°C; IR (λmax(KBr): 3445, 2919, 2850, 1837, 1733, 1705, 1675, 1640, 1543, 1522, 1461, 1380, 1341, 1250, 1155, 974, 845, 831, 759 and 730 to 720 cm^{-1}; ^{1}H- NMR (300 MHz,δppm, $CDCL_3$, TMS): and ^{13}C- NMR (δppm, $CDCL_3$, 75MHz) (Table 3). FAB-MS(m/z,rel,int):M^+486(6.94), 471(5.56), 469(5.56), 459(5.56), 443(9.72), 441(8.33), 429(69.44), 427(36.11),

Table 3. Crystallization of Compound S1.

Carbon No.	Type of carbon	H1-NMR delta ppm ($CDCL_3$, 300 MHz)	C13- NMR, delta ppm ($CDCL_3$, 75 MHz)
1	CH2	1.49(d, j = 5.1 Hz)	37.52
2	CH	3.51 (m)	72.10
3	C	-	231.0
4	CH	5.82 (S)	122.0
5	C	-	161.7
6	C	-	223.0
7	CH2		34.23
8	CH		31.95
9	CH	1.83	50.42
10	C		36.40
11	CH2	1.09	21.35
12	CH2	1.49, 1.13	42.60
13	C		46.13
14	CH		56.35
15	CH2		24.60
16	CH2	1.95 (m)	28.50
17	C		57.04
18	CH3	0.65 (S)	12.24
19	CH3	1.01 (S)	19.65
20	CH		32.18
21	CH3	0.98	19.04
22	CH2	2.32(d, J = 5.7 Hz)	40.10
23	CH2	1.26	26.40
24	CH	2.23	24.60
25	CH	2.33 (m)	29.50
26	CO		183.4
27	CH3	0.82 (d, j = 7.5Hz)	19.30
28	CH2		23.35
29	CH3	0.95 (t)	12.12
30	OCH3	3.75 brs	64.87

423(11.11), 411(100), 397(43.05), 395(43.05), 383(15.28), 367(9.72), 353(6.94), 327(5.56), 311(6.94), 299(8.33), 287(6.94), 269(15.28), 239(9.72), 227(12.5), 213(15.28), 199(13.89), 185(18.06), 175(19.44), 159(41.67), 145(43.61), 133(40.28), 107(45.8), 105(48.61), 95(50.0).

The TLC examination of the isolated compound from benzene:acetone extract of the leaves of *A. odoratissimus* showed a single spot upon exposure to iodine vapour. It was readily soluble in $CHCL_3$. To the best of our knowledge, this compound has not been previously isolated or separated from any other sources. 2β- hydroxy-stigmasta-4-en-3,6-dione-methoxy ester was isolated from this plant for the first time in our laboratory. The IR spectrum showed peaks corresponding to enedione pattern (1837 and 1733 cm^{-1}), the carbonyl group (1718 cm^{-1}), C=C unsaturation and isopropyl group (1639 and 1380 to 1341 cm^{-1}), respectively (Ali, 2001; Dewick, 1997; Song et al., 2004; SAS Institute SAS/STAT user's Guide, 1997; Branco et al., 2001). The peak at 3445 cm^{-1} corresponds to hydroxyl group. The enedione pattern is further confirmed by the absorption at 238 nm in the UV spectrum. The peak at 1705 cm^{-1} showed the presence of $COOCH_3$ group in the molecule (Silverstein et al., 1984). The absorption peaks at 759, 831 and 845 cm^{-1} are characteristic of Δ^4- unsaturated and skeletal vibrations of steroidal moiety (Fujita et al., 1992; Bellamy, 1975). The ^1H spectrum showed the presence of two singlets, integrating three protons each at δH 0.65 and 1.01 were assigned to two angular methyl groups at C-18, C-19 and a doublet at δH 0.98 (J = 6.5Hz) to the methyl protons at C-21. A doublet was assigned to 27- Me at δH 0.82 (J = 7.5 Hz). The singlet at δH 5.32 for one proton revealed the presence of Olefinic proton at C-4 in the ring usually observed in the steroids having Δ^4 un-saturation with enedione (Peres and Nagem,

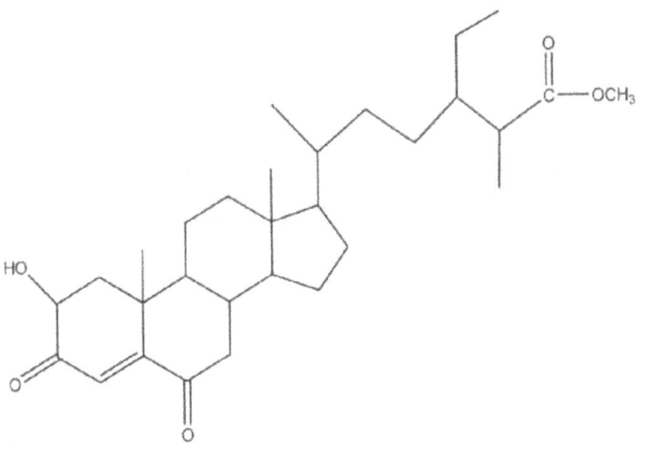

Scheme 1. 2β-hydroxy-stigmasta-4-en-3,6-dione-methoxy ester.

1997; Chung et al., 2006; Li et al., 1999). A multiplet at δH 3.50 showed the presence of 2α- methine proton (axial) interacting with C-1 equatorial, C-1 axial for hydroxyl group in the Ring A. There is a broad singlet at δH 3.75 for methoxy group at C-30 of methyl ester at C-26. The attachment of this group to C-25 is confirmed by the deshielding of H-25 at δH 2.33 as multiplet. A triplet at δH 0.95 was assigned to methyl proton at C-29. Most of the other peaks corresponded to those of stigmast-4-en-3,6- dione and other similar compounds of this class reported from different sources (Akihisa et al., 1992; Tsuda and Schroepter, 1979). The multiplicity of carbon atoms in the ^{13}C-NMR ($CDCL_3$, 75 MHz) spectrum confirmed the suggested structure. The methyl carbons appeared at δc 12.24 (C-18), and δc 19.65 (C-19), 19.04 (C-21), 19.30 (C-27, and 12.12 (C-29). The upfield resonances of C-18 and C-19 at δc 12.24 and 12.12 suggested the location of the methoxy ester group at C-26 (Peres and Nagem, 1997; Chung et al., 2006; Li et al. (1999). The most deshielded carbonyl carbons at C-3, C-6 was assigned to the peak at δc 231 and 223, respectively. It shows the presence of Ene-di-One system in the molecule. The downfield shift of these carbonyl molecules compared to reference compound is possible only when the C-2 is attached with hydroxyl group. The peak at δc 72.1 for C-2 was assigned to Carbinolic carbon. The deshielded carbonyl carbon signal at δc 183.4 was for – $COOCH_3$ at C-26 and the peak at δc 64.87 for OCH_3 group. The deshielded signals at δc 161.7 and 122.0 were assigned to vinylic carbons at C-5 and C-4 of the Ring A, respectively (Akihisa et al., 1992; Tsuda and Schroepter, 1979; Panico et al., 1994). The molecular ion peak of S1 in FAB-MASS was found at m/z 509(M^+Na), suggesting the molecular formula as $C_{30}H_{46}O_5$ (Cal. 486.6972). The fragmentation pattern is characteristic of a steroid as it showed peaks at m/z 468(M-H_2O, 429(M+2H-$COOCH_3$), 395[M+H-Me-H_2O-COOH), 471(M-CH_3), 427(M-$COOCH_3$). 287(M-side chain), thus, confirming the number of carbon atoms in the side chain and the $COOCH_3$ group at C-26 carbon. Peaks at m/z 345,288 and 274 were formed due to (M-side chain), [M-(side chain+ cleavage of C_{16} and C_{17} bond)] and [M-(side chain+ ring D cleavage)], respectively (Chung et al., 2006; Li et al., 1999; McLafferty, 1973).

Thus, the IR, ^{1}H-NMR, ^{13}C-NMR and Mass-Spectral analysis along with physical properties established the identity of Compound S1 as 2β- hydroxy-Stigmasta- 4-en-3, 6-dione- methoxy ester (Scheme 1).

Conclusion

The results of the present investigation constitute the occurrence of 2β- hydroxy- Stigmasta- 4-en-3,6-dione methoxy ester type compound in plant kingdom. The title compound has been isolated from this plant for the first time.

ACKNOWLEDGEMENT

The authors are thankful to the Head, School of Study in Chemistry and Biochemistry, Vikram University Ujjain for providing laboratory and library facilities.

REFERENCES

Akihisa T, Hayashi Y, Petterson GW, Shimizu N, Tamura T (1992). 4-alpha methylvernosterol and other Sterols from Vernonia anthelmintica seeds. Phytochemistry 31(5):1759-1763.

Ali M (2001). Techniques in Terpenoid Identification. Birla Publications Delhi, India.

Bellamy LJ (1975). The Infrared Spectra of Complex Molecules. Chapman and Hall London.

Bheemasankara rao ch, Suseela K, Subba rao PV, Gopala KP, Subba raju GV (1984). Chemical examination of some Indian Medicinal Plants. Ind. J. Chem. 23B:787-788.

Chakabati B, Chaudhuri A, Choudhury PR (1968). Analysis of the essential oils of Artabotrys odoratissimus fruits and leaves. J. Indian Med. Assoc. 51(5):227-229.

Chakarabarti B, Chaudhuri A, Chowdhury PR J (1968). Antifertility effect of green leaves of *Artabotrys odoratissimus*. Indian Med. Assoc. 51:227.

Chopra RN, Nayar SL, Chopra IC (1956). Glossary of Indian Medicinal plants, antifertility effect of green leaves of *Artabotrys odoratissimu.* (C.S.I.R. Publication, New Delhi) 25.

Chung Ill-M, Ali M, Khan TD, Choung MG, Park HJ, Ahmed A (2006). Novel anthracene derivatives isolated from Rice Hulls of *Oryza sativa*. Bull. Korean Chem. Soc. 27(1):93-98.

Connoly JD, Haquc ME, Hasan CM, Haider SS (1994). Constituents of the stem bark of *Artabotrys odoratissimus* . Fitoterapia, 65(1):92-93.

Dewick P (1997). Medicinal Natural Products - A Bio-synthetic Approach. John Wiley and Sons Ltd. Chichester, England.

Fujita T, Liu D, Veda S, Takeda Y (1992). Xanthones from *Polygala tenuifolia*. Phytochemistry. Phytochemistry 31:3997-4000.

Haider SS, Hasan CM, Hussain CF (1991). Chemical constituents of stem bark of *Artabotrys odoratissimus*. J. Bangla Acad. Sci. 15(1):59-62.

Li W, Chan C, Leung H, Yeung H, Xiao P (1999). Xanthones from *Polygela caudata*. Phytochemistry. 51:953-958.

McLafferty FW (1973). Interpretation of Mass Spectra. Benjamin Inc. Massachusets. pp. 118-119.

Panico R, Powell WH, Richer JC (1994). IUPAC Commission on the Nomenclature of Organic Chemistry: A guide to IUPAC Nomenclature of Organic Compounds. Blackwell Scientific Publications Ltd., Oxford UK. Supersedes in part ref. 8.pp. 71-73.

Peres V, Nagem TJ (1997). Trioxygenated naturally occurring Xanthones. Phytochem. 44:191-214.

Perold GW, Rosenber MEK, Howard AS, Huddle PA (1978) Anthraquinones from *Artabotrys odoratissimus* leaves. J. Chem. Soc Perkin Trans. 1:239.

SAS (1997). Institute SAS/ STAT user's Guide, SAS Institute, Cary, NC 6.12.

Sharma M, Desiraju S, Chaurey D, Mehta BK (2002). GC-MS study o *Artabotrys odoratissimus* fatty oil (leaves). Grasas Aceites 53(2):187-189.

Silverstein RM, Bassler GC, Morill TC (1984). Spectrometri Identification of Organic Compounds. John Wiley and Sons, Nev York 4th Edn. 181.

Singh JP, Singh AK, Ranjan R, (2009). Chemical constituents o *Artabotrys odoratissimus* (seeds). Rasayan J. Chem. 2(1):156-158.

Song HK, Ahn JK, Ahmad A, Hahn SJ, Kim SJ, Chung IM (2004) Identification of allelochemicals in rice root exudates at variou phonological phases and their influence on barnyard grass Allelopathy J. 13(2): 173-188.

Trivedi CP, Saxena SP, Emmanuel J (1971). Preliminary phytochemica and pharmacological studies on *Artabotrys odoratissimus*. Indian J Med. Res. 59:635.

Evaluation of drinking water quality

Krishna Vaidya[1]* and Mohini Gadhia[2]

[1]Department of Biology, K.B.S. College, Vapi, Gujarat, India.
[2]Veer Narmad South Gujarat University, Surat, Gujarat, India.

Navsari District is located in the south eastern part of Gujarat State in the coastal lowland along Purna River in India. It's geographical coordinates are 20°51' 0" North, 72°55' 0" East. In the present study, the physico-chemical parameters of Navsari District (Gujarat, India) have been analyzed regarding their suitability for drinking purpose. The study was carried out by collection of water samples from six sampling sites. These samples are analyzed for turbidity, pH, total solids, total suspended solids, total dissolved solids, total hardness, magnesium hardness, calcium hardness, phenolphthalein alkalinity and total alkalinity. The analyze results is compared with permissible limits as prescribed by World Health Organization (WHO), Gujarat pollution control board (GPCB) for drinking water quality.

Key words: Drinking water, hardness, total solids, pollution, Navsari.

INTRODUCTION

Water is the most beautiful and precious gift of nature without which no life could survive on earth (Dara, 1998; Kumar and Tripathi, 2000). Water takes many different shapes on earth and to study water a new science evolved named as "Hydrology" which is the science to know the properties, distribution and behavior of water in nature (Fair and Geyer, 1958). Among the various needs of water, the most essential need is drinking. Surface water and ground water are two major sources for the supply of drinking water. Surface water comes from lakes, reservoirs and rivers. Groundwater comes from wells that the water supplier drills into aquifers (Park, 1997). Maintaining the quality of water is the most important one for human being since it is directly linked with his daily life (Gosh, 2002). Thus, proper and managed study of water, especially freshwater is essential to understand the relationship and interdependence of various constituents of any habitat.

The town of Navsari is approximately about 2000 years old. The city is situated at southeastern Gujarat State, west-central India. It is situated in the coastal lowland along the Purna River. District covers an area of 2,211 km^2 and has population of 1,229,463 of which 27.36% is urban. It lies between 20°51' 0" North, 72°55' 0" East. Weather is pleasant almost all the year around, sunny from September to May, rainy from June to August. There are two lakes in the city namely; Dudhiya Talao and Sarbatiya Talao. The main source of Nagarpalika water works supply in Navsari City comes from Kakarapar through a canal and is stored in a small reservoir called "Dudhiya Talao" (Patel et al., 2000). The kakrapar dam is constructed across the river Tapi and down stream of Ukai dam. To monitor the potable water quality, total selected six sampling sites shown in Plate 1 are untreated water of Dudhiya Talao (Site 1), treated water of Navsari water works (Site 2), Station Area (Site 3), Lunsikui Area (Site 4), Chhapra Village (Site 5) and Viraval Village (Site 6).

MATERIALS AND METHODS

In the present study, six sampling sites were selected. The sampling was done on seasonal pattern. Composite sampling method was particularly adopted in (Site 1). The taps were kept open for 2 to 3 min while collecting samples from pipeline supply to remove the possible impurities in water through pipes. Water samples were collected at fixed time to maintain the consistency in the results. Care was also taken for collection timing depending on

*Corresponding author. E-mail: krish_bio@yahoo.co.in.

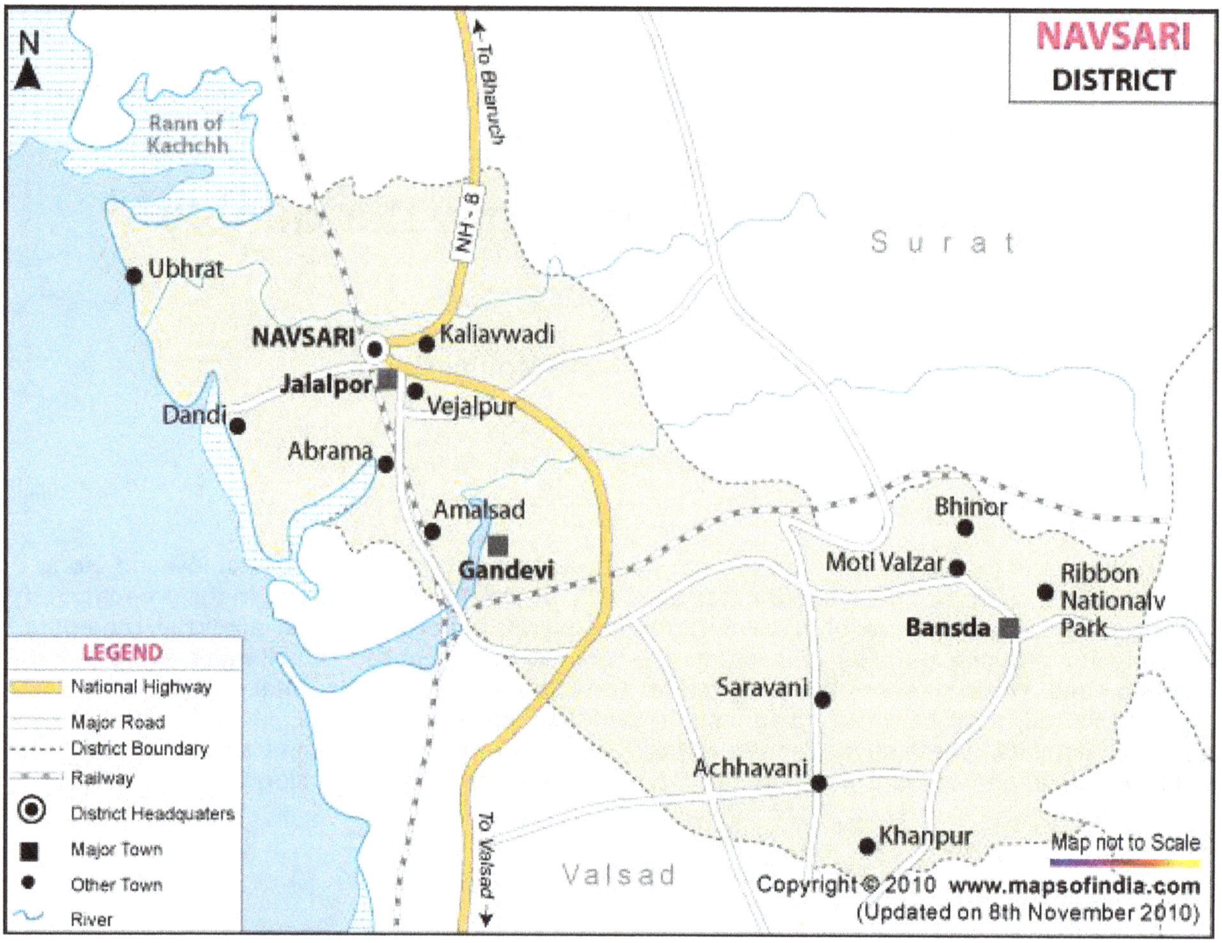

Plate 1. Map of Navsari.

water supply from Navsari Nagar Palika.

The methods of APHA (1995) and Trivedi and Goel (1986) were followed for water analysis. The parameters such as turbidity, pH, total solids, total suspended solids, total dissolved solids, total hardness, magnesium hardness, calcium hardness, phenolphthalein alkalinity and total alkalinity were brought for further analysis.

RESULTS

The results of physico-chemical parameters of average six sampling sites are shown through Graph 1. Turbidity of drinking water of Navsari District was higher from Site 1 and reduced drastically after Site 2. More or less same value was recorded from remaining Sites 3 to 6 in all the three seasons except in summer season from Site 3.

The pH was predominantly alkaline ranged between 7.48 to 7.83 in monsoon season, 7.41 to 8.10 in winter season and 7.41 to 8.08 in summer season throughout the study from all the six sampling sites of Navsari district. pH of drinking water was found highest in Site 1 whereas lowest in Site 6 during all the three seasons. The pH of Sites 2, 3, 4 and 5 were more or less same in all three seasons.

Total solids were recorded maximum in the range of 664.47 to 990.50 mg/l from Sites 1, 5, 6 and was found to be minimum 488.16 mg/l from Site 4 in monsoon season whereas it was recorded more or less same from Sites 2 and 3 during all the three seasons. Total dissolved solids of untreated water from Sites 1, 5, 6 were recorded in range of 649.19 to 972.50 mg/l in all three seasons whereas in Sites 2, 3 and 4 were more or less same in range between 472.31 to 749.38 mg/l.

The results of total suspended solids were found to be higher from Site 1 and reduced drastically after the Site 2. More or less same value was recorded from Sites 5 and 6 whereas in Sites 3 and 4 results were close to Sites 5 and 6.

Total hardness of Site 1 was found to be minimum during all three seasons. It was recorded in the range 306.50 to 486.47 mg/l from Sites 2, 3 and 4 whereas it was found to be in similar range of 426.85 to 565.36 mg/l from Sites 5 and 6. The results of calcium hardness, in Site 1 was far below than the Site 2 in summer and monsoon seasons whereas it was found in same range between 241.82 to 499.66 mg/l from Sites 3, 4, 5 and 6.

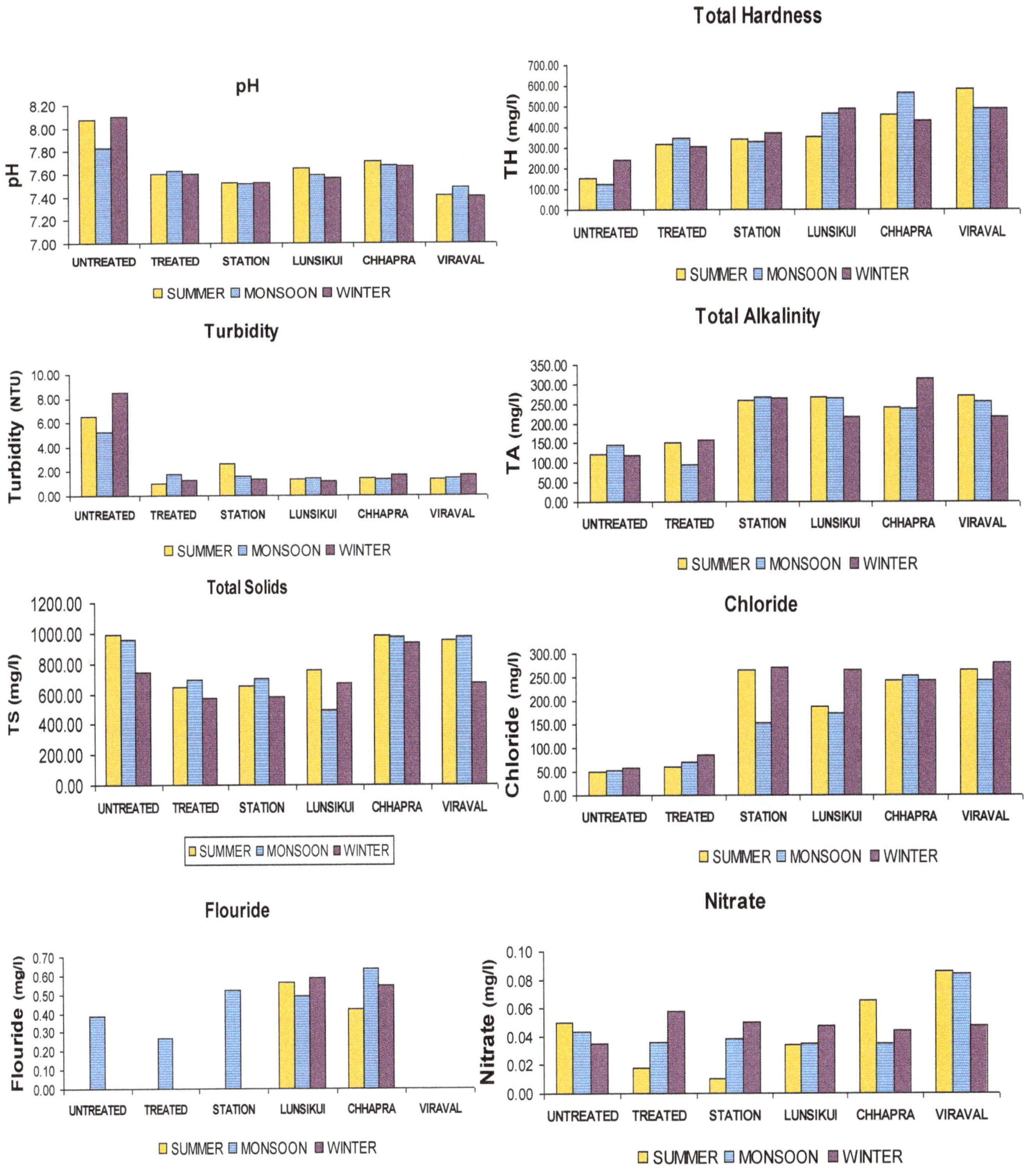

Graph 1. The PH, turbidity, total solid, flouride, total hardest, total alkalinity, chloride and nitrate of Navsari District.

Magnesium hardness was recorded in range 58 to 67.78 mg/l from Site 1 and was minimum from Site 2 whereas it was more or less similar range from 67.09 to 95.63 mg/l from Sites 3, 4, 5 and 6 during all the three seasons.

Total alkalinity of Sites 1 and 2 was found minimum during all the three seasons in range 93.50 to 156.50 mg/l whereas it was more or less similar range from 217.69 to 315.03 mg/l from Sites 3, 4, 5 and 6 during all the three seasons as shown in the Graph 1. Phenolphthalein alkalinity were recorded in the range of

1.66 to 4.06 mg/l from Sites 2 to 6 except during monsoon season from Site 1 it was highest in range of 7.25 mg/l was depicted in the graph.

DISCUSSION

In the present study, turbidity was observed highest in untreated water and was reduced in all the sites and maintained well. The highest turbidity in untreated water was due to presence of clay and silt brought with runoff of water from Kakrapar canal though it has not crossed the standard limits.

pH is the measure of the intensity of acidity or alkalinity and measures the concentration of hydrogen ions in water (Mackee and Wolf, 1963). pH value with 7 is considered to be the best and most ideal (Sawyer and Mc Carty, 1967). During the present study, pH was found in alkaline range between 7.4 to 8.10 which was under the desirable limit.

The survey regarding the taste threshold level of TDS was done by Bruvold and Ongerth (1969) and was concluded that the range between 658 to 758 mg/l was good enough and the range between 1283 to 1333 mg/l was unpalatable for drinking. So, water with presence of high level of TDS was not used by the consumers. In the present study, TDS was found in the range 472.31 to 972.50 mg/l which was within the desirable limit.

Hardness is defined as the concentration of calcium and magnesium ions content of water (Kumar and Kakrani, 2000). Most probably natural water supplies contain at least some hardness due to dissolved calcium and magnesium salts (Fulvio and Olori, 1965). Hardness was higher from sampling Sites 3 to 6 compared to untreated and treated. However, the value did not cross the limits.

Calcium is important as a nutrients, its deficiency causes rickets (Trivedi and Goel, 1986). High concentrations of calcium are not desirable in washing, laundering and bathing. Scale formation in boilers takes place by high calcium along with magnesium (Park, 1997). In the present study, calcium was found highest from Sites 5 and 6 due to bore well water.

Magnesium also occurs in all kinds of natural waters with calcium, but its concentration remains generally lower than the calcium (Purohit and Saxena, 1990). So, if calcium and magnesium is high in water, it may cause kidney disease (Taylor, 1958). In the present study, magnesium was found below the desirable limit.

Alkalinity in natural waters is due to free hydroxyl ions and hydrolysis of salts formed by weak acids and strong bases. Water with low alkalinity is more likely to be corrosive, which could cause deterioration of plumbing and an increasing chance for lead in water if present in pipe, solder or plumbing fixtures (Frank, 1987). In the present study, alkalinity was high from Sites 3 to 6; this may be due to corrosion in distributing pipes and the bore well supply but were found in normal range.

Conclusion

The research work is an attempt to assess the drinking water quality. The physico-chemical analyses reveal that the present status of drinking water quality is suitable for drinking purposes. The quality of physic-chemical falls within the standard limits of WHO standards. This indicates that the water of Navsari District and its vicinity is suitable for drinking purpose.

REFERENCES

American Public Health Association, American Water Works Association, Water Environment Federation, 1995. In: Standard Methods for the Examination of Water and Wastewater 19th Edition Washington D.C., New York.

Bruvold WH, Ongerth HJ (1969). In: The taste of Water, Public Health Reports, Rawat Publication, Jaipur, pp. 110-135.

Dara SS (1998). In: A Textbook of Environmental Chemistry and Pollution Control, S. Chand Publication, New Delhi, pp. 64-69.

Frank N (1987). In: Water quality hand book, Mc Graw hill publication, New York, 2nd Edition, pp. 13-19.

Fulvio DE, Olori L (1965). In: Hardness of Drinking Water and Public Health, Pergamon Publication, New York, p. 95.

Gosh GK (2002). In: Water of India: Quality and Quantity, A.P.H. Publication, New Delhi, pp. 9, 13, 19, 97, 191.

Kumar A, Tripathi G (2000). In: Water Pollution Assessment and Management, Daya Publication, New Delhi, pp. 1-Fair, GM, Geyer JC (1958). In: Elements of Water Supply and Wastewater Disposal, John Wiley Publication, USA, p. 1.

Kumar V, Kakrani B (2000). In: Water-Environment and Pollution, Agro Bios Publication, New Delhi, pp. 1-26.

Mackee JE, Wolf HW (1963). In: Water Quality Criteria, Mc Graw Hill Publication, New York 2nd Edition, pp. 136-247, 270, 275-277.

Park K (1997). In: Text book of Preventive and Social Medicine, Banarsidas Publication, Jabalpur, 15th Edition, pp. 468-479.

Patel AM, Malik MM, Parikh SR (2000). In: Ground water fluctuation and quality in Surat branch of Kakrapar command, pp. 1-3.

Purohit SS, Saxena MM (1990). In: Water life and pollution, Agro Botanical Publication, New Delhi, pp. 61-64.

Sawyer CN, Mc Carty PL (1967). In: Chemistry for Sanitary Engineers, McGraw Hill, Toronto, 2nd Edition, pp. 117-132.

Taylor EW (1958). In: The examination of waters and water supplies, Churchill publication, London, pp. 27-48.

Trivedi RK, Goel PK (1986). In: Chemical and Biological method for water pollution standard, Ashish publication, New Delhi, pp. 1-25; 100-105.

N-doped carbon aerogels for carbon dioxide (CO_2) capture

Aimé Serge Ello*, Jacques Aboua Yapo and Albert Trokourey

Université Félix Houphouët-Boigny de Cocody, Laboratoire de Chimie physique 22 bp 582 Abidjan 22, Cote d'ivoire.

The N-doped carbon aerogel materials were investigated to explore their potential use in carbon dioxide (CO_2) capture. Carbon aerogels were prepared from a nitrogen-containing polymer precursor using urea as nitrogen source into the polymer matrix through the sol-gel method. CO_2 capture performances were evaluated from 273 to 298K at 1 bar. CO_2 adsorption is influenced by the micropores and amount of a nitrogen-containing the samples. N2 adsorption isotherms have showed different structures of these carbon materials. CO_2 adsorption capacities up to 3.6 mmol/g (298K) and 4.5 mmol/g (273K) were achieved. Both texture and surface chemistry affect the CO_2 capture performance of the adsorbents.

Key word: Adsorption, carbon dioxide (CO_2), urea, carbon, aerogel, N-doped.

INTRODUCTION

Global warming is caused mainly by greenhouse gas emissions such as methane (CH_4) and carbon dioxide (CO_2). These gases are gaining more attention due to their potential environmental and socio-economic negative consequences. Among the technologies proposed to reduce CO_2 emissions, adsorption is considered as a very promising technology for CO_2 capture since adsorbents have been discovered (Choi et al., 2009). These materials have a high adsorption capacity, great selectivity, good mechanical properties and remain stable over repeated adsorption–desorption cycles (Choi et al., 2009). Different types of solid sorbents have been, or are currently being investigated as potential adsorbents for CO_2 capture, including carbon-based sorbents (Zou et al., 2002), supported carbonates (Gregory et al., 2006) and zeolites (Jerome et al., 2008; Simone et al., 2004). Studies have shown some limitations of these materials either because of their low adsorption capacity. The introduction of new functional groups leading to several different configurations of these sorbents, including amine-functionalized porous carbons, organic resins and polymers offer promising alternatives (Jason et al., 2008; Zhijian et al., 2008). The incorporation of basic nitrogen groups into the carbon framework ensures an improved adsorption/absorption for acidic gases (Foad et al., 2000; Pels et al., 1995).

Porous carbon materials used for CO_2 capture are mostly prepared by post-synthetic amine modification or ammonia treatment (Plaza et al., 2007) which leads to materials lacking stability and, in addition, the reagents are corrosive, with the same disadvantages as for modified silica and zeolites. Alternatively, nitrogen containing porous carbons can be prepared directly from nitrogen-rich precursors. Porous nitrogen-enriched carbons from melamine-formaldehyde resins using fumed silica had been prepared (Pevida et al., 2008). Porous carbons monolith using L-lysine as nitrogen source showed maximum CO_2 adsorption capacity of 3.13 mmol/g under 1 atmosphere at 298 K (Hao et al., 2010). Highly porous N-doped carbon activated by KOH (Sevilla et al., 2011) adsorbed CO_2 from 2.7 to 3.9 mmol/g at 298K. The structural characterizations and CO_2 adsorption on the resulting carbons indicated that CO_2 uptake was sometimes independent of the specific surface area but closely related to the micropore volume and the N content of the carbons. It was also observed that direct pyrolysis of these precursors usually results in bulk carbon products with less porosity (Takashi, 2000).

In the present work, we synthesized monolith N-doped carbon aerogels with sol-gel method using urea and resorcinol-formaldehyde as nitrogen and carbon precursor, respectively. Carbon aerogels are highly

*Corresponding author. E-mail: elloserge@gmail.com.

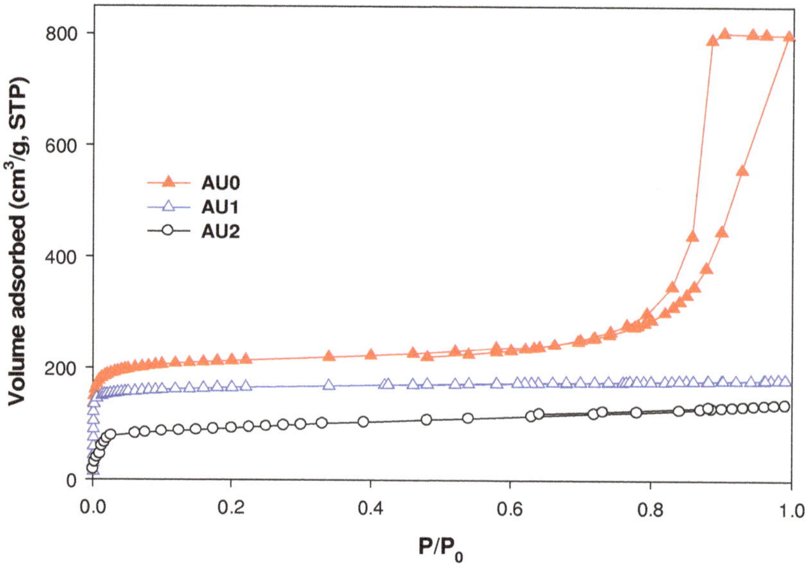

Figure 1. Adsorption-desorption isotherms of N-doped and carbon aerogels.

porous materials with a network structure consisting of interconnected sphere-like nodules whose sizes are regulated by the synthesis conditions (Elkhatat and Al-Muhtaseb, 2011). Here, we investigate the use of urea as nitrogen-containing precursors to introduce nitrogen groups on the surface of standard carbon aerogels. Urea can be hydroxymethylolated by the addition of formaldehyde to the amino groups. This occurs by a series of reactions that lead to the formation of mono- and di-methylolureas (Park et al., 2011; Dalin et al., 2011). The complex surface nitrogen functionalities can be obtained after pyrolysis when those precursors are incorporated into sol–gel polymerization. This is part of the first report on the N-doped aerogel material by sol gel method and can lead to more applications.

MATERIALS AND METHODS

Sample preparation

3.7 g of resorcinol and different amounts of urea 1 to 2 g were dissolved in 5 g of water and catalyst was added (weight ratio: Resorcinol / sodium Carbonate, R/C = 500). After stirring for 20 min, 5.1 ml of formaldehyde was added while stirring for 30 min at room temperature. Sol–gel polymerization of the mixture was carried out in a sealed glass cylinder by holding the mixture at 358K for 72 h in the oven. Acetone was selected as the drying solvent for ambient drying because of its low surface tension (20.66 dyn / cm) and its low boiling point (329K) which reduces the shrinkage of Resorcinol-Fromaldehyde-Urea (RFU) gels during drying (Jun et al., 2006). The wet gelling was performed at room temperature for 3 days, and then RFU gels were dried completely at ambient conditions for 72 h. After drying, RFU aerogel particles were placed in a quartz boat and heated to 1073K, ramp 278K / min under inert nitrogen (N_2) atmosphere in a tubular furnace for 2 h. The samples were denoted, respectively as AU0; AU1 AU2, where the numbers indicate the molar ratio of resorcinol/urea in the sample.

Characterization

Nitrogen adsorption isotherms were measured at 77K on ASAP 2010 volumetric analyzers (Micromeritics, Inc., GA). The BET (Brunauer-Emmett-Teller) specific surface area was calculated from nitrogen adsorption isotherms in the relative pressure (P/P_0) range of 0.05-0.2. The total pore volume was estimated from the amount adsorbed at a relative pressure (P/P_0) of ~0.99. The CO_2-adsorption isotherms of the AU-i series were measured using a Micromeritics ASAP 2020 static volumetric analyzer at 298 and 273K. Before adsorption measurements, all samples were degassed at 473K ensuring that the residual pressure fell below 3 μmHg. It was then cooled down to 273 and 298K, followed by the introduction of CO_2 into the system. The CO_2-adsorption capacity in terms of adsorbed volume under standard temperature and pressure (STP) was then recorded.

RESULTS AND DISCUSSION

Characterization

According to the International Union of Pure and Applied Chemistry (IUPAC), six types (Type I to VI) isotherms are used to describe different porous and materials interactions and four types of hysteresis loops (Types H1-H4) for texture of the adsorbent (Thommes, 2010). The N_2 adsorption and desorption isotherms of carbon aerogel and N-doped carbon aerogel displayed in Figure 1, show different types of isotherms. The sample AU0 type IV with H2 hysteresis typically observed with more complex pore structure in which network effect is important (Thommes, 2010). Unfortunately, we do not have images from scanning electron microscopy.

Adsorption isotherms of Type I (AU1 and AU2) are observed when the pores size of the carbon materials are in micro order and the potential of interaction between the

Table 1. Physical and chemical characteristics of carbon and N-containing carbonaerogel.

Samples	Physical properties				Chemical properties			
	S_{BET} (m^2/g)	V_{micro} (cm^3/g)	V_{meso} (cm^3/g)	V_t (cm^3/g)	N (wt%)	O (wt%)	H (wt%)	C (wt%)
AU0	746.7	0.19	1.05	1.24	0	10.82	0.81	88.37
AU1	576.5	0.18	0.10	0.28	0.51	19.35	0.60	79.54
AU2	321.8	0.13	0.08	0.21	1.44	17.68	0.50	80.38

S_{BET}: Surface area calculated by BET (Brunauer-Emmett-Teller) method. V_t = total volume at P/P_0 = 0.995;V_{meso} = volume of mesopore (V_t-V_{micro}).

surface of the carbon and N_2 is high. Thus, N_2 isotherms exhibited a high nitrogen uptake at low relative pressures (P/Po < 0.1), corresponding to the filling of micropores.

The N-doped carbon aerogels AU1, AU2 showed the lowest nitrogen uptake and also, their surface areas reduced considerably from 746.7 m^2/g (AU0) to 576 (AU1) and 314 m^2/g (AU2) when the molarratio of resorcinol/urea (r/u) increased corresponding to less developed porosity.

Table 1 shows the physical and chemical characteristics of N-containing carbon aerogels. The nitrogen content increased from 0.51 to 1.44% with the ratio of resorcinol/urea (r/u). We did not expect to get more residual nitrogen-content in sample AU2 than AU1. This very low range of value was also observed at this high temperature (1073K) because nitrogen was consumed which was in good agreement with the literature (Hongwei et al., 2012; Zhangxiong et al., 2012). We have not been able to use other techniques such as; Fourier transform infrared (FTIR) and x-ray photoelectron spectroscopy (XPS) to properly investigate bonding of nitrogen. Concerning theporosity of the structure, the pore size distribution of the samples were made up mostly of 0.7 nm for AU0 and AU1 but larger fractions of mesopores with pore diameters of up to 2 nm for AU0 and 1.2 nm for AU2 were obtained. The total pore volume decayed with addition of urea from 1.24 to 0.28 cm^3/g together with the volume of micropores from 0.19 to 0.13 cm^3/g.

The same trend was observed with the volume of the mesopores which decreased significantly from 1.05 to 0.08 cm^3/g. We therefore concluded that urea inhibited the development of mesopores. The percentage of other atoms such as oxygen, hydrogen and carbon did not change considerably between carbon aerogel and N-doped carbon aerogel.

CO_2 adsorption

Figures 2, 3 and 4 show the isotherms for CO_2 adsorption at 298 and 273K for the aerogel and N-doped carbon aerogels. The adsorption increased from 2.7 to 3.6 mmol/g at room temperature and 4 to 4.5 mmol/g at 273K. The highest value obtained at the ambient temperature was superior to what we observed in the literature (3.2 mmol/g) for N-doped monolith carbon material (Hao et al., 2010) but slightly inferior for other N-doped porous carbon material (3.9 mmol/g) with 1700 m^2/g surface area after activation with KOH (Sevilla et al., 2011). Our samples were prepared without anyactivation, just a conventional sol-gel recipe with Na_2CO_3 catalyst in few amounts (R/C = 500). The N-doped aerogel carbon AU1 had a better adsorption compared to the N-doped carbons AU2. Despite its high surface area and total pore volume AU0 adsorbed lesser than N-doped carbon. Table 1 has showed that N-content in AU2 was higher than AU1 but no CO_2 capacity adsorption. The difference of adsorption of N-doped carbon aerogel would not be attributed only to amount of nitrogen containing the sample but also to the micropore and pore size.

AU2 had more nitrogen content but lesser micropore and pore size (1.2 nm) than AU1 and AU0 though it is known that narrow pores are favorable to CO_2 adsorption. The isosteric heats of adsorption (Qst) for CO_2 (Figure 4) as calculated from the adsorption isotherms at 273 and 298K using the Clausius–Clapeyron equation confirmed the nitrogen content of the samples. The initial Qst values of 24.8 and 33.57 kJ/mol were shown for AU1 and AU2, respectively which decreased with the amount adsorbed. The values were not so high but indicated a little strong adsorbent–adsorbate interaction between the N-containing carbon framework and CO_2 molecules.

In this work, we confirmed also the theory observed by several authors (Plaza et al., 2007; Pevida et al., 2008; Hao et al., 2010; Sevilla et al., 2011; D'Alessandro et al., 2010; Zhangxiong et al., 2012) that the presence of basic N groups leads to an improvement of the adsorption of acidic gases such as CO_2. This could be due to the strong pole–pole interactions between the large quadrupole moment of CO_2 molecules (Lifeng and Ralph, 2011) and the polar sites associated with N groups coupled with the sample porosity (Sevilla et al., 2011). By comparing the properties of AU2 and AU0, we noted that the higher value of adsorption of AU2 was at low pressure due to the presence of nitrogen and not to its micropore until 300 mmHg, while AU1 gave a lower nitrogen content and micropore near to AU0. Table 2

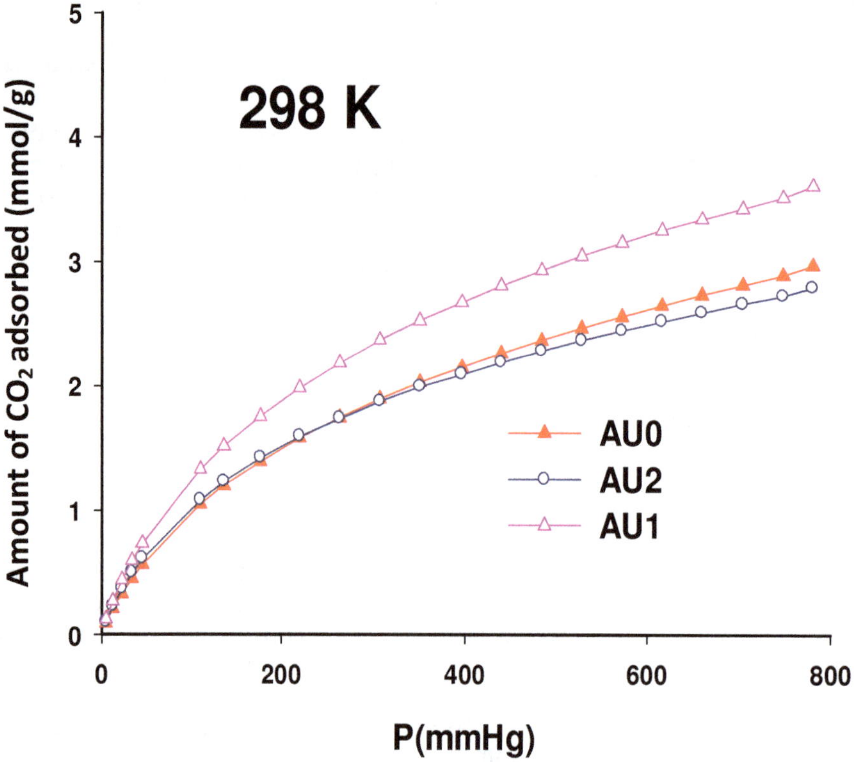

Figure 2. CO_2 adsorption capacity of N-doped and carbon aerogel at 298K.

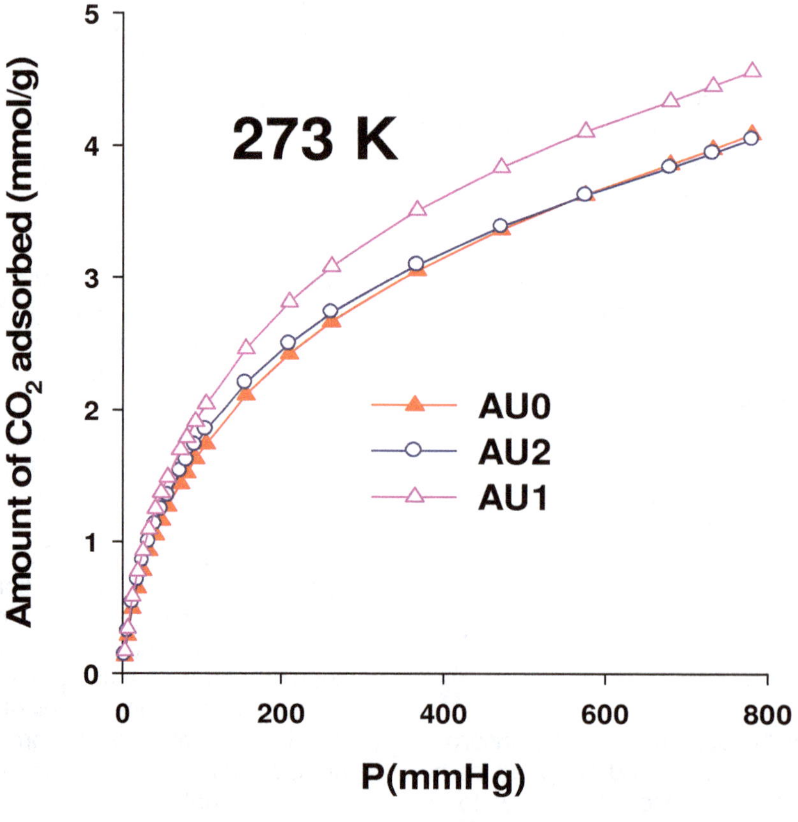

Figure 3. CO_2 adsorption capacity of N-doped and carbon aerogel at 273K.

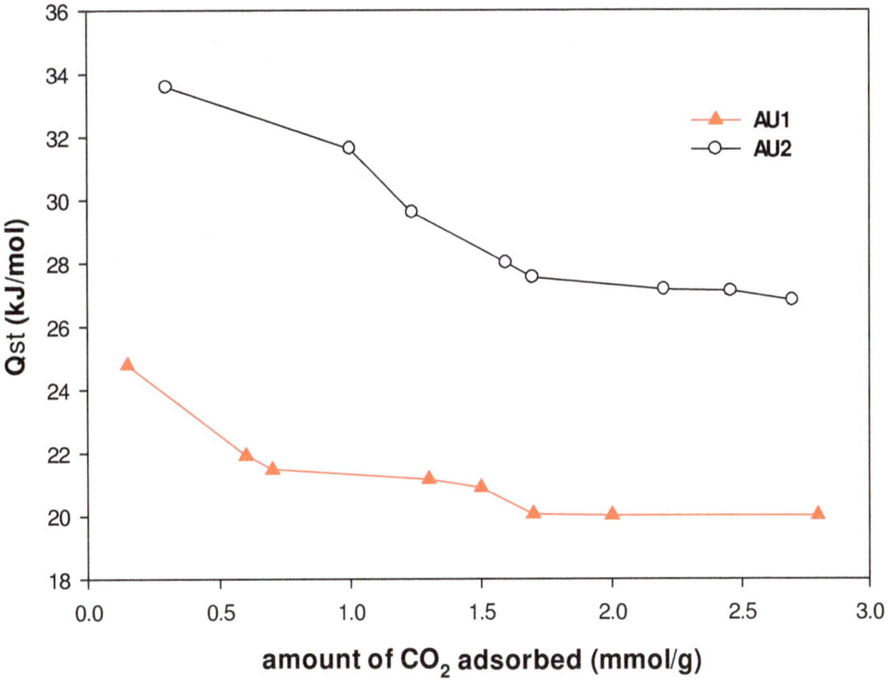

Figure 4. Isosteric heat of CO_2 adsorption (Qst) as function of amount of CO_2 adsorbed.

Table 2. CO_2 adsorption capacities of nitrogen-containing porous carbons observed in the literature at 25°C and 1 Atm.

Origin and N-doped carbon materials	Nitrogen content wt (%)	Capacity (mmol/g)	References
Melamine-formaldehyde	3.15	2.25	(Pevida et al., 2008)
Phenol-formaldehyde-lysine	1.91	3.13	(Hao et al., 2010)
Polypyrrole-Based Porous	10.1	3.9	(Sevilla et al., 2011)
Urea-formaldehyde-resorcinol	0.51	3.6	This work

shows a summary of some values of CO_2 adsorbed cited in the text.

Conclusion

N-doped carbon aerogel materials with different structures have been synthesized by a classical sol-gel method, with urea as a doped precursor. N_2 isotherms showed a difference of pores development. CO_2 adsorption experiment was affected by the nitrogen content and the micropore structure. The CO_2 adsorption and pore structures increased the amount of urea and the N-doped structure adsorbed more than the non-doped carbon aerogel.

ACKNOWLEDGEMENT

This work was supported by the Merit Scholarship Program for High Technology (MSP) from Islamic Bank of Development (IBD).

REFERENCES

Choi S, Drese JH, Jones CW (2009). Adsorbent Materials for Carbon Dioxide Capture from Large Anthropogenic Point Sources. Chem. Sus. Chem. 9(2):796-854.

D'Alessandro DM, Smit B, Long JR (2010). Carbon Dioxide Capture: Prospects for New Materials. Angew. Chem. Int. Ed. 35(49):6058-6082.

Dalin W, Mingming C, Chengyang W, Jie B, Jiaming Z (2011). Synthesis of carbon microspheres from urea formaldehyde resin. Mater. Lett. 7(65):1069-1072.

ElKhatat AM, Al-Muhtaseb SA (2011). Advances in Tailoring Resorcinol-Formaldehyde Organic and Carbon Gels. Adv. Mater. 26(23):2887-2903.

Foad A, Andrey B, Teresa JB (2000). Adsorption/Oxidation of Hydrogen Sulfide on Nitrogen-Containing Activated Carbons; Langmuir 4(16):1980-1986.

Gregory PK, Seamus WD, Alan LC (2006). Diethylenetriamine[propyl(silyl)]-Functionalized (DT) Mesoporous Silicas as CO_2 Adsorbents. Ind. Eng. Chem. Res. 8(45):2626-2633.

Hao GP, Li WC, Qian D, Lu AH (2010). Rapid Synthesis of Nitrogen-Doped Porous Carbon Monolith for CO_2 Capture. Adv. Mater.

7(22):853-857.
Hongwei Y, Youzhu, Y, Shik CET (2012). Nitrogen-enriched carbonaceous materials with hierarchical micro-mesopore structures for efficient CO_2 capture. Chem. Eng. J. pp. 185-186, 374-379.
Jason CH, Jeffrey HD, Daniel JF, McMahan LG, Genggeng Q, Christopher WJ (2008). Designing Adsorbents for CO_2 Capture from Flue Gas- HyperbranchedAminosilicas Capable of Capturing CO_2 Reversibly. J. Am. Chem. Soc. 10(130):2902-2903.
Jerome M, Marc C, Francis M (2008). Experimental Investigation on CO_2 Post-Combustion Capture by Indirect Thermal Swing Adsorption Using 13X and 5A Zeolites. Ind. Eng. Chem. Res. 1(47):209-215.
Jun L, Xianyou W, Qinghua H, Sergio GPJS (2006). Studies on preparation and performances of carbon aerogel electrodes for the application of supercapacitor. J. Power Sour. 1(158):784-788.
Lifeng W, Ralph TY (2011). Increasing Selective CO_2 Adsorption on Amine-Grafted SBA-15 by Increasing Silanol Density. J. Phys. Chem. C 43(115):21264-21272.
Park BD, Jeong HW, Lee SM (2011). Morphology and Chemical Elements Detection of Cured Urea–Formaldehyde Resins. J. Appl. Polym. Sci. 3(120):1475-1482.
Pels JR, Kapteijn F, Moulijn JA, Zhu Q. Thomas KM (1995). Evolution of nitrogen functionalities in carbonaceous materials during pyrolis. Carbon 11(33):1641-1653.
Pevida C, Drage TC, Snape CE (2008). Silica-templated melamine–formaldehyde resin derived adsorbents for CO_2 capture. Carbon 11(46):1464-1474.
Plaza MG, Pevida, C, Arenillas A, Rubiera F, Pis JJ (2007). CO_2 capture by adsorption with nitrogen enriched carbons. Fuel 14(86):2204-2212.
Sevilla M, Valle-Vigón P, Fuertes AB (2011). N-Doped Polypyrrole-Based Porous Carbons for CO_2 Capture. Adv. Funct. Mater. 14(21):2781-2787.
Simone C, Carlos AG, Alírio ER (2004). Adsorption Equilibrium of Methane, Carbon Dioxide, and Nitrogen on Zeolite 13X at High Pressures. J. Chem. Eng. Data 4(49):1095-1101.
Takashi K (2000). Control of pore structure in carbon. Carbon 2(38):269-286.
Thommes M (2010). Physical Adsorption Characterization of Nanoporous Materials. Chem. Ing. Tech. 7(82):1059-1073.
Zhangxiong W, Paul AW, Dongyuan Z (2012). Post-enrichment of nitrogen in soft-templated ordered mesoporous carbon materials for highly efficient phenol removal and CO_2 capture. J. Mater. Chem. 22(22):11379-11389.
Zhijian L, Bandar F, Caspar JS, Alan LC (2008). Stepwise growth of melamine-based dendrimers into mesopores and their CO_2 adsorption. Microporous Mesoporous Mater. 1-3(111):536-543.
Zou Y, Vera M, Alírio ER (2002). Adsorption of carbon dioxide at high temperature a review. Separ. Purif. Tech. 2-3(26):195-203.

Groundwater quality around new cement factory, Ibese, Ogun State, Southwest Nigeria

Ayedun, H.*, Oyede, R. T., Osinfade, B. G., Oguntade, B. K., Umar, B. F. and Abiaziem, C. V

Department of Science Laboratory Technology, Federal Polytechnic Ilaro, P. M. B. 50 Ilaro, Ogun State, Nigeria.

Assessment of physicochemical parameters and trace metals were carried out in groundwater of Ibese town and environs, Southwest Nigeria. The mean value of pH, electrical conductivity (EC), total dissolved solid (TDS) and alkalinity (HCO_3^{2-}) are 5.90 ± 0.32, 1328.00 ± 676.26 μScm^{-1}, 664.00 ± 353.58 mgL^{-1} and 94.00 ± 18.55 mgL^{-1}. The chloride concentration was found to be higher than World Health Organization (WHO) recommended limit of 250 mgL^{-1} in drinking water in majority of the samples. The suitability of the water for irrigation was established using sodium adsorption ratio (SAR) which ranges from 2.61 to 11.56. Concentration of heavy metals [Lead (Pb), Manganese (Mn), Nickel (Ni) and cadmium (Cd)] was determined using atomic absorption spectrophotometer and most of them are below detection limit of the instrument. It was concluded that the alkalinity was as a result of water-rock reaction, and treatment to remove high chloride concentration is required before consumption. The establishment of Dangote cement factory in the area should facilitate the provision of potable water for the inhabitants by the company as part of community service.

Key words: Ibese, groundwater, cement, heavy metals, sodium adsorption ratio.

INTRODUCTION

Groundwater and surface water are important source of water supply in the world which is needed for human survival and industrial development. The ground and surface water chemistry is controlled by the composition of its recharge components as well as geological and hydrological variations within the aquifers (Shahnawaz and Singh, 2009). Polluted groundwater and surface water sources are the cause for the spread of epidemic and chronic disease in human beings. Industrialization and increase in population are responsible for depletion of our groundwater sources (Khodapanah et al., 2009). Improved knowledge is required for understanding and evaluating the suitability of groundwater for different purposes. The knowledge of water-rock interaction as well anthropogenic influence is necessary for eventual utilization and management (Todd, 1995; Kelly, 1940).

Groundwater quality comprises physical, chemical and biological qualities of groundwater (Oluseyi et al., 2011). Temperature, turbidity, colour, taste and odour make up the list of physical water quality parameters. Since most groundwater are colourless, odourless and without specific taste, the concern is the chemical qualities. Naturally, groundwater contains mineral ions and these ions slowly dissolve from soil particles, sediments and rocks as the water travels along mineral surfaces in the pores or fractures of the unsaturated zone and the aquifer.

Some dissolved solids may have originated in the precipitation water or river water that recharges the aquifer. However, human activities can alter the natural composition of groundwater through mining activities, disposal or dissemination of chemicals and microbial matter at the land surface and into soils or through injection of wastes directly into groundwater.

Cement industries are generally associated with high dust emissions into atmosphere. Emitted dusts are naturally eliminated as deposits to the earth surface through dry or wet deposition in rainfall (Olaleye, 2005;

*Corresponding author. E-mail: hayedun2000@yahoo.com.

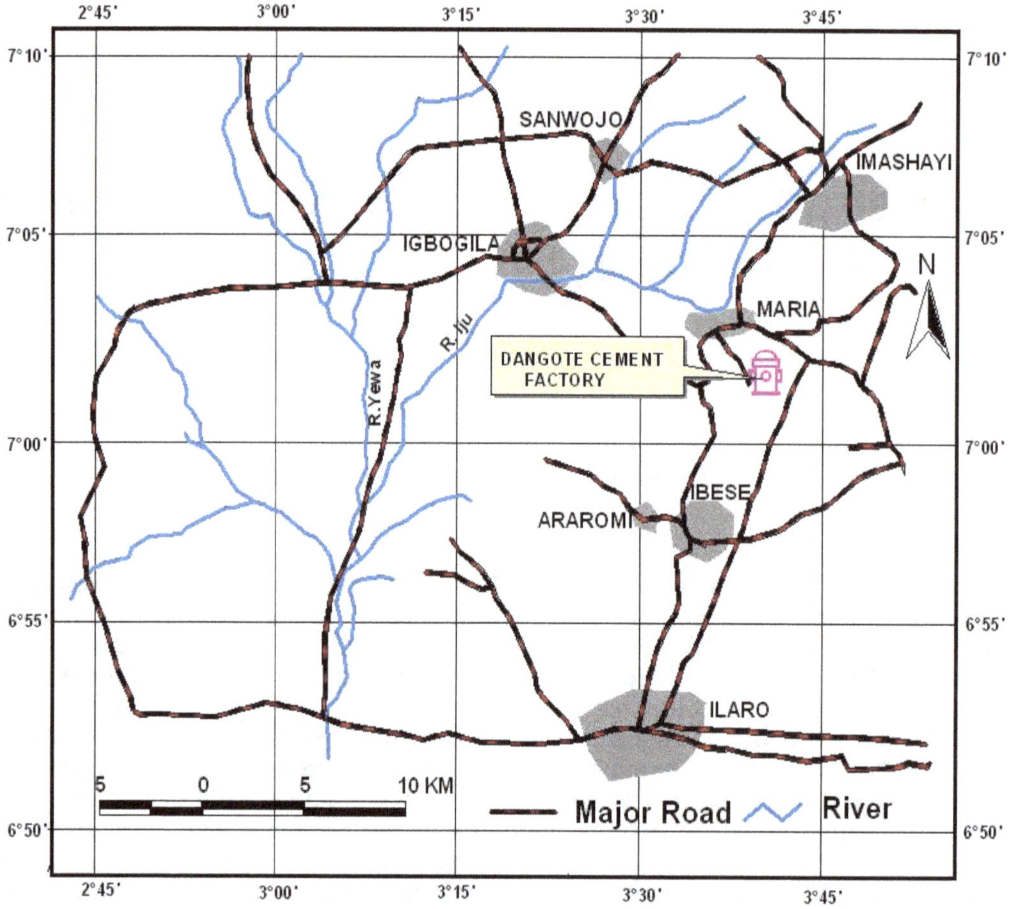

Figure 1. A map showing the sampling locations.

Asubiojo et al., 1991). The damaging consequences of the released dust particles for the soil, flora and fauna of the cement factory neighborhood could be considerable (Akeredolu, 1989) and the damaging effects of dust fall, which is characterized by enriched toxic heavy metals such as Arsenic (As), Lead (Pb), Nickel (Ni), Chromium (Cr), Copper (Cu), Zinc (Zn), Manganese (Mn) and Cadmium (Cd) (Adejumo et al., 1994).

A cement factory (Dangote Cement) was situated at Ibese but under construction as at the period of sampling (October/November, 2010) and commenced production by the end of 2011 (Figure 1). The aim of the present study is to determine physicochemical parameters and heavy metal concentration in groundwater of Ibese and environs to serve as baseline data against which future changes will be compared.

MATERIALS AND METHODS

Study area

Ibese is located in Egbado (Yewa) North of Ogun state Southwest Nigeria. It lies on latitude 6° 58′ 0″ N and longitude 3° 2′ 0″ E. The inhabitants are mainly farmers and traders. A cement factory (Dangote Cement) was situated at Ibese but under construction as at the period of sampling (October/November 2010) and commenced production by the end of 2011. The study area consists of Ibese, Imasayi, Sawonjo, Maria and Igbogila which are located few kilometers downstream to the cement factory.

Geology of the study area

The geology of Ibese and environs consists of Ewekoro formation which is marine and Paleocene age. It consists of a limestone unit several meters in thickness which is overlain by a shale unit almost three times as thick as the lime stone.

Method

Water samples were collected from 25 wells around Dangote cement factory, Ibese, Ogun State, Southwest Nigeria in September, 2010. Plastic bottles were used for the collection and kept in the ice, while samples meant for metal analysis were preserved by the addition of concentrated nitric acid (5 ml to 1 L of water). Parameters like pH, temperature, electrical conductivity (EC) and depth were measured on the field. The pH, conductivity and total dissolved solid (TDS) were measured with pH-conductivity-TDS meter (COMBO HI model 98130). The redox potential (RP) and temperature were measured using RP-meter

Table 1. Range of physicochemical parameters from all locations.

Parameter	Ibese	Sawonjo	Maria	Imasayi	Igbogila	Mean ± SD	WHO
pH	6.42	6.1	5.5	5.7	5.8	5.90 ± 0.32	6.5 - 8.5
Temp (°C)	28.06	30.1	29.3	28.7	29.2	29.07 ± 0.68	-
EC (μScm^{-1})	1130	2570	520	1320	1100	1328.00 ± 676.26	-
TDS (mgL^{-1})	560	1320	260	660	520	664.00 ± 353.58	-
RP (mV)	39.53	70	20.8	25.2	58.5	42.81 ± 18.93	-
Depth (m)	22	9.5	10.7	30	8.5	16.14 ± 8.47	-
HCO_3^{2-} ($mgCaCO_3L^{-1}$)	120	90	110	70	80	94.00 ± 18.55	-
DO (mgL^{-1})	1.14	5.2	6.1	3.4	2.2	3.61 ± 1.84	-
Cl^- (mgL^{-1})	350	270.5	275.2	120.92	280.71	259.47 ± 75.12	250
NO_3^- (mgL^{-1})	6.49	5.2	7.2	4.2	5.1	5.64 ± 1.07	50
PO_3^{2-} (mgL^{-1})	0.41	0.32	0.75	0.62	0.42	0.50 ± 0.16	-
SO_4^{2-} (mgL^{-1})	9.03	10.3	15.21	10.52	10.63	11.14 ± 2.12	250

Table 2. Concentration of metals in groundwater of the study area.

Parameter	Ibese	Sawonjo	Maria	Imasayi	Igbogila	Mean	WHO
Na (mgL^{-1})	7.65	35	7	30	35	22.95 ± 12.87	200
K (mgL^{-1})	0.34	11.7	5.45	2.26	11.62	6.27 ± 4.69	-
Ca (mgL^{-1})	3.9	17.3	9.4	12.7	20.1	12.68 ± 5.73	-
Mg (mgL^{-1})	0.29	6.23	4.96	4.51	4.79	4.16 ± 2.02	-
SAR	5.28	10.2	2.61	10.24	9.92	7.65 ± 3.14	-
MH	6.92	26.48	34.54	26.21	19.24	22.68 ± 9.25	-
%Na	65.59	0.66	46.44	65.21	65.19	48.62 ± 25.07	-
Pb (mgL^{-1})	ND	0.01	0.01	ND	ND	0.01 ± 00	0.01
Mn (mgL^{-1})	0.03	0.01	0.05	0.01	0.01	0.02 ± 02	0.5
Ni (mgL^{-1})	ND	ND	ND	0.01	ND	0.01 ± 00	0.02
Cu (mgL^{-1})	ND	0.01	ND	0.01	ND	0.01 ± 00	2
Cd (mgL^{-1})	ND	ND	ND	ND	ND	ND	0.03

(Thermo Orion model), while the dissolved oxygen (DO) was measured using DO-meter (HACH model). Anions like SO_4^{2-}, PO_4^{3-} and NO_3^{2-} were determined using ultraviolet (UV)-Visible Spectrophotometer (Camspec model). Turbidimetric method was used for SO_4^{2-} determination, Vanado-Molybdo-Phosphoric acid method was used for PO_4^{3-} determination, while Salicylate method was used for NO_3^{2-} determination. The Cl^- concentration was determined by Mohr's method, while HCO_3^{2-} was determined by titration against 0.01 M H_2SO_4 using mixed indicator (Bromocresol green-Methyl red solution). Ethylene diamine tetraacetic acid (EDTA) titration was used for Mg^{2+} and Ca^{2+}, while Flame Photometer was used for Na^+ and K^+ concentration determination. The heavy metals (Cd, Pb, Ni, Mn and Cu) were determined using atomic absorption spectrophotometer (Bulk Scientific Model).

RESULTS

The results of the physicochemical parameters from Ibese, Sawonjo, Maria, Imasayi and Igbogila are shown in Table 1. The concentration of heavy metals (Pb, Mn, Ni and Cd), essential elements (Sodium (Na), Potassium (K), Magnesium (Mg) and Cu) as well as calculated percent (%) Na, sodium adsorption ratio (SAR), Magnesium hazard (MH) are shown in Table 2.

DISCUSSION

The range of pH values (5.5 to 6.42),which are below 6.5 to 8.5 recommended by World Health Organization (WHO) in drinking water as shown in Table 1 indicated that, all the groundwater collected from the study areas are slightly acidic. Acidic water as recorded from groundwater of the study area can cause corrosion of water pipes and can affect gastrointestinal tract when consumed which can lead to diarrhea. Longe and Balogun (2010) reported pH range of 5.30 to 7.07 in groundwater of Lagos, Nigeria, while Obiefuna and Orazulike (2011) reported pH range of 6.5 to 7.8 in groundwater from Yola area, Northwestern, Nigeria. The closeness of the value may be due to similarity in geology

Table 3. Salinity and alkali standard (Sadashivaiah et al., 2008; Richards, 1954).

EC	SAR	Class
100 - 250	<10	Excellent
250 - 750	10 - 18	Good
750 - 2250	18 - 26	Doubtful
>2250	>26	Unsuitable

of Lagos and the study area which are all sedimentary and coastal in nature. Water temperature is the most important factor of the water which has a great deal of influence on various chemical and biological reactions taking place in water. The temperature ranges recorded in groundwater in the present study are 28.06 to 30.1°C. Rahim and Hussain (2011) reported temperature ranges of 28 to 30°C in groundwater of Beed city, Maharashtra, India. The sulphate concentration ranges from 9.03 to 15.21 mgL^{-1} which are within normal range recommended for drinking water by WHO. Oluseyi et al. (2011) reported mean value of 84.63 mgL^{-1} in groundwater of Ewekoro cement factory environment.

Close relationship exists between groundwater quality and land use (Orebiyi et al., 2010). Various land use activities can result in groundwater contamination. Potential sources of groundwater pollution include solid waste landfills, on-site excreta disposal systems, cemetery and animal wastes resulting from human activities among others. In Ibese and environs, the sources previously mentioned could be responsible for concentration of sulphate, phosphate and nitrate recorded in the area because the environment is rural and agricultural based. Nitrate concentration ranged between 4.2 to 7.2 mgL^{-1} which were all within WHO acceptable limit of 50 mgL^{-1} in drinking water. This implies that all the agricultural and anthropogenic activities in the area do not affect the groundwater sources. High nitrate concentration causes met-hemoglobinemia with the symptoms of paleness, bluish mucous membranes, digestive and respiratory problems (McCasland et al., 2007). If leachate is released from landfill or open dumpsites to the surrounding soil without proper collection and treatment, it could contaminate groundwater resources (Somjai and Suporn, 1993). Studies have shown that leachates cause an increase in dissolved inorganic substances such as chloride sulphate, bicarbonate, sodium and potassium of groundwater (Zanoni and Fungaroli, 1973).

The chloride content ranges from 120.92 to 350 mgL^{-1} with the values higher than 250 mgL^{-1} recommended by WHO recorded in Ibese, Sawonjo, Maria and Igbogila. The values recorded from Imasayi may be due to location which is less water-logged like others. Similarly, wells in Imasayi are deeper than those from other locations. Excess salt increases the osmotic pressure of the soil solution that can result in a physiological drought condition. Shahidullah et al. (2000) reported no chloride toxicity in groundwater from Phulpurthana, Mymensingh district of Bangladesh. Ramkumar et al. (2009) reported chloride in the range of 5.3 to 51.9 mgL^{-1} in groundwater of Vedaraninyan, India and reported the groundwater to be of NaCl types which indicate the influence of sea water in the wells. Similarly, the high level recorded in Ibese environs is likely due to saline intrusion because of the coastal nature of the area. The concentration of HCO_3^{2-}could result from dissolution of CO_2 gas likely formed by anoxic biodegradation of organic matters in domestic sewage, wastes water and landfills. Jun-Yan (2010) reported higher value in groundwater of Nandong, China. The EC ranges from 520 to 1320 μScm^{-1}. Gbadebo et al. 2010 reported the mean value of 349 ± 9.74 μScm^{-1} in groundwater of Agbara area of Ogun State, Nigeria, while Khodapanah et al. (2009) reported conductivity value of 350 to 23600 μScm^{-1}. The value found in the present study indicated that 20% of all the wells were above 2,250 μScm^{-1} as unsuitable according to Sadashivaiah et al. (2008). The large variation in EC is mainly attributed to lithologic composition and anthropogenic activities prevailing in the area. Normally, irrigation water with an EC of < 700 μScm^{-1} causes little or no threat to most crops, while EC > 3000 μScm^{-1} may limit their growth (Tijani et al., 2005).

The sequence of mean concentration values of cations in groundwater of the study area is Na > Ca > K > Mg as shown in Table 2. The Ca^{2+}, Mg^{2+}, HCO_3^{2-} and pH are primarily controlled by water-rock interaction in the area (Adedeji and Ajibade, 2005). The suitability of groundwater for irrigation is contingent on the effects on the mineral constituents of the water on both the plant and soil. Salts can be highly harmful as they can limit the growth of plants physically, by restricting the uptake of water through modification of osmotic processes. Salinity, sodicity and toxicity generally need to be considered for evaluation of the suitability for irrigation (Todd, 1995). Parameters such as EC, %Na, MH and SAR, were used to assess the suitability of water for irrigation.

The Na and alkali hazard in the use of water for irrigation is determined by the absolute and relative concentration of cations and is expressed as the SAR. The SAR was calculated from: SAR = Na / (Ca+ Mg) / 2. The values obtained from this study range from 2.61 to 11.56. Richards (1954) stated that if SAR <10 (Excellent), 10 to 18 (Good), 18 to 26 (Doubtful), >26 (unsuitable) as shown in Table 3. Fifty-seven percent (57%) of the groundwater samples in the study area are excellent for irrigation, while 43% are good for the same purpose. Na^+ was the most dominant among the major cations with mean values of 19.04 ± 12.62, while K^+ has the least mean values of 4.72 ± 4.66 mgL^{-1}. This trend is similar to what was reported by Khodapanah et al. (2009). Similarly, Mitra et al. (2007) reported an average Na concentration of 18.8 mgL^{-1} in groundwater of Northwest

Honshu Island, Japan. The Na in irrigation waters is expressed using the following equation: %Na = (Na + K) / (Ca + Mg + Na + K) * 100 where all the ionic concentrations are expressed in mg/L. High percentage of Na^+ with respect to Ca^{2+}, Mg^{2+} and Na^+ in irrigation water, causes deflocculating and impairing of soil permeability (WHO, 2008). The range of $\%Na^+$ in groundwater of the study area is 0.66 to 65.49 %. Locations with %Na greater than 60% are not suitable for irrigation. Majority of the heavy metals were not detected in groundwater of the study area. Pb concentrations are below detection limit in Ibese, Imasayi and Igbogila, while other heavy metals (Pb, Mn, Ni and Cd) are within the range recommended by WHO in drinking water.

Conclusion

Most of the physicochemical parameters examined in groundwater of Ibese and environs showed that they are suitable for drinking. The concentration of Ca^{2+}, Mg^{2+}, HCO_3^{2-} and pH are due to water-rock reaction and most of the heavy metals analyzed were not detected before the commencement of Dangote cement factory in the area. The chloride content which was higher in concentration than WHO recommended limit in drinking water can be removed by using reverse osmosis because direct removal may be too expensive. Also, for agricultural purposes, like irrigation, groundwater samples from the study area are equally suitable. We advise that the company should provide boreholes for the people as part of community service as soon as possible to guide against unforeseen situation of contamination in order to safeguard the health of people living in the area.

REFERENCES

Adedeji A, Ajibade IT (2005). Quality of well water in Ede area Southwestern Nigeria. J. Hum. Ecol. 17:223-228.

Adejumo JA, Obioh OOJ, Akeredolu FA, Olaniyi HB, Asubiojo OI (1994). The atmospheric deposition of major, minor and trace elements within and around three cement factories. J. Radioanal. Nucl. Chem. 179:95-204.

Akeredolu FA (1989). Atmospheric Environment Problems in Nigeria: An overview. Atmos. Environ. 23(4):783-792.

Asubiojo O I, Aina PO, Oluwole AF, Arshed W, Akanle OA, Spyrou NM (1991). Effects of Cement production on the elemental composition of soils in the neighborhood of two cement factories. Water Air Soil Pollut. 58:819-828.

Jun-Yan YJ (2010). Effects of land use on hydrochemistry and contamination of Karst groundwater from Nandong underground river system, China. Water Air Soil Pollut. 210:123-141.

Kelly WP (1940). Permissible composition and concentration of irrigation waters. Proceedings ASCE 66:607.

Khodapanah L, Sulaiman WNA, Khodapanah N (2009). Groundwater quality assessment for different purposes in Eshtehard distric, Tehran, Iran. Eur. J. Sci. Res. 36(4):543-553.

Longe EO, Balogun MR (2010) Groundwater quality assessment near a municipal landfill, Lagos, Nigeria. Res. J. Appl. Sci. Eng. Technol. 2(1):39-44.

McCasland M, Trautman NM, Robert RJ, Porter KS (2007). Nitrate: Health effects in drinking water. http:// pscp.ccc.cornell.edu/facts slides-self/facts/nit-heef-grw85.aspx.

Olaleye VF (2005). Chthofaun survey of rivers within Ewekoro cement manufacturing operational areas. Biodivers. Conserv. 14:2171-2184.

Oluseyi T, Olayinka K, Adeleke I (2011). Assessment of groundwater pollution in the in the residential areas of Ewekoro and Shagamu due to cement production. Afr. J. Environ. Sci. Technol. 5(10):786-794.

Orebiyi EO, Awomeso JA, Idowu OA, Martins O, Oguntoke O, Taiwo AM (2010). Assessment of Pollution hazard of shallow well water in Abeokuta and Environs, Southwest, Nigeria. Am. J. Environ. Sci. 6(1):50-56.

Rahim A, Hussain S (2011). Assessment of groundwater quality of Beed city, Maharashrata India. Orient. J. Chem. 27(3):1273-1275.

Ramkumar T, Venkatramanon S, Mary AI, Tamilselvi M, Ramesh G (2009). Hydro-geochemical Quality of Groundwater in Vedaranniyan Town, Tamilnadu, India. Res. J. Earth. Sci. 1(1):28-34.

Richards LA (1954). Diagnosis and improvement of saline alkali soils: Agriculture. Handbook 60, US Department of Agriculture, Washington D.C. Vol. 160.

Sadashivaiah C, Ramakrishnaiah RC, Ranganna G (2008). 'Hydrochemical Analysis and Evaluation of Groundwater Quality in Tumkur Taluk, Karnataka State, India". Int. J. Environ. Res. Public Health 5(3):158-164.

Shahnawaz MD, Singh KM (2009). Groundwater quality of Piro and Jagdishpur Blocks Bhorps district, a middle gangatic plain. Intl. J. Pharma. Qual. Assur. 1(1):9-10.

Somjai K, Suporn KTI (1993). Monitoring and evaluation of shallow well water quality near a waste disposal site. Environ. Int. 19:579-587.

Todd DK (1995). Grounddwater Hydrology. John Wiley and Sons Inc. New York.

Tijani MN, Balogun SA, Adeleye MA (2005). Chemical and Microbiological Assessment of water and Bottom-sediments Contaminations in Awba Lake (U.I), Ibadan, SW-Nigeria. Mater. Geo-environ. 52(1):123-126.

WHO (2008). Guidelines for drinking water quality. 3rd Ed. Health criteria and supporting information, Geneva.

Zanoni AE, Fungaroli AA (1973). Potential for groundwater pollution from the land disposal of solid wastes. Crit. Rev. Environ. Sci. Technol. 3:225-260.

Accessing the potential of *Lonchocarpus laxiflorus* roots (LLR) plant biomass to remove Cadmium (II) ions from aqueous solutions: Equilibrium and kinetic studies

A. H. Santuraki[1] and A. A. Muazu

Department of Chemical Sciences, Federal University Kashere, Gombe State – Nigeria.

The potential of *Lonchocarpus laxiflorus* roots (LLR) for the removal of Cd (II) ions from aqueous solutions has been investigated. The adsorption of Cd (II) ions was found to be affected by solution pH, contact time, adsorbent dosage, initial metal ion concentration, and temperature. The equilibrium was analysed using Langmuir and Freundlich isotherm models. The data was found to have a closer correlation with the Lamgmur isotherm as evidenced by a higher correlation coefficient (R^2). The adsorption capacity for *L. laxiflorus* was found to be 19.35 mg g^{-1} and the removal efficiency of the LLR was found to be 98% under optimized conditions. The kinetics data was also subjected to pseudo-first-order and the pseudo-second-order kinetic models. The data could be explained better using the pseudo-second-order kinetic model. The selected biosorbent has been analyzed by FT-IR for identification of contributing functional groups, x-ray powder diffraction (XRD) for its structural properties and scanning electron microscope (SEM) for the structural morphology analysis.

Key words: Biosorption, cadmium, isotherms, kinetics, *Lonchocarpus laxiflorus*.

INTRODUCTION

Despite the requirement for healthy environment, the environment has been found to be contaminated with various pollutants. This has now posed a great challenge to human wellbeing. Such pollutants may be found in air, water soil, coastal erosion, overfishing and deforestation as well as disposal of waste, which constitute several heavy metals. Contamination of water by heavy metals is one of the most challenging environmental issues currently. Cadmium is one of the most toxic metals apart from lead and mercury. It has been reported to cause renal dysfunction, hypertension, lung insufficiency, bone lesions and cancer (Feng et al., 2010) which is a leading cause of death. The cadmium drinking water guidelines value recommended by WHO is 0.003 mg L^{-1} (WHO, 2008). Cadmium accumulates both in the environment and the body causing long term damage to life (Nida et al., 2012). Cadmium is one of the heavy metals with a greatest potential hazard to humans and environment

(Fouad et al., 2012).

The principal sources of Cd into the environment are electroplating, smelting, alloy manufacturing, pigments, plastic, battery, mining and refining processes (Gupta and Nyaka, 2012). Once released into the environment cadmium is toxic to plants animals and microorganisms (Bailey et al., 1999). The metal is non-biodegradable, persistent and bioaccumulate mainly in the kidneys and liver of vertebrates, invertebrates and also in algae (Ajay et al., 2005). A number of methods have been employed to remove Cd (II) ions from the environment including, ion exchange, reverse osmosis (Gupta and Nyaka, 2012) membrane filtration, electrochemical treatment, and adsorption. etc where each of them have limitations such as high cost and production of hazardous by- products are found expensive and sometimes ineffective, especially when metals are present in solution at very low concentration within the range 1 to 100 mg/cm^3 (Santhi and Manonmani, 2012).

Consequently, it is essential to find new methods for effective removal of cadmium from water and wastewater. Compared with other traditional methods, adsorption is quite popular due to its simplicity and high efficiency, as well as the availability of a wide range of adsorbents (Orhan and Buyukgungor, 1993; Babel and Kurniawan, 2003). Activated carbon is the best useful adsorbent of heavy metals from waste water. However, the cost of activated carbon is high; its regeneration also requires additional expense. Therefore there is need to come up with other alternative that provide an easy, feasible, reliable, low cost adsorbents especially those of biological origin commonly referred to as biosorbents to improve the water quality. Biosorption has gained a lot of credibility currently because of its eco- friendly nature, excellent performance and cost- effectiveness (Davis et al., 2003). Equilibrium isotherm models and kinetic models were applied to the data obtained for a better understanding of the adsorption process. Thus the objective of the present study is to investigate the binding of metal ion cadmium (II) by *Lonchocarpus laxiflorus* roots (LLR) in its immobilized form from aqueous solutions and to study the effect of various factors affecting the efficiency of the process.

MATERIALS AND METHODS

Biomass preparation

Plant collection and treatment

The roots of *L. laxiflorus* plant were collected from a tree behind Modibbo Adama Federal University of Technology Yola, Nigeria. The plant was wash thoroughly under running water to remove dust and any adhering particle and then rise with distilled water. The sample was air dry for 2 weeks and the dry roots was grinded in analytical mill and sieve to obtain adsorbent of known particle size range. The biomass powder was kept in an air tight bottle for further study (Igwe and Abia, 2006). Our studies indicated that though there was no big difference in the adsorption rates of the various parts of plant, the roots were better. Therefore roots were used in this adsorption experiments.

Chemicals

All chemicals used in the present work were of analytical grade. The stock solution of Cd^{2+} ions was prepared in 1.0 g L^{-1} concentration. Cadmium solution of 1000 mg/cm^3 concentration was prepared by dissolving 2.103 g of $Cd(NO_3)_2$ in 250 cm^3 of distilled water and make up to 1000 cm^3 in a volumetric flask. The pH of the solutions was adjusted using 0.1 mol L^{-1} HCl and NaOH solutions.

Analysis of metal ions

Atomic absorption spectrophotometer equipped with an air acetylene flame, controlled with computer was used to investigate the concentration of cadmium metal ion. The hallow cathode lamp of Cd, was used at 283.3 nm wavelength, while slit is 0.2 nm, operated at 8 mA (Suleiman et al., 2007).

Reparation of sodium alginate and calcium chloride stock solution

Sodium alginate was prepared by weighing 4.00 g and making it up to 100 cm^3 mark with distilled water in a volumetric flask and left overnight for complete dissolution. 0.12 M of calcium chloride was prepared by weighing 26.28 g in to 1000 cm^3 volumetric flask and making up to mark with distilled water according to a standard procedure described by Osemeahon and Esenowo, (2012).

Immobilization of the roots of *L. laxiflorus* plant

Sodium alginate was used for immobilization of the roots of *L. laxiflorus* plant. 50 cm^3 of sample solution prepared by dissolving 4 g of each test sample in 100 cm^3 of distilled water and mix with 50 cm^3 of 4% stock solution of sodium alginate and stir vigorously in 250 cm^3 beaker, to obtain a homogenous mixture. After mixing, the solution the solution was drawn through hyperdemic needles and was added drop wise to a stirred solution of 1 M $CaCl_2$. A retention time of 1 h was allowed for the reaction to obtain complete precipitation of the immobilized leave powder of *L. laxiflorus* plant. The beads thus formed that is, sodium alginates were kept in fresh $CaCl_2$ solution. Before sorption studies, the beads were removed and allowed to dry at room temperature. The dried solid mass was stored in a polythene bag for further use (Mishra, 2013).

Biosorption experiments

The experiments were carried out in the batch mode for the measurement of adsorption capacities. From 100 ppm of cadmium metal ion solution, 50 ml was taken into a 250 ml conical flask and 0.2 g of the LLR was added corked with a rubber bung and shaken with a flask shaker for 2 h at room temperature (30°C) at 180 rpm. The separation of the adsorbents and solutions were carried out by filtration with whatman filter paper No 42 and the filtrates were stored in sample cans for use. The residual metal ions concentrations were determined using atomic absorption spectrophotometer (AAS). Pyeunicam Model SP. For studies on effect of temperature the adsorption studies were carried out at 25, 30, 40, 50, 60 and 70°C. The percentage adsorption was calculated using the following equation:

$$\%Adsorption = [(C_i - C_f/C_i)] \times 100 \quad (1)$$

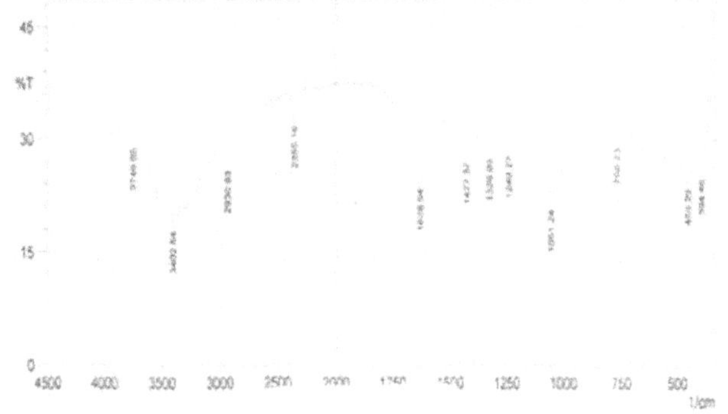

Figure 1. FTIR spectrum of *L. laxiflorus* root.

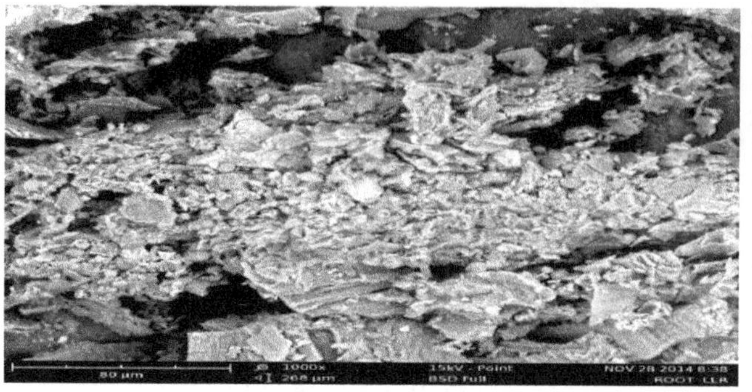

Figure 2. SEM of LLR powder in raw form.

Where C_i= Initial metal ion concentration and C_f = Equilibrium metal ion concentration (mg/L).

The amount of metal absorbed by the biosorbent was also calculated as

$$q_e = V(C_o - C_e)/M \quad (2)$$

Where, q is the amount of metal ion adsorbed in mg/g; C_o is the initial metal ion concentration in mg/cm^3; C_e is the final concentration in mg/cm^3, V is the volume of metal ion solution in liters; M is the mass of the root of *L. laxiflorus* powder used in gram.

RESULTS AND DISCUSSION

The FTIR is measured in the range of 400 to 4000 cm $^{-1}$ wave number. The FTIR of the adsorbent displays a number of adsorption peaks, indicating the complex nature of the studied adsorbent. The IR bands consisted of four regions, the broad OH band (3200-3600 cm^{-1}), C-H stretching region (2800-3000 cm^{-1}), carbonyl group stretching region (1550 -1750 cm^{-1}), and finger print bands (below 1550 cm^{-1}) (Shin et al., 2007).

From Figure 1, the absorption of peak at approximately 3400 were due to stretching vibrations of hydroxyl groups which are one of the main components of cell wall polysaccharides of the plants (Suantak et al., 2011). The adsorption peak at 2930 cm^{-1} is likely due to the presence of C-H asymmetric stretching vibration in –COOH group (Kumar et al., 2012) of methylene groups on the surface. The absorption band at 1630 cm^{-1} may be assigned to Amide I and II (protein) respectively (Pradhan et al., 2007). While the peaks in the range of 1427 to 1328 cm^{-1} could be attributed to carboxylate group (Pradhan et al., 2007). The broad peak at 1240 could be due to (C- C) or (C- H) or (C-O) stretching of the carboxyl groups (Singh et al., 2010). The peak in the region 1051 cm^{-1} is due to the presence of C- C Stretching of the polysaccharides (Singh et al., 2010). In conclusion, the FTIR spectroscopic analysis of the plant biomasses indicated the presence of hydroxyl, amide and carboxylate groups as the main functional groups involved in the complexation of metal ions for biosorption processes.

The adsorption capacity of the adsorbent depends upon porosity as well as chemical reactivity of the functional groups at the adsorbent surfaces. It seems that these functional groups participate in metal binding process.

Scanning electron microscope (SEM) analysis is another important tool used in the determination of the surface morphology of an adsorbent. The SEM image and micro- analysis of LLR plant is shown in Figure 2. The external surface of LLR is full of cavities with well developed porous structure. The external surfaces show a rough area having different pore diameters distributed over the surface of the biomass which may be responsible for metal removal. This analysis reveals a highly porous structure for LLR biomass. It could also be seen that more uneven and rough surface morphology exist in all the adsorbent.

X-ray diffraction (XRD) analysis was carried out by using XRD system with Cu-Kα radiation. The XRD patterns for LLR in raw powdered form is shown in Figure 3. These gives information about the changes in the crystalline and amorphous nature of the adsorbents. Sharp intensity XRD peaks have been observed at typical scanning angles of 2θ = 20. The sharp peaks present in the figures indicated the crystalline nature of the material. In addition, the presence of other weak intensity peaks in the spectra indicates the amorphous nature of the three adsorbents. The amorphous nature of the adsorbents suggests that metal ions can easily penetrate the surface which is desirable for an effective removal. These results are in good agreement with those reported Kugbe et al. (2009).

Effect of pH on metal biosorption

Hydrogen ion concentration is one of the important factors that influence the adsorption behavior of metal ions in aqueous solutions. It affects the solubility of metal ions in solution, replaces some of the positive ions found in active sites and affects the degree of ionization of the adsorbate during the process of biosorption. This is because it affects solution chemistry and also the

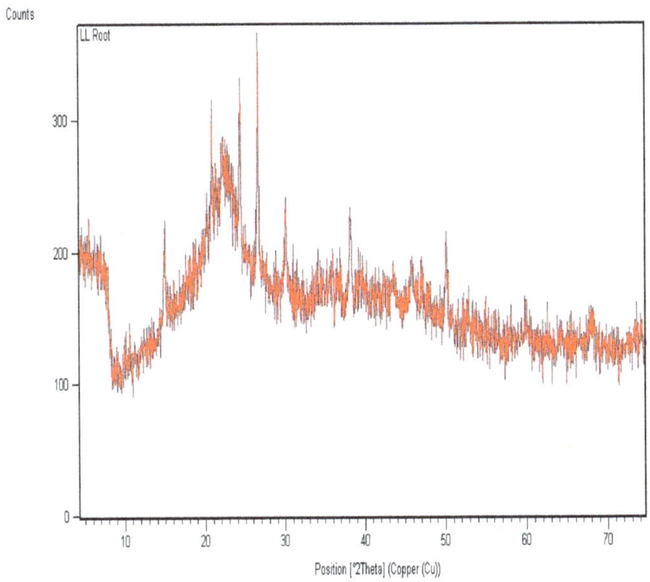

Figure 3. XRD image of *L. laxiflorus* roots raw powdered.

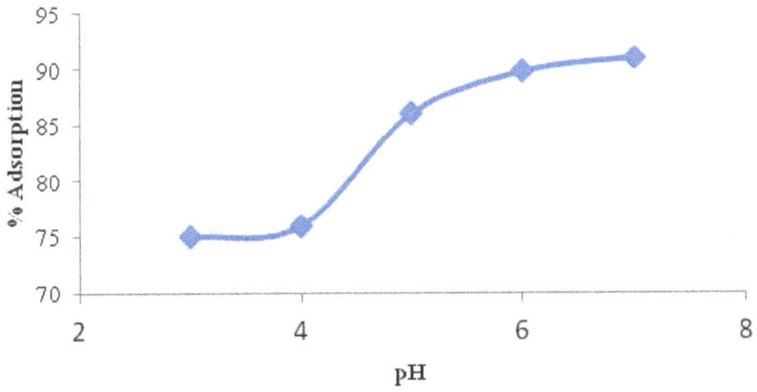

Figure 4. Effect of pH on % adsorption of Cd (II) ions on LLR biomass.

speciation of the metal ions. The effect of initial pH on biosorption of Cd (II) ions onto *L. laxiflorus* was evaluated in the pH range of 3.0 to 7.0. Studies in pH range above 7.0 were not attempted as there is precipitation of cadmium (II) hydroxides.

From Figure 4 it could be seen that Cd (II) ions adsorption increased as the pH increased. At low pH values, protons occupy the biosorption sites on the biosorbent surface and therefore less Cd (II) ions can be adsorbed because of electrostatic repulsion between the metal cations and the protons occupying the binding sites. When the pH was increased, the biosorbent surface became more negatively charged and the biosorption of the metal cations increased drastically until equilibrium was reached at pH 6.0 to 7.0. At pH of >7.0 there is formation of hydroxylated complexes of the metal ions and these complexes compete with the metal

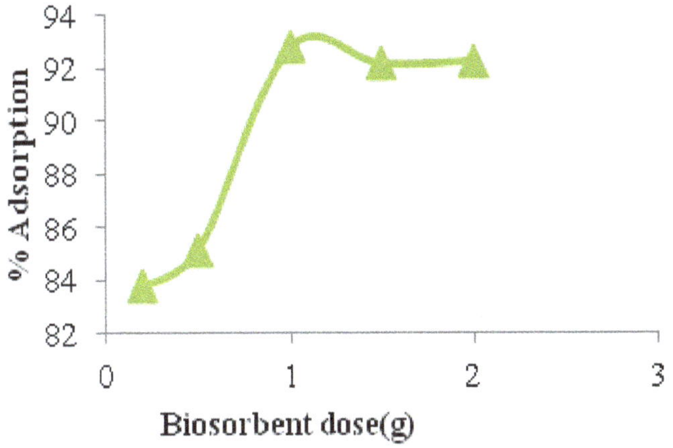

Figure 5. Effect of dosage on % adsorption of Cd (II) ions on LLR biomass.

cations for the adsorption sites hence a reduction in the effective metal cations removal. Therefore adsorption experiments at pH above this were not considered.

Effect of biosorbent dosage

The effect of biomass dosage on adsorption of Cd (II) ions is indicated in Figure 5. The number of available binding sites and exchanging ions for the biosorption depends upon the amount of biosorbent in the biosorption system. This is attributed to the fact that it determines the number of binding sites available to remove the metal ions at a given concentration. The dosage also determines the adsorption capacity of the biosorbent with an increase in mass reducing the biosorption capacity as the mass increase from 0.1 to 2.5 g per 20 ml of adsorbate. An increase in the % adsorption is attributable to an increase in the number of binding sites for the metal cations. Similar results were recorded in the literature for other adsorbents. However, the mass could not be increased infinitely as at some point all the solution is sequestered leaving no residual solution for concentration determination. Similar trend have been found by Mahajan and Sud (2011).

Effect of initial metal concentration

The initial concentration remarkably affected the uptake of Cd (II) ions in solution. The efficiency of Cd (II) ions adsorption by LLR at different initial concentrations (10 to 80 mg L^{-1}) was investigated as shown in Figure 6. At a lower concentration, the adsorption sites take up the available metal ions much quickly due to less competition among the metal ions for the available binding sites which are fixed in this case. However, as the concentration increases the competition for the limited

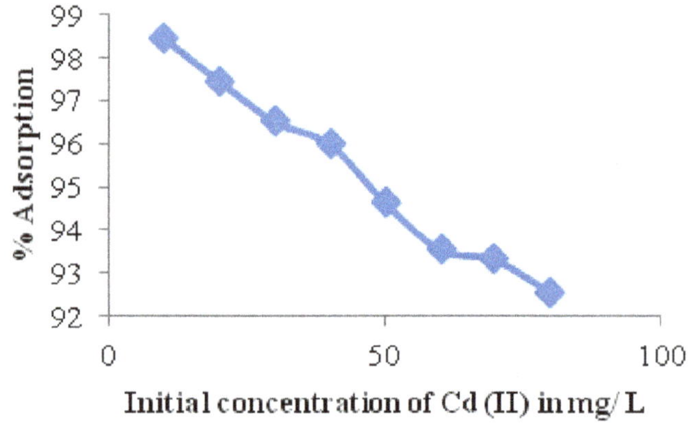

Figure 6. Effect of initial cadmium concentration on adsorption of LLR biomass.

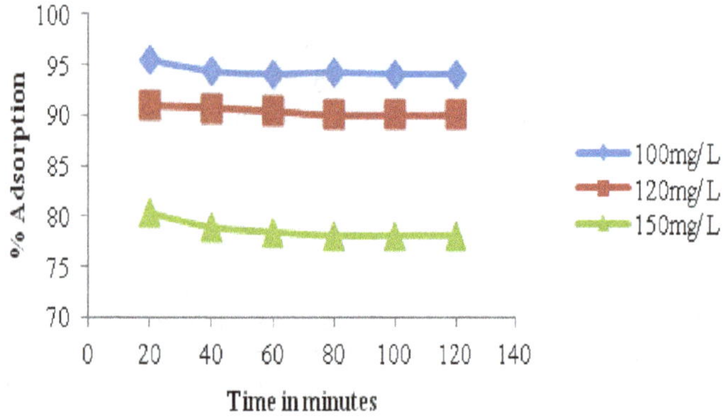

Figure 7. Effect of contact time on adsorption of Cd (II) ions by LLR.

binding sites sets in as the binding sites become saturated (Mahajan et al., 2013).

Effect of contact time

Contact time is an important parameter for any successful use of the biosorbents for practical purposes. Effect of contact time on adsorption of Cd (II) ions was investigated keeping the biomass in contact with the metal ion solution for different time periods between 0 to 60 min. It was noted that as adsorption proceeds, the sorbent reaches saturation state, at this point the sobbed solute tends to desorbs back into solution (Figure 7). Eventually, the rate of adsorption and desorption are equal at equilibrium. When the system attains equilibrium, no further net adsorption occurs. The time taken to attain equilibrium is very important for process optimization. The rate of adsorption is very fast at first and over 95% of total biosorption of Cd (II) ions occurs in the first 5 min and thereafter it proceeds at a slower rate

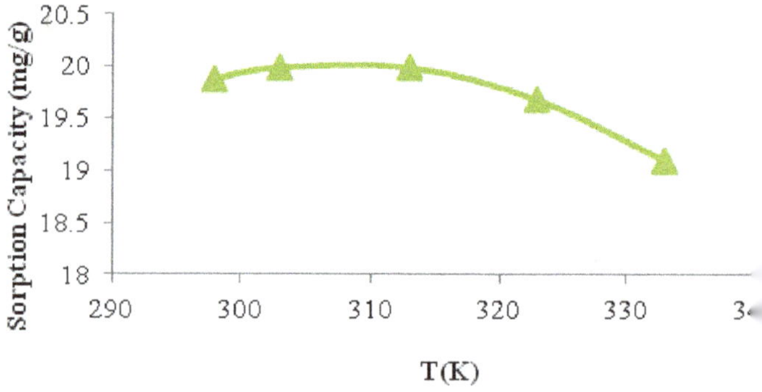

Figure 8. Effect of temperature on % adsorption of Cd (II) ions by LLR.

and finally no further significant adsorption is noted beyond 20 min of contact time. The very fast adsorption makes the material suitable for continuous flow water treatment systems (Sarin and Paint, 2006).

Effect of temperature

Temperature of the medium affects the removal efficiency of pollutants in aqueous solutions. This is because a change in temperature in turn affects the solubility of pollutants and also the kinetic energy of the adsorbing ions. Therefore the effect of temperature on adsorption of Cd (II) ions was investigated and the data is shown in Figure 8. The results indicate that the percentage adsorption increases with increase in temperature up to 40°C, after that any increase in temperature is accompanied by a reduction in % adsorption. This can be attributed to the fact that with increase in temperature of the solution, the attractive forces between the biomass surface and Cd (II) ions are weakened thus decreasing the sorption efficiency. This could be due to increase in the tendency for the Cd (II) ions to escape from the solid phase of the biosorbent to the liquid phase with increase in temperature. Finally increased temperature beyond 40°C could have destroyed some of the binding sites on the biosorbent surface due to bond rupture (Meena et al., 2005).

Biosorption kinetics

Kinetic study provides useful information about the mechanism of adsorption and subsequently investigation of the controlling mechanism of biosorption as either mass transfer or chemisorption. This helps in obtaining the optimum operating conditions for industrial-scale batch processes. A good correlation of the kinetic data explains the biosorption mechanism of the metal ion on the solid phase (Garima and Dhiraj, 2013). In order to

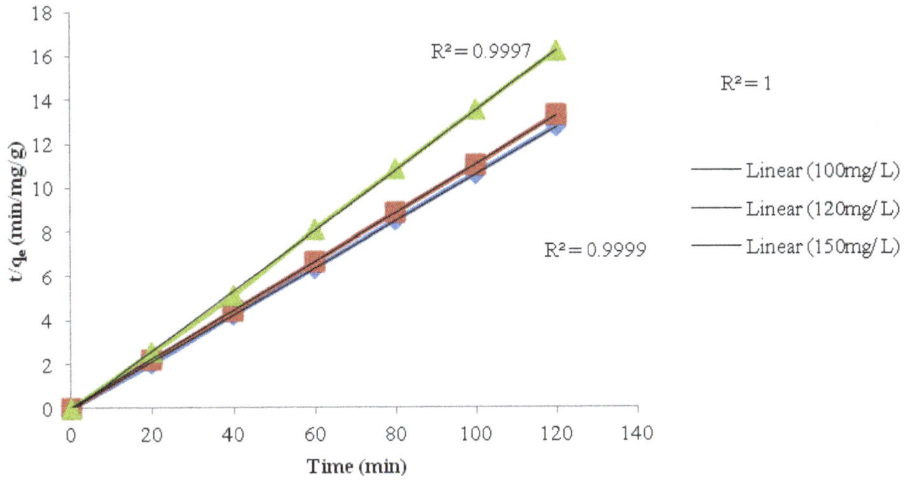

Figure 9. Pseudo-second -order- plots of Cd (II) adsorption onto LLR.

evaluate the kinetic mechanism that controls the biosorption process, the pseudo -first-order models (Lagergren, 1898) were applied for biosorption of Cd (II) ions on the biosorbent. The Lagergren pseudo- first – order rate model is represented by the equation:

$$\log qe - qt = \log qe - \frac{K1}{2.303}t \quad (3)$$

Where q_e and q_t are the amounts of metal adsorbed (mg g^{-1}) at equilibrium and at time t respect ively, and K_1 is the rate constant of pseudo-first-order biosorption (min $^{-1}$). The q_e and rate constant were calculated from the slope and intercept of plot of log (q_e- q_t) against time t.

The pseudo-second-order equation (Ho and Mckay, 1999) assumes that the rate limiting step might be due to chemical adsorption. According to this model metal cations can bind to two binding sites on the adsorbent surface. The equation can be expressed as shown below:

$$\frac{t}{q_t} = \frac{1}{K_2 q_e^2} + \frac{1}{q_e^t} \quad (4)$$

Where k_2 is the rate constant of the pseudo-second-order adsorption (g/mg/min). If the adsorption kinetics obeys the pseudo-second-order model, a linear plot of t/q t versus t can be observed as shown in Figure 9.

Biosorption isotherms

For optimization of the biosorption process design, it is imperative to obtain the appropriate correlation for the equilibrium data. Biosorption isotherms describe how adsorbate interacts with the biosorbent and the residual metal ions in solution during the surface biosorption. The isotherms also help in determination of adsorption capacity of the biosorbent for the metal ions. The data on Cd (II) ions biosorption was fitted with the Langmuir (1918) and Freundlich (1906) isotherms (Figures 10 and 11).

The Langmuir isotherm assumes monolayer coverage of the adsorbate onto a homogeneous adsorbent surface and the biosorption of each cation onto the surface has equal activation energy. The Langmuir isotherm can be exppressed as:

$$\frac{C_e}{q_e} = \frac{1}{q_{max}b+} + \frac{C_e}{q_{max}} \quad (5)$$

Where qmax is the monolayer capacity of the biosorbent (mg g^{-1}), and b is the biosorption constant (L mg $^{-1}$). The plot of ce/qe versus ce should be a straight line with a slope of 1/qmax and intercept of 1/q max b when the biosorption follows Langmuir equation. The Freundlich equation can be expressed as:

$$\log q_e = \log K_f + \frac{1}{n} \log C_e \quad (6)$$

Where K_f and 1/n are the Freundlich isotherm constants related to biosorption capacity and biosorption intensity respectively. If the equation applies then a plot of log qe versus log Ce will give a straight line of slope 1/n and intercept as K_f

LLR has a higher value for n suggesting multilayer sorption on the surface of the plant biomass, which is due to its various chemical functional groups. The values of b and K obtained for LLR also confirms a higher sorption capacity and superior performance of LLR adsorbents for Cd (II) ions adsorption from aqueous solution. Similar results are reported by Mahajan et al. (2013), Iqwe et al. (2005) and Kurniawan and Thiam (2010).

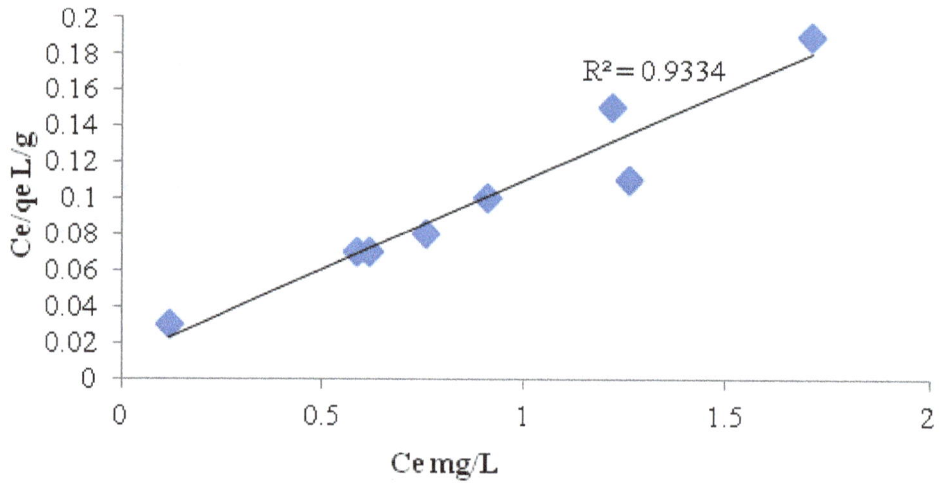

Figure 10. Langmur isotherm for the sorption of cd(II) using LLR.

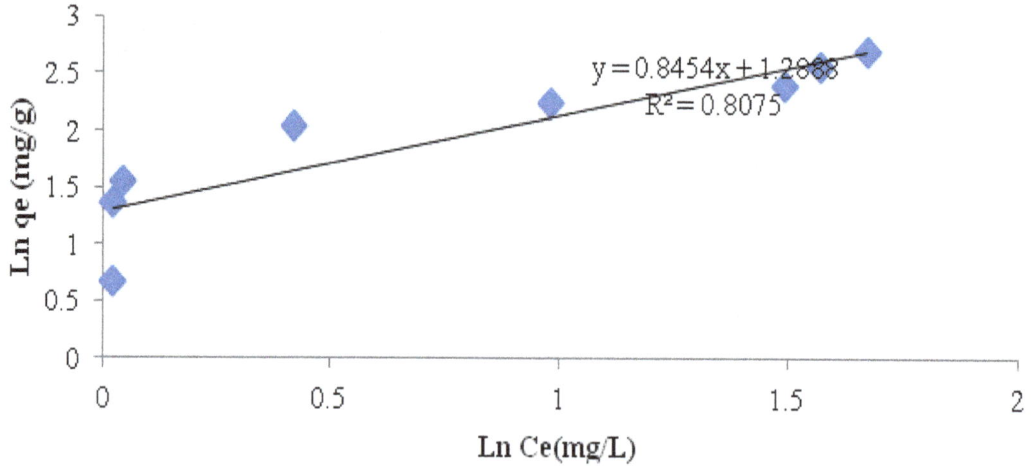

Figure 11. Freundlich adsorption isotherm for cadmium by LLR.

Conclusion

Adsorption of cadmium ions from aqueous solution using the *Lonchocarpus laxiflorus* plant roots material was investigated. Various contributing parameters such as contact time, initial metal ion concentration, solution pH, and adsorbent dose was optimized for maximum removal efficiency. The sorption data fitted well with Langmuir isotherm with high R^2 values. The kinetic studies indicated that the pseudo second order model was the best one in describing the kinetics of cadmium (II) adsorbed onto roots powder. A large number of carbonyl and hydroxyl groups were observed in the FTIR analysis, XRD studies reveal the crystalline structure of the biosorbent and SEM studies showed the presence of various moieties that enhances the adsorption phenomenon. Excellent removal efficiency in its encapsulated form explores the utilization of the biomass at the commercial scale for small scale industries, making it of potential commercial use.

Conflict of Interest

The authors have not declared any conflict of interest.

REFERENCES

Ajay KM, Mishra KJ, Rai PK, Chitra R, Nagar PN (2005). Removal of heavy metal ions from aqueous solution using carbon aerogel as an adsorbent. Indian J. Chem. Technol. 10:72-78.

Babel S, Kurniawan TA (2003). Low-cost adsorbents for heavy metals uptake from contaminated water: A review. J. Hazard. Mater. 97(1-3):219–243.

Bailey SE, Olin TJ, Bricka RM, Adrian DD (1999). A review of potentially low-cost biosorbents for heavy metals. Water Res. 33:2469-2479.

Davis TA, Volesky B, Mucci A (2003). A review of the biochemistry of heavy metals biosorption by brown algae. Water Res. 37(18):4311-4330.

Feng N, Guo X, Liang S, Zhu Y, Liu J (2010). Biosorption of heavy metals from aqueous solutions by chemically modified orange peel. J. Hazardous Mater.185:49-54.

Fouad K, Noureddin A, Mohammed CN (2012). Adsorptive removal of cadmium from aqueous solution by cork biomass. Equilibrium, dynamic and thermodynamic studies. Arab. J. Chem. Doi:10.1016/J. arabjc.2011.12.013

Freundlich HMF (1906). Uber die adsorption in lasugen. J. Phys. Chem. 57:385–470.

Garima M, Dhiraj S (2013). Accessing the Potential of Lingocellulosic Agricultural Waste Biomass for Removal of Ni (II) Metal Ions from Aqueous Streams. Int. J. Sci. Eng. Res. 4(40):1713-1720.

Gupta VK, Nayak A (2012). Cadmium removal and recovery from aqueous solutions by novel adsorbents prepared from orange peel and Fe_2O_3 nanoparticles. Chem. Eng. J. 180:81–90.

Ho YS, McKay G (1999). Pseudo—second order model for sorption processes. Process Biochem. 34:451–465.

Igwe JC, Abia AA (2006). A bioseparation process for removing heavy metals from wastewater using biosorbent. Afr. J. Biotechnol. 5(12):1167-1177.

Kugbe J, Matsue N, Henmi T (2009). Synthesis of Linde Type A Zeolite-Goethite Nanocomposite as an Adsorbent for Cationic and Anionic Pollutants. J. Hazardous Mater. 164(2-3):929-935.

Kumar KK, Prasad MK, Sarada B, Murthy CVR (2012). Studies on biosorption of Nickel using immobilized fungus. *Rhizomucor tauricus*. Bioresour. 7(4):5059-5073.

Kurniawan K, Thami Larabu P (2010). Removal of iron (111) from aqueous solution using Ricinus communes Seed Shell and polypyrole coated Ricinus communes seed shell activated carbon. Int. J. Chem. Tech. Res. CODEN (USA): IJCRGG. 2(1):26-35.

Lagergren S (1898). Zur theorie der sogenannten adsorption gelöster stoffe, K. Sven. Vetenskapsakad. Handl. 24:1-39.

Langmuir I (1918). The adsorption of gases on plane surfaces of glass, mica and platinum. J. Am. Chem. Soc. 40:1361–1403.

Mahajan G, Umesh G, Vinod G (2013). Utilization Properties of Jatropha De- oiled cake for removal of Nickel (ii) from aqueous solutions. BioResources 8(4):5596-5611.

Mahajan G, Sud D (2011). Kinetics and equilibrium studies of chromium (vi) metal ion remediation by Arachis hypogeal shells. Agreen approach. BioResources 6:3324-3338.

Meena AK, Mishra GK, Rai PK, Rajagopal C, Nagal PN (2005). Removal of heavy metal ions from aqueous solutions using carbon aerogel as an adsorbent. J. Harzadous Mater. 122:161-170.

Mishra SP (2013). Adsorption of Cu and Zn on calcium alginate immobilized penicillium sp. Indian J. Chem. Technol. 20:21-25.

Nida MS, Ahlam MF, Aklam A (2012). Biosorption of Cadmium(11) from Aqueous Solution by *Prunus avium* Leaves. Am. J. Environ. Eng. 2(5):123-127.

Orhan Y, Buyukgungor H (1993). The removal of heavy metals by using agricultural wastes. Water Sci. nigricans. Water Res. 32:1437-1444.

Osemeahon SA, Esenowo DI (2012). Evaluation on the application of immobilized Konkoli (*Maesopsis eminii*) immobilized leaves for the removal of Pb^{2+} Cd^{2+} and Zn^{2+} in aqueous solution. Afr. J. Pure Appl. Chem. 6(5):59–64.

Pradhan J, Das SN, Thaker RS (2007). Adsorption of hexavalent chromium from aqueous solution by using activated red mud. J. Colloid Interf. Sci. 217(1):137–141.

Santhi IT, Manonmani S (2012). Adsorption of Methylene Blue from aqueous Solution onto a waste aquacultural Shell Powders (prawn waste). Sustain. Environ. Res. 22(1):45-51.

Sarin V, Paint KK (2006). Removal of chromium from indusrial waste water by using eucalphus bark. Bioresource Technol. 97: 15–20.

Shin EW, Karthikeyan KG, Tshabalala MA (2007). Adsorption mechanism of cadmium on juniper bark and wood. Bioresour. Technol. 98:588-594.

Singh R, Chadetric R, Kumar R, Bhatia D, Kumar Bishnoi NR, Singh N (2010). Biosorption Optimization of lead II, Cadmiun II and Copper II using response surface methodology and applicability in isotherm and thermodynamics modeling. J. Hazard. Mater. 174:623-634.

Suantak KC, Baloma J, Shri C (2011). Removal of arsenic (iii) from aqueous solution by biosorption on to maize (*Zea mays*) leaves surface: parameters optimization sorption isotherms, kinetics and thermodynamic studies. Res. J. Chem. Sci. 1(5):73–79.

World Health Organisation (Ed). (2008). Guidelines for drinking water quality, world health organization, ISBN 9241544743, Geneva.

Emamectin benzoate residue analysis in ground and terrestrial field water in Lahore suburban area

Sheraz Ahmed[1], Fahim A. Qureshi[1], Amjad Islam[1,2]* and Ahmed Adnan[1]

[1]Department of Chemistry, Government College University, Katchery Road, Lahore-54000, Pakistan.
[2]Industrial Chemicals Division, Guides Corporation, Davis Road, Lahore-54000, Pakistan.

High performance liquid chromatography with diode array detector was used to determine the ememactin residue in ground water in Lahore suburban area. The samples were extracted with chloroform through solvent extraction by using a separating funnel. The chemicals used were chloroform, methanol (HPLC grade) and acetonitrile (HPLC grade). Mobile phase used was acetonitrile (98%) and methanol (2%). 10 samples were analyzed and emamectin benzoate was detected in 5 samples. Emamectin benzoate residue was detected in terrestrial and ground water. It was present in trace amount. The concentration range varies from 0.04 to 1.4 ppm of terrestrial field water. With respect to toxicity, it is class II pesticide by W.H.O. Presence of pesticides in large amounts can create a lot of diseases in human body.

Key words: New York codes rules and regulations (NYCRR), reference dose (RfD), United States environmental protection agency (USEPA).

INTRODUCTION

Emamectin benzoate belongs to the avermectin group and is administered to the fish through feed. It is a non-systemic insecticide which penetrates leaf tissue by transmular movement. It paralyses the *Lepidoptera*, which stops feeding within hours of ingestion and die. Emamectin benzoate is effective against several stages of the salmon louse, copepodite, chalimus, and mobile pre-adult and adult stages (Lees et al., 2008). It is a highly efficient insecticide against many larvae of Lepidoptera and other insects. It is not ovicidal and can penetrate into plant cuticula, is effective even at a very low dosage and is not disruptive to beneficial arthropods during integrated pest management. Emamectin benzoate is a very effective agricultural insecticide that is applied to various crops in a low concentration (Yoshii et al., 2000). It is used as an acaricides or parasites for animals and plants.

Emamectin benzoate is the 4′-deoxy-4′-epi-methylamino benzoate salt of avermectin B_1 (abamectin), which is similar structurally to natural fermentation products of *Streptomyces avermitilis*. Emamectin is a mixture of approximately 90% emamectin B_{1a} and approximately 10% emamectin B_{1b}. Emamectin benzoate is being developed as a newer broad spectrum insecticide for vegetables. The mechanism of action involves stimulation of high affinity gamma-amino butyric acid (GABA) receptors and a consequent increase in membrane chloride ion permeability. Animal studies point out a wide margin of safety because mammalian species are much less sensitive due to lower GABA receptor affinities and relative impermeability of the blood brain barrier (Sung et al., 2009).

It is a white to off-white powder. Its density is 1.20 (23°C), vapour pressure is 3×10^{-8} mm Hg (torr) and solubility in water is 0.024 g/L (pH 7, 25°C). The metabolites of emamectin are 8, 9-Z isomer (defined as 8, 9-Z), 4”-epi-(N-formyl-N-methyl) amino-4”-deoxy-avermectin B_1 (methyl formyl amino form, defined as MFA) and 4”-epi-amino-4”-deoxy-avermectin B_1 (amino form, defined as Amino) (Yoshii et al., 2004). Its melting point is 141 to 146°C. It is non-corrosive, posing a slight fire hazard if exposed to flame. It is soluble in acetone and methanol but hardly soluble in hexane. The half life of emamectin in water was found to vary between 0.7 days during summer conditions to 35.4 days during winter conditions. The difference is due to light sensitivity not temperature. Emamectin benzoate decomposes in a

*Corresponding author. E-mail: aiqchemist@gmail.com, sherazahmed83@gmail.com.

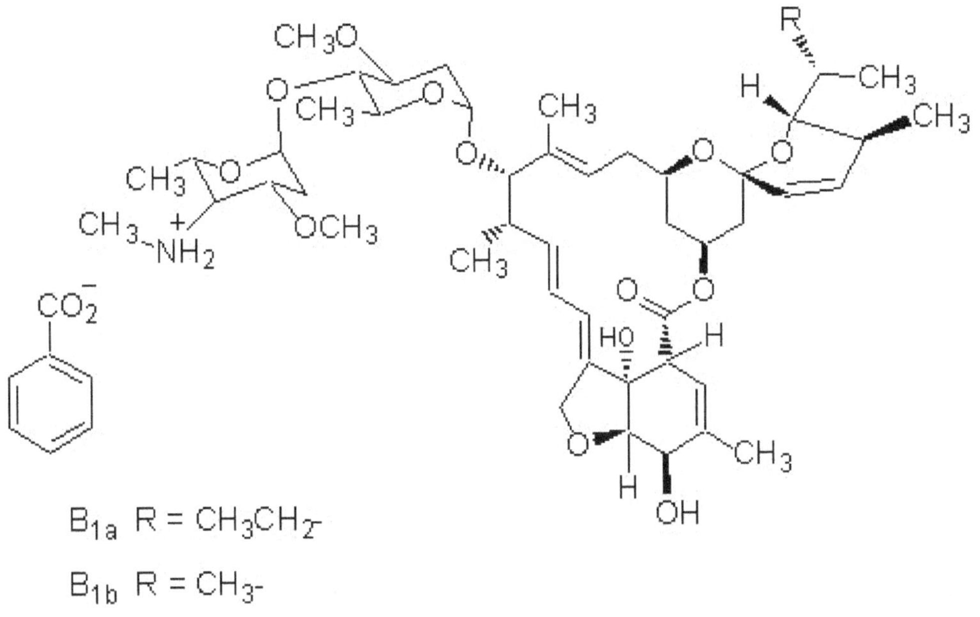

Figure 1. Structure of emamectin.

kinetic phase with a half-life, which includes aerobic and anaerobic metabolism, of 174.2 days. The anaerobic half-life of emamectin benzoate is 427.4 days. Emamectin benzoate exhibits no mobility. It tightly bounds to soil and does not move readily in environment. Terrestrial field dissipation studies indicate a maximum half-life of 15 to 17 days. No quantifiable residues were found at 6 to 12 inches, and no detectable residues were found below 12 inches. Emamectin benzoate is highly toxic to marine/estuarine crustaceans. The leaching studies show that depending upon the type of soil, 0.1 to 0.82% of the applied material can be desorbed.

The acceptable daily intake (ADI) has been set at 1 µg/kg body weight by WHO/FAO. But the United States Environmental Protection Agency (USEPA) calculated an oral reference dose (RfD) for emamectin benzoate of 0.00025 mg/kg/day based on the no-observed-effect level (NOEL) from the sub chronic neurotoxicity in mice (0.075 mg/kg/day) and an uncertainty factor of 300. Environment Conservation Department in New York does not register this product as a "restricted use pesticide". Emamectin benzoate is "not likely" to be a human carcinogen.

In plants, metabolism has been investigated in cabbage, lettuce, sweet corn and in other vegetables (Crouch et al., 1997; Crouch and Feely, 1995). Maximum emamectin benzoate residues established by USEPA in or on head and stem Brassica vegetables, celery and head lettuce each at 0.025 parts per million (ppm). The USEPA estimated that chronic dietary exposure to these residues would be no more than five percent of the population adjusted dose (PAD) for infants and children and no more than four percent of the reference dose (RFD) for adults. This chronic exposure analysis is based on the assumption that 25% of crops are treated and contain tolerance level residues.

Emamectin has neither antibacterial nor fungicidal properties. It is meant to kill salmon lice, which is a crustacean. It is the second class pesticide by WHO/FAO. Abamectin emamectin benzoate is an upgrade to improve the product, its activity is much higher than the avermectin and such as the stomach poisoning toxicity is avermectin 2146 times. In normal cases, the effect of emamectin benzoate was significantly better than abamectin, emamectin benzoate control diamondback moth should be better than the effect of avermectin (Yoshii et al., 2001) (Figure 1).

Several methods have been developed for the determination of emamectin residues in soil, crops and water (Hicks et al., 1997; Chukwudebe et al., 1997). Liquid chromatographic method using a UV detector was reported for the determination of emamectin residues in water (Blackwell et al., 2004).

This work is different because it involves a diode array detector. High performance liquid chromatography is more favorable than conventional liquid chromatography for pesticide residue analysis. In Pakistan, emamectin is in use for the last four to six years and there is not much work reported dealing with residue analysis.

MATERIALS AND METHODS

Organic solvents (residue analysis grade) for dissolving and extracting were methanol (HPLC), chloroform and acetonitrile (HPLC grade) purchased from Merck (Germany). Standard material (95%) was obtained from Shanghai Taizhao Chemicals (China). Purity of the standard after recrystallization was more than 95%.

Table 1. Comparison of ground water and terrestrial water samples.

S/N	Place of sampling	Ground water	Terrestrial field water
1	Cotton field (Kasoor)	Traces	Detected
2	Bitter goard (Changa Manga)	Traces	Detected
3	Patato field (Raiwind)	Traces	Detected
4	Wheat field (Patooki)	Traces	Detected
5	Brassica field (Sialkot Road)	Traces	Detected

Table 2. Concentration and retention time data of the samples.

S/N	Place of sampling	Terrestrial field water					
		B_{1a} x = y/9320.9+1692/9320.9			B_{1b} x = y/50446+714.48/50446		
		RT	y = area	x = (mg/L)	RT	y = area	x = (mg/L)
1	Cotton field (Kasoor)	4.24	5186	0.74	3.70	4859	0.11
2	Bitter goard (Changa Manga)	4.23	2262	0.42	3.70	1686	0.047
3	Patato field (Raiwind)	4.25	7437	0.97	3.71	4164	0.09
4	Wheat field (Patooki)	4.18	11522	1.4	3.66	5828	0.12
5	Brassica field (Sialkot Road)	4.24	8798	1.12	3.69	3690	0.08

Stock standard solutions were prepared by accurately weighing 100 mg of standard emamectin benzoate and dissolved in acetonitrile and methanol (98:2), after which the volume is made to 100 ml. Working standard solutions of 1, 5, 10 and 20 ppm were prepared by dilution with acetonitrile and methanol (98:2). For each standard solution took 0.25, 1.25, 2.5 and 5 ml of stock standard solution and 25 ml solvent, acetonitrile and methanol (98:2) for each working standard. Stock standard solution and working standard solutions were stored under refrigeration (4°C).

High-performance liquid chromatography (HPLC) is one of the premier analytical techniques widely used in analytical laboratories. High-performance liquid chromatography or high pressure liquid chromatography (HPLC) is a form of column chromatography used frequently in biochemistry and analytical chemistry to separate, identify, and quantify compounds. HPLC utilizes a column that holds chromatographic packing material (stationary phase), a pump that moves the mobile phase(s) through the column, and a detector that shows the retention times of the molecules. Retention time varies depending on the interactions between the stationary phase, the molecules being analyzed, and the solvent(s) used.

A high performance liquid chromatograph (HPLC), Agilent 1100, equipped with diode array detector (DAD) was used for the analysis of residues of emamectin benzoate. An HPLC column, Thermo scientific (4.6 mm ID; 250 mm length; 5 µm particle size; pore size 120 A, stainless steel) was used for emamectin residue analysis. Mobile phase was a combination of acetonitrile (98%) and methanol (2%). Injection volume was 100 µl. The solvent was pumped at a flow rate of 1.2 ml/min with gradient elution. Solvents used for samples preparation were methanol and acetonitrile. Wave length of light was 246 nm. Analysis was carried out at room temperature (25°C).

Sampling

10 samples of ground water and terrestrial field water (5 terrestrial and 5 ground water samples) were collected from the different field crops in different villages near Lahore. These water samples were collected from Kasoor (Ganda Sing border), Changa Manga, Patooki (Hunjaran Kalan), village Vinya and Raiwind. These samples were analyzed after 24 h being stored at 4°C until analyses. These samples were collected from December 2008 to February 2009 at temperature 12°C.

Extraction

All samples (1000 ml of water for each sample) were extracted one by one from chloroform through solvent extraction using a separating funnel and were filtered through 0.4 µm filter paper one by one separately. Extracted samples were then allowed in the beakers to dry at room temperature after covering the beakers with aluminium foil with holes, after evaporation of chloroform from all samples, the residues were deposited at the bottom of the beakers. The residue from each beaker was dissolved in 25 ml solvent (98 ml acetonitrile : 2 ml methanol) and then was filtered once again through 0.4 µm filter paper one by one separately for making sample solution and analyzed by HPLC.

RESULTS AND DISCUSSION

10 samples of ground water and terrestrial field water (5 terrestrial and 5 ground water samples) were analyzed for pesticide residue. Emamectin was identified in these samples from its retention time by comparing it with authentic standard. Comparison of these samples is shown in Table 1. Concentrations of these samples are shown in Table 2. Maximum residue limit (MRL) value of emamectin is 0.5 ppm by WHO/FAO in ground water. Terrestrial toxicity effect concentration from technical report of environment agency, Rio house (England) is 570 to 1318 (mg kg^{-1}).

Emamectin is a mixture of emamectin B_{1a} and emamectin B_{1b}. Emamectin B_{1a} is a major component and emamectin B_{1b} is the minor component. Each component has its own peak. Two peaks were obtained

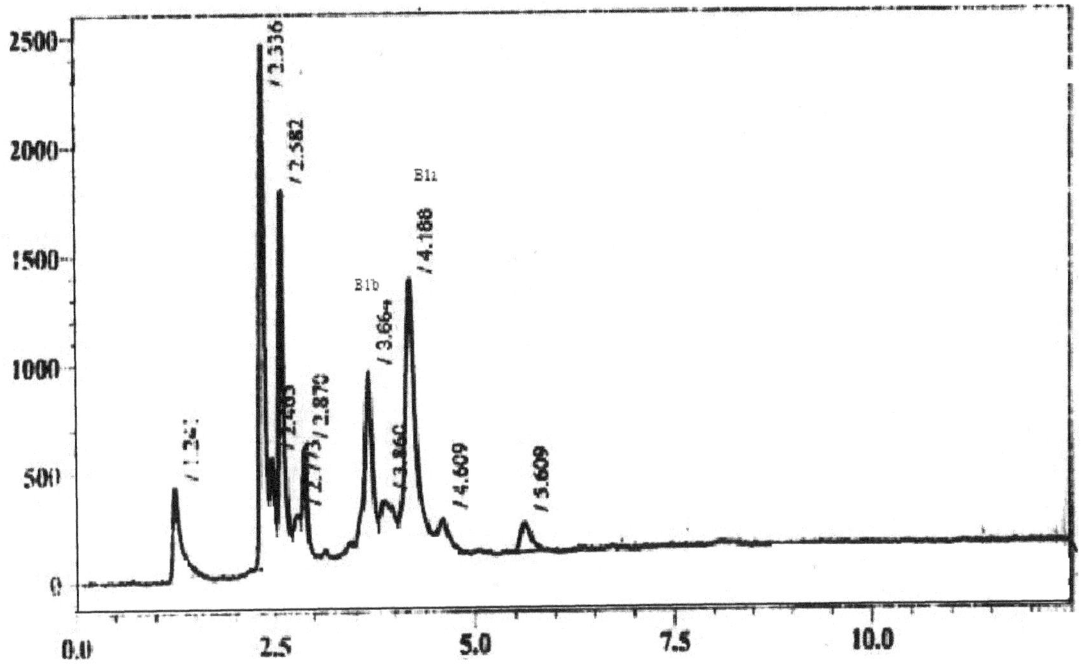

Figure 2. Chromatogram of a water sample.

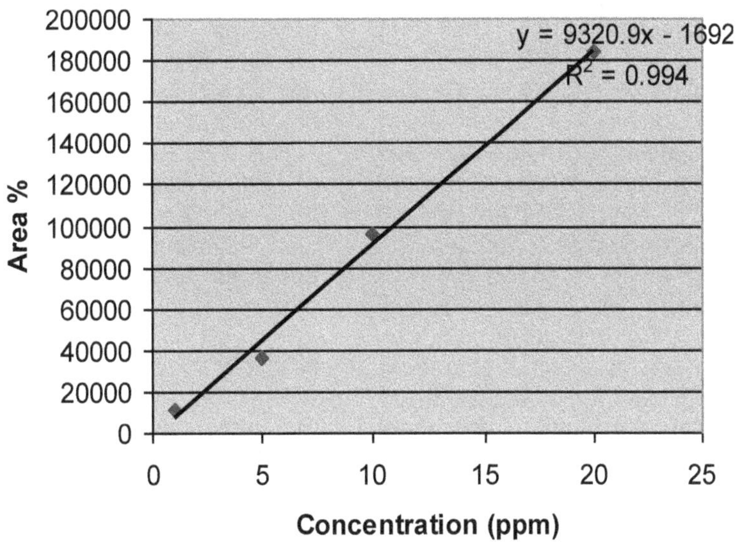

Figure 3. Calibration curve of a sample for emamectin B_{1a}.

in the samples also. The concentration range varies from 0.04 to 1.4 ppm in which emamectin benzoate B_{1a} concentration ranges from 0.42 to 1.4 ppm whereas emamectin benzoate B_{1b} concentration ranges from 0.04 to 0.12 ppm.

Chromatogram of a water sample is shown in Figure 2. Figure 3 shows a calibration curve for emamectin B_{1a} and it is a graph between concentration (on x-axis) and area (on y-axis). Figure 4 shows a calibration curve for emamectin B_{1b}.

At the time of sampling, the temperature was 12°C. With the temperature increase, the toxicity of avermectin against pests (Particularly virulent stomach poison) increases. However, emamectin benzoate and abamectin toxicity with rising temperature rate of increase is inconsistent.

Emamectin benzoate is poorly soluble in water. The hydrolysis half-life of emamectin benzoate is 19.5 weeks at pH 9 and stable at pH 5.2, 6.2, 7.2 and 8.2. At the time of sampling, pH of water samples varies from 6.2 to 7.8.

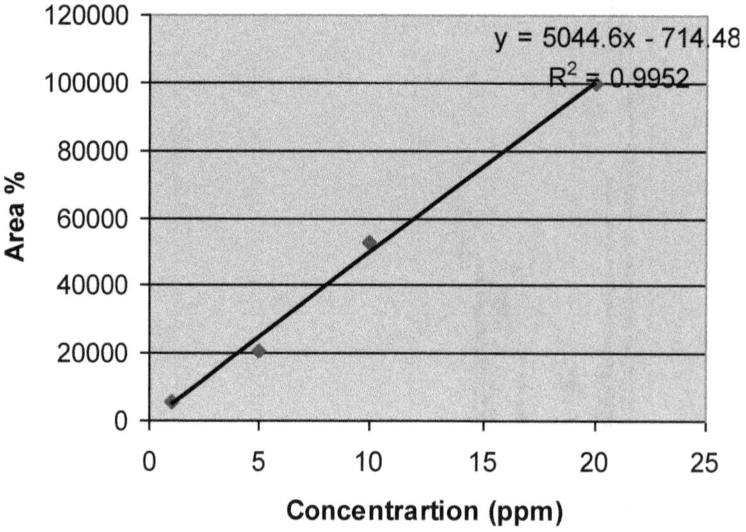

Figure 4. Calibration curve of a sample for emamectin B_{1b}.

There are no chemical specific Federal or State drinking water standards for emamectin benzoate. Based on its chemical structure, emamectin benzoate falls under the 50 µg/L general New York State drinking water standard for "unspecified organic contaminants" (10 NYCRR Part 5, Public Water Systems), based on the USEPA Office of Pesticide Programs RfD of 0.00025 mg/kg/day and the Department of Environmental Conservation Water Quality Regulation procedures for deriving surface water and groundwater standards and guidance values from non-oncogenic effects (6 NYCRR - Part 702.5), an ambient water quality value for emamectin benzoate would be 1.8 µg/L.

Use of avermectin production levels are generally higher than that of emamectin benzoate (emamectin benzoate production, mostly 1 or 0.5%, while abamectin is generally 1.8 or 2%), the dilution factor are generally higher than the avermectin emamectin benzoate (avermectin 1000 to 1500 times, emamectin benzoate 2000 to 3000 times).

Due to its low application rate and the number of target crops limited crops, the labeled use of proclaim insecticide does not appear to pose significant risks to the workers or to the general public through dietary exposure.

Conclusions

The results show that traces of emamectin benzoate residue are found in ground water. Emamectin benzoate residue is found in terrestrial water in detectable amount. It is not much toxic for human health in low doses. Emamectin benzoate was very stable when stored in a refrigerator. This holds also protection from light which is also a prerequisite for stability. Because of the high potency of the emamectin benzoate against parasites and insects very small amounts are applied, resulting in extremely low concentrations in samples obtained during field water and ground water studies. No analytical problems were encountered in the processing of samples. The results indicate that emamectin benzoate is not a very dangerous pesticide like carbamates and organophosphates.

This is quite a beneficial study on the basis of local use of the pesticide that recommends emamectin benzoate as a high potency and low dose pesticide. This may be a good alternate to high dose pesticides and thus may be a better application economically and environmentally.

ACKNOWLEDGEMENTS

The authors are grateful to Dr. Fahim Ashraf Qureshi and Dr. Ahmad Adnan from the Department of Chemistry, Government College (GC), University Lahore, Pakistan, for their critical review on this manuscript.

REFERENCES

Blackwell P, Lutzhoft H, Halling-Sorensen HP, Boxall ABA, Kay P (2004). Fast and robust simultaneous determination of three veterinary antibiotics in groundwater and surface water using a tandem solid phase extraction with HPLC-UV detection. J. Chromatogr. A, 1045(1-2): 111-117.

Chukwudebe CA, Atkins HR, Wislocki GP (1997). Metabolic fate of emamectin benzoate in soil. J. Agric. Food Chem., 45(10): 4137-4146.

Crouch SL, Feely FW (1995). Fate of [14C] emamectin benzoate in Head Lettuce. J. Agric. Food Chem., 43(12): 3075-3087.

Crouch SL, Wrzesinski LC, Feely FW (1997). Fate of [$^{14}C/^{3}H$] emamectin benzoate in cabbage extractable residues. J. Agric. Food Chem., 45(7): 2744-2757.

Hicks MB, Payne LD, Prabhu SV, Wehner TA (1997). Determination of emamectin benzoate in freshwater and seawater at pictogram-per-milliliter levels by liquid chromatography with florescence detection. J. AOAC Int., 80: 1098-1103.

Lees F, Baillie M, Gettinby G, Revie WC (2008). The efficacy of emamectin benzoate against infestations of *Lepeophtheirus salmonis* on Farmed Atlantic Salmon (*Salmo salar* L) in Scotland, 2002-2006. *PLoS ONE*, 3(2): 1549.

Sung FY, Huang TI, Fan KC, Lin HC, Lin PS (2009). Avermectin intoxication with coma, myoclonus, and polyneuropathy. "Clinical Toxicology", 47(7): 686-688.

Yoshii K, Ishimitsu S, Tonogai Y, Arakawa K, Murata H, Mikani H (2004). Simultaneous determination of emamectin, its metabolites, milbimectin, ivermectin and abamectin in Tomato, Japanese Radish and Tea by LC/MS. J. Health Sci., 50(1): 17-24.

Yoshii K, Kaihara A, Tsumura Y, Ishimitsu S, Tonogai Y (2000). Liquid chromatographic determination of emamectin, milbemectin, ivermectin and abamectin in crops and confirmation by liquid chromatography–mass spectrometry. J. Chromatogr. A, 896: 75-85.

Yoshii K, Kaihara A, Tsumura Y, Ishimitsu S, Tonogai Y (2001). Simultaneous Determination of residues of emamectin and its metabolites, and milbemectin, avermectin, and abamectin in crops by liquid chromatography with fluorescence detection. J. AOAC Int., 84: 910-917.

Status and distributions of pesticides buried at five sites in Arusha and Mbeya regions, Tanzania

John A. M. Mahugija

Chemistry Department, University of Dar es Salaam, P. O. Box 35061 Dar es Salaam, Tanzania.

This study investigated the compositions and distributions of pesticides buried at five sites in Tanzania. Samples were collected from various depths up to 220 cm. The analytes in samples were extracted by pressurized fluid extraction. Labelled internal standards were used for identification and quantification. Clean-up of extracts was conducted by using column chromatographic techniques. Determination of the analytes was performed by using a high resolution GC–MS. The dominant compounds detected were 4,4'-DDT (dichlorodiphenyltrichloroethane), 2,4'-DDT, 4,4'-DDE, 2,4'-DDE, 4,4'-DDD, 2,4'-DDD, hexachlorocyclohexane (HCH) isomers (α-, β-, γ-, δ- and ε-HCH), aldrin, dieldrin, endrin, pentachlorobenzene, hexachlorobenzene, pentachloroanisole, endosulfans, chlordanes, heptachlor and heptachloroepoxides. The highest concentrations of total DDT and total HCH ranged 5.2 to 5410 mg/kg dry weight (dw) and 1.4 to 42200 mg/kg dw, respectively, while the highest concentrations for other compounds varied from 1.1 to 7200 mg/kg dw. The results indicated aged contaminants but the rate of degradation was generally very slow. The concentrations of the compounds decreased with increase in distance from the burial points. There were variations in the distribution of the compounds among the sampling sites and depths; some showed even distribution or accumulation downwards indicating some risks for groundwater contamination.

Key words: Buried pesticides, soil, Tanzania.

INTRODUCTION

The banning of pesticide introductions into certain countries, poor storage and lack of store management, unawareness, overstocking, inadequate assessment of needs, difficulties in forecasting outbreaks of pests, inappropriate formulations and containers supplied by pesticide distributors, excessive or uncoordinated donations are among the factors which contribute to the accumulation of obsolete pesticides in developing countries (FAO, 1998, 2001). Common types of obsolete pesticides include organochlorine pesticides such as dichlorodiphenyltrichloroethane(DDT),hexachlorocylohexanes (HCHs), aldrin, dieldrin, endrin, hexachlorobenzene, chlordane and heptachlor, which have been withdrawn or banned for public health and environmental reasons (FAO, 1996, 2001). Organochlorine pesticides are described as persistent organic pollutants (POPs) which are organic compounds that, to a varying degree, resist photolytic, biological and chemical degradation in natural environments. POPs are often characterized by low water solubility and high lipid solubility, leading to their bioaccumulation in fatty tissues and toxicity. They are also semi-volatile, enabling them to move long distances in the atmosphere before deposition occurs (Ritter et al., 1995).

There are no easy disposal methods that are safe, cheap and generally applicable under circumstances prevailing in developing countries. As a result, the pesticides are improperly disposed, which often leads to serious environmental contamination. The common disposal methods used include open burning, discharge to sewer, burying or landfill disposal, which are likely to cause severe damage to public health and the environment (FAO, 1996).

The problem of pesticide burial sites exists in most African, East European and Caucasian countries. For example, significant contamination has been found in soil around sites where tonnes of obsolete pesticides were buried in Belarus (PAN UK, 2005), Armenia (Manvelyan et al., 2006), Poland (Gałuszka et al., 2011) and other countries. Typically, the content of buried pesticides is dominated by organochlorine pesticides. Other pesticide types, such as organophosphates, carbamate pesticides, dinitrophenols, phenoxyacids, and inorganic compounds are stored in smaller quantities (Gałuszka et al., 2011).

Soil is an important component of the terrestrial ecosystem. It generally acts as a primary sink and emission source for contaminants (Rauf et al., 2002). The concentrations and distribution features of pesticides in soil profiles are subject to physical, chemical and biological processes. The fate processes of pesticides in the environment are divided into three major types: (i) Adsorption, which is the binding of pesticides to mineral or organic matter; (ii) Transfer processes that move pesticides in the environment; and (iii) Degradation processes that break down pesticides (Harrison, 1990; Sun et al., 1997).

The tendency of different pesticides to adsorb to soil varies with their chemical and physical characteristics. Some pesticides, bind very tightly while others bind only weakly and are readily desorbed, that is, released back into the soil solution. Adsorption influences whether a pesticide is free to enter into other biological and chemical pathways. Since a strongly adsorbed pesticide is less likely to be dissolved in water, it is also less likely to be absorbed into plant roots and soil organisms or to leach beyond the root zone (Harrison, 1990; EXTOXNET, 1993; Waldron, 1992; Mansour, 1993). Many soil characteristics influence pesticide adsorption, including structure, texture, organic matter content, pH, and moisture content. Soil texture affects movement of water through soil and, therefore, affects the movement of dissolved pesticides. The coarser the soil, the faster the movement of the percolating water, and the less opportunity for adsorption of dissolved chemicals. Soils with more clay tend to hold water and dissolved chemicals longer. These soils also have far more surface area on which pesticides can be adsorbed (McEwen and Stephenson, 1979; Harrison, 1990). Soil organic matter influences how much water a soil can hold and how well it will adsorb pesticides. Increasing fractions of organic matter increases the adsorptive capacity of soils. This characteristic is due in part to particles of organic matter providing soils with a large number of sites, onto which pesticides can bind. The chemical nature of a pesticide determines sorption equilibria by influencing its direct affinity for the clay or organic matter or by influencing its solubility or affinity for the soil solution (McEwen and Stephenson, 1979; Harrison, 1990; Mansour, 1993). Soils with highly developed structure may encourage the movement of pesticide-laden water between the structural units, known as peds, without contacting the individual soil particles within the units. This has the effect of reducing the potential for adsorption (Harrison, 1990). The fate of pesticides in soils varies with pH differences, or the acid/base balance of the soil solution, primarily because of the influence of pH on sorption phenomena (Mansour, 1993). Soil pH also affects chemical degradation. Soil pH impacts the chemical's reactivity and certain soil functions such as microbial metabolism (Buyuksonmez et al., 1999). Wet soils tend to adsorb fewer pesticides than dry soils, either because water molecules compete for the binding sites or because some of the pesticide molecules remain dissolved in the soil water (Harrison, 1990). Soil temperature also affects pesticide adsorption because when the temperature increases, some bonds break and cause the desorption of some pesticide molecules. At high temperatures the greater solubility of pesticides also results into more pesticides being available in the soil solution (McEwen and Stephenson, 1979; Waldron, 1992).

Pesticide transfer refers to the movement of pesticides from their sites of origin. Too much movement can lead to injury of nontarget species (including humans), and surface and groundwater contamination. Five processes that can move pesticides are volatilization, runoff, leaching, absorption, and physical removal. All these are influenced by management practices such as the methods used to store, apply or dispose the pesticides (Harrison, 1990; Waldron, 1992).

Pesticide degradation processes break down pesticide molecules into simpler and, generally, less toxic compounds, although in some cases may be more toxic. The three major types of pesticide degradation are microbial degradation, chemical degradation, and photodegradation (Harrison, 1990; Buyuksonmez et al., 1999).

The aim of this study was to investigate the compositions and distributions of pesticides buried at the five sites located in Arusha and Mbeya regions in Tanzania. There were no clear records as to when the pesticides were buried at the studied sites, but it was generally indicated that the pesticides were buried more than ten years before this study.

MATERIALS AND METHODS

Study areas

The study areas were located in Arusha and Mbeya regions (Figure 1). At the Plant Health Services Office at Tengeru in Arumeru

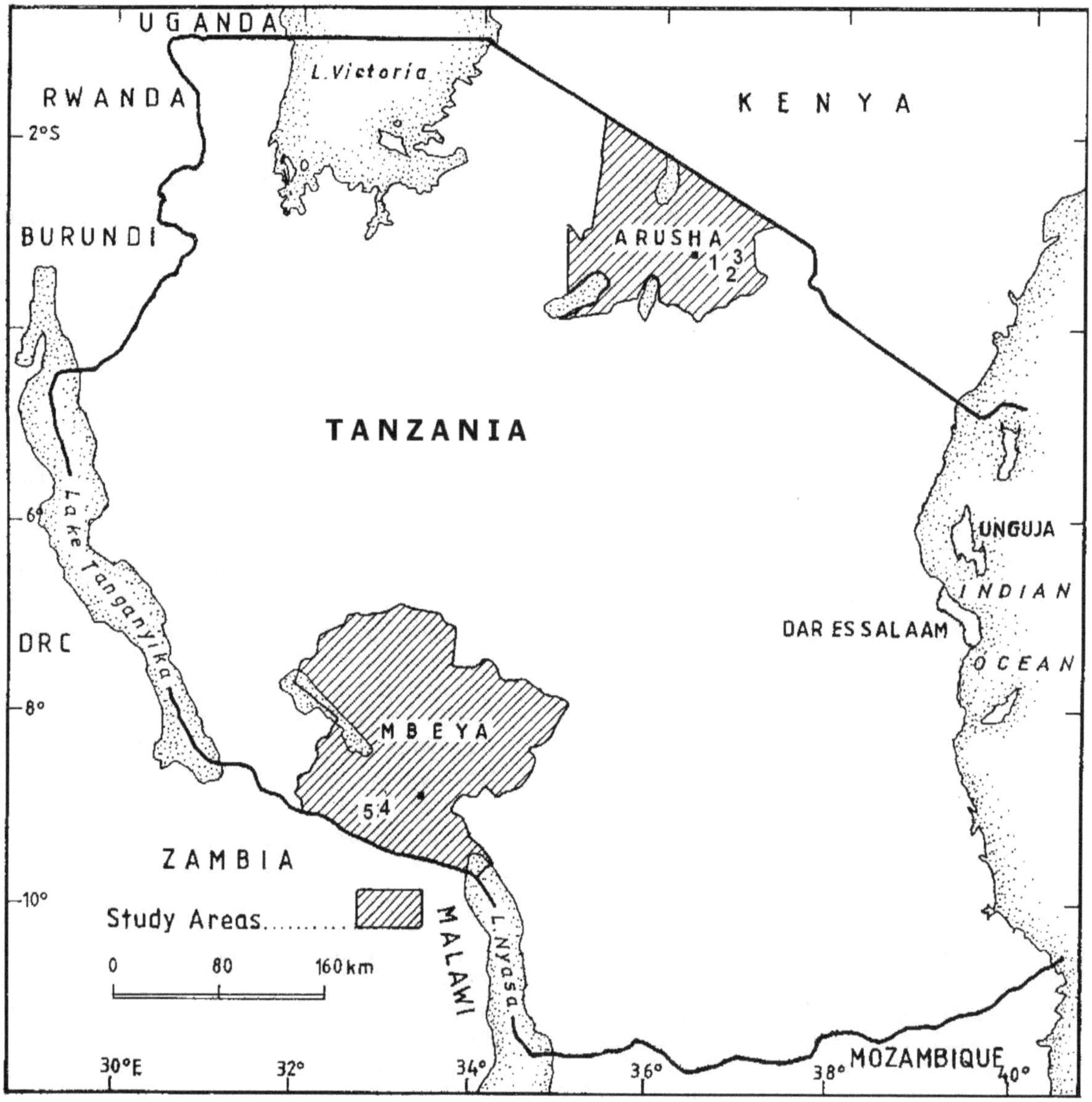

Figure 1. Map showing the sampling sites: *1 = Tengeru, 2 = ACU maize farm (Usa river), 3 = Akheri vegetable garden, 4 = TaCRI-Mbimba, 5 = MBOCU.*

district, Arusha region (locust control centre for Tanzania), pesticides contained in drums were kept outside the office at the backyard area. The drums rusted and the contents leaked into the soil. The drums were removed to another unknown location while the contaminants were buried in soil, covered with a polyethylene sheeting and fresh soil added on top to avoid the odour, but that did not prevent the smell from permeating the area. The location is beside the Plant Health Services Offices and the workers there often complained of the strong smell invading the offices, especially during dry season. The location is at the top of a sloping bank of approximately 3 m. Cultivation activities were being carried out very close to the contaminated area.

Pesticides collected from some stores by the Arusha Co-operative Union (ACU) were buried at another site located in a maize farm owned by ACU at Usa River Chemchem maize farm in Arumeru district, covering an area approximately 160 m^2. No vegetation was found to grow at the area where the pesticides were buried.

At Akheri village in Arumeru district in Arusha region, pesticides collected from a nearby former storage building owned by Meru Co-operative Union, were buried at a small area in a vegetable garden. This was a working garden with continuous planting and harvesting of vegetables such as spinach, onions and cabbages. The pesticides were buried directly in the vegetable garden right in the irrigation path, which implied the possibility of being dispersed and introduced into the food chain.

Other sites where pesticides were buried are Tanzania Coffee Research Institute (TaCRI)-Mbimba and Mbozi Cooperative Union (MBOCU). The TaCRI-Mbimba site is located at Mbimba village in Mbozi district, Mbeya region within the coffee research farm at the Tanzania Coffee Research Institute. At this site, pesticides were dumped and buried in a pit of an area approximately 25 m^2. At MBOCU headquarters in Vwawa town, Mbozi district in Mbeya region, pesticides and empty containers were dumped in a pit of about 42 m^2. The site is located very close to residential areas and office buildings. Cultivation was being carried out in the vicinity of this site.

Sampling

Soil samples were collected from five sites located in Arusha and Mbeya regions (Figure 1). Soil samples from Tengeru, ACU maize farm and Akheri were collected in January 2009. Samples from Mbimba and MBOCU were collected in April 2009. Samples were collected at points within the burial areas and approximately 5 to 60 m from those points down and up slope. Soil samples were collected from 19 points at Tengeru and the sampling depths were 30, 75, 150 and 220 cm. At ACU maize farm, samples were collected at 10, 30 and 75 cm depths and were from 13 points. The sampling depths at Akheri were 10, 30 and 50 cm and the samples were collected from 8 points. The samples were collected from 50 and 100 cm at Mbimba, while 10, 50 and 100 cm were the sampling depths for MBOCU.

Sampling was conducted applying the standard guidelines (Åkerblom, 1995). The tools used for sampling of soil included clean spades, hoes, clean aluminium foil, clean buckets, measuring tapes, folding rulers, insulating boxes, solvents for cleaning, bottles for waste solvents, kitchen roll papers, polyethylene bags, waste plastic bags and stainless steel spoons. Using a hoe and spade, a test pit was prepared a little deeper than the desired depth, then using a clean spade a slice about 5 cm thick was made along the vertical wall of the pit at the desired depth and the soil was thrown away. Another clean spade or spoon was used to take the sample. Samples at deeper points and presumably less contaminated points were collected first before the surface or presumably contaminated depths or points. A sample was obtained by collecting at least five subsamples from different points at the same depth within the pit. The stones, sticks, plant roots and other unwanted materials were removed by using a clean spoon. The sample was ground and mixed very thoroughly on aluminium foil. The sample was immediately wrapped in an aluminium foil and placed in a polyethylene bag then put in an insulated box. The samples were transported to the laboratory and stored in a freezer at -28°C until extraction.

Extraction, clean up and gas chromatographic analysis

Extraction, clean up and analysis of the soil samples were conducted at the Institute of Ecological Chemistry, German Research Centre for Environmental Health in May – October 2009. The procedures by Schramm et al. (2008) were adopted with modifications. Soil samples were extracted by pressurized fluid extraction using an Accelerated Solvent Extractor (ASE 200 Dionex). A cellulose filter was inserted into the inner bottom of the extraction cell, then sea sand dried at 550°C (ca. 1 g) was added. The sample (0.5-5 g) mixed with hydromatrix for drying and dispersing was added into the cell and a filter placed on top. Prior to extraction, the sample was spiked with ^{13}C–labelled and deuterated internal standards (10 µl of a mixture containing 333-1000 pg/µl of organochlorine compounds in nonane). All the internal standards for the compounds determined were ^{13}C–labelled except for 4,4′-DDD, which was a deuterated standard. Highly contaminated samples were not spiked with internal standards prior to extraction, but the diluted extracts were spiked during clean-up. The samples were quantitatively extracted by an accelerated solvent extractor at a temperature of 120°C and pressure of 120 bar and with *n*-hexane:acetone (75:25) as the extraction solvent mixture. Two static cycles of 10 min were applied for a complete extraction. Another sub-sample of each sample was dried for 24 h at 105°C and then weighed for moisture and dry weight determination. The extracts were passed over anhydrous sodium sulfate to remove water. The extracts were concentrated using vacuum rotary evaporation and the solvent was reconstituted to hexane:dichloromethane (1:1).

Clean-up of the concentrated extracts (0.5-1 ml for undiluted extracts or 50-100 µl of diluted extracts for highly contaminated samples) to remove interferences was conducted by using 10 g silica gel (grade 60), 5 g alumina with 3% H_2O and 5 g anhydrous sodium sulfate, packed in that order in glass column (30 cm long with an internal diameter of 2.5 cm). The extracts were eluted with 100 ml of a mixture of hexane and dichloromethane (1:1) at a flow rate of *ca.* 2 drops per second (about 0.1 ml/s) and concentrated to 1 ml using a rotary evaporator, then using a very gentle stream of nitrogen to ca. 0.2 ml. The solvent was changed to acetonitrile and concentrated using nitrogen to ca. 0.2 ml. Further clean-up was performed through a C18 SPE cartridge using 1 g C_{18}-modified silica gel and the eluting solvent used was acetonitrile (5 ml). The extracts were concentrated by blowing a very gentle stream of nitrogen to ca. 0.2 ml. The concentrated extracts were transferred into clean vials containing a recovery standard (20 µl of a 1 ng/µl solution of ^{13}C-pentachlorotoluene and ^{13}C-1,2,3,7,8,9-hexachlorodibenzo-*p*-dioxin in nonane) and extracts were concentrated with a gentle flow of nitrogen to 20 µl ready for analytical determination.

Instrumental analysis of the organochlorine pesticides and metabolites was performed by using a high resolution gas chromatograph coupled to a high resolution mass spectrometer (HRGC–MS). An Agilent 6890 GC equipped with a capillary column (Rtx-Dioxin2, 40 m, 0.18 mm ID, 0.18 µm film thickness, Restek) was used. The temperature program was 60°C (1.5 min), 25°C/min to 140°C (0 min), 8°C/min to 300°C (20 min). 0.5 µl was injected using an autosampler (A200S, CTC) in pulsed splitless mode by a cold injection system CIS 4 (Gerstel). The temperature programme for the injector was: 120°C, 12°C/s, 280°C, 5 min. The carrier gas was helium in a constant flow of 1.3 ml/min. The temperature at the transferline was 300°C. The measurement was conducted with a Finnigan MAT 95S mass spectrometer (Thermo) with a resolution of 10 000. The ionisation mode was EI at 50 eV and 260°C and the detection was by using the selected ion monitoring (SIM) mode. The two most intense ions of the molecular ion cluster were monitored for the analytes and labelled standards. The identification and quantification criteria included confirmation of retention times, relative retention times and isotope ratios for the labelled standards and respective analytes. The mass fragment with the highest intensity of the molecular ion was used for quantification while the other was used as a ratio mass (Schramm et al., 2008).

Analytical quality assurance and control

Separate tools were used to collect different samples from different depths and points. Tools to be reused were thoroughly cleaned with water and soap and rinsed with dichloromethane and acetone. The labelled pesticide standards were of over 99% certified purity (obtained from Dr. Ehrenstorfer, Augsburg, Germany). The standard pesticide solutions and samples were stored in glass-stoppered flasks or vials in deep freezers at –28°C. All organic solvents were of picograde quality (obtained from LGC Promochem, Wesel, Germany). After use, all glassware and tools were rinsed with a technical mixture of toluene, acetone and hexane, and washed with water and detergent in a washing machine. Thereafter, the glassware were dried in an oven overnight at programmed temperatures up to a maximum temperature of 450°C. The silica gel was heated overnight at 550°C to reduce background levels. Analysis of blanks, certified reference materials and recovery tests were used to check contamination and performance of the method. No significant peaks appeared in the chromatograms of the blanks. Recoveries of labelled internal standards varied between 51 and 120% with average recoveries ranging from 76 to 112% and the

Table 1a. Minimum, maximum and mean concentrations and detection frequencies of pesticides and degradation products in soil at Tengeru -Plant Health Services Office.

Compound	Concentration (mg/kg dw), n = 54		Detection frequency (%)
	Min–Max	Mean	
4,4'-DDT	0.0014–4800	110	100
2,4'-DDT	0.0004–435	11	100
4,4'-DDD	0.0002–88	3.3	100
2,4'-DDD	nd–9.0	0.3	94.4
4,4'-DDE	0.00014–110	3.3	100
2,4'-DDE	nd–10	0.24	98.1
Total DDT	0.0024–5410	124	100
α-HCH	0.0001–5000	92.4	100
β-HCH	0.0001–18000	324	100
γ-HCH	nd–13000	232.2	98
δ-HCH	0.0001–3200	59.1	100
ε-HCH	nd–3011	56	98
Total HCH	0.0004–42200	764	100
Pentachlorobenzene	nd–52	1.0	92.6
Hexachlorobenzene	nd–2.0	0.041	98
Pentachloroanisole	nd–0.082	0.002	74.1
trans-Chlordane	nd–0.061	0.002	50
cis-Chlordane	nd–0.019	0.0007	32
oxy-Chlordane	nd	nd	0
Heptachlor	nd–0.094	0.0018	42.6
cis-Heptachloroepoxide	nd–0.00004	0.00001	59.3
trans-Heptachloroepoxide	nd–0.0002	0.00002	27.8
Aldrin	nd–2.2	0.04	90.7
Dieldrin	0.00004–20.2	0.4	100
Endrin	nd–0.09	0.002	61.1
Endosulfan-I	nd–0.31	0.01	70.4
Endosulfan-II	nd–0.21	0.008	74.1
Methoxychlor	nd–0.0003	0.00001	64.8
Mirex	nd	nd	0

nd = not detected, dw = dry weight.

coefficients of variation were in the range of 4 to 19%. The detection limit was defined as three times the average noise value measured. The limits of detection ranged from 0.001 to 0.005 ng/g. The limit of quantification was three times the limit of detection and every signal below this limit was treated as not detectable.

RESULTS AND DISCUSSION

Composition of pesticides and degradation products

The concentrations of the compounds detected in soil samples are presented in Tables 1 to 4. The major contaminants detected in soil samples included DDT isomers and their major degradation products (dichlorodiphenyldichloroethane (DDD) and dichlorodiphenyldichloroethylene (DDE), HCH isomers (α-HCH, β-HCH, γ-HCH, δ-HCH and ε-HCH), aldrin, dieldrin, endrin, endosulfans, pentachlorobenzene, hexachlorobenzene, chlordanes, and heptachlors.

DDT, DDE and DDD

The highest concentrations of total DDT (4,4'-DDT + 2,4'-DDT + 4,4'-DDD + 2,4'-DDD + 4,4'-DDE + 2,4'-DDE) in samples varied from 5.2 mg/kg dw (ACU maize farm) to 5410 mg/kg dw (Tengeru). The highest concentrations of DDT isomers and degradation products ranged as follows: Tengeru 9.0 mg/kg dw (2,4'-DDD) to 4800 mg/kg dw (4,4'-DDT), ACU maize farm 0.035 mg/kg dw (2,4'-DDE) to 3.0 mg/kg dw (4,4'-DDT), Akheri vegetable garden 1.1 mg/kg dw (2,4'-DDE) to 70 mg/kg dw (4,4'-DDT), Mbimba 4.0 mg/kg dw (2,4'-DDE) to 45 mg/kg dw (4,4'-DDT) and MBOCU 6.1 mg/kg dw (2,4'-DDD) to

Table 1b. Distribution of the concentrations of DDT, DDD and DDE in soil samples at Tengeru.

Compounds	Concentrations (mg/kg dw)						
	Burial point (2 sampling points)						
	30 cm, n = 2		75 cm, n = 2		150 cm, n = 2		220 cm
	Min–Max	Mean	Min–Max	Mean	Min–Max	Mean	n = 1
4,4'-DDT	0.015–47	24	900.4–4800	2850	0.41–0.83	0.62	3.022
2,4'-DDT	0.001–4.4	2.201	150–435	293	0.018–0.075	0.047	0.4
4,4'-DDD	0.0002–2.520	1.3	87–88	87.5	0.018–0.0553	0.037	0.046
2,4'-DDD	0.00003–0.634	0.32	4.3–9.0	6.7	0.0004–0.0231	0.012	0.005
4,4'-DDE	0.005–3.51	1.8	59–110	85	0.0683–0.0934	0.081	0.078
2,4'-DDE	0.0001–0.12	0.06	2.81–10	6.4	0.0022–0.003	0.003	0.0053
Total DDT	0.021–58	29	1210–5410	3310	0.542–1.10	0.821	4.0
(DDE+DDD)/DDT	0.1–0.3	0.2	0.04–0.2	0.12	0.2–0.3	0.3	0.04
	Down slope (10–60 m from burial point, 8 sampling points)						
	30 cm, n = 8		75 cm, n = 8		150 cm, n = 8		220 cm
	Min–Max	Mean	Min–Max	Mean	Min–Max	Mean	
4,4'-DDT	0.026–1.3	0.240	0.008–0.464	0.13	0.006–0.900	0.18	na
2,4'-DDT	0.0023–0.12	0.026	0.002–0.056	0.019	0.0013–0.11	0.0252	na
4,4'-DDD	0.0007–0.083	0.016	0.0011–0.19	0.031	0.0006–0.02	0.0042	na
2,4'-DDD	0.0001–0.0053	0.0011	0.0003–0.032	0.0052	0.0003–0.0023	0.0007	na
4,4'-DDE	0.0057–2.0	0.533	0.0011–0.0272	0.0132	0.002–0.014	0.0058	na
2,4'-DDE	0.0001–0.053	0.0092	0.0001–0.0012	0.0005	0.0001–0.0006	0.0003	na
Total DDT	0.041–3.41	0.80	0.013–0.56	0.20	0.011–1.0	0.21	na
(DDE+DDD)/DDT	0.1–6.9	1.8	0.1–0.9	0.4	0.03–0.5	0.2	na
	Up slope (5–20 m from burial point, 9 sampling points)						
	30 cm, n = 9		75 cm, n = 9		150 cm, n = 9		220 cm
	Min–Max	Mean	Min–Max	Mean	Min–Max	Mean	
4,4'-DDT	0.0014–0.083	0.023	0.0052–0.0454	0.0223	0.003–0.063	0.0174	na
2,4'-DDT	0.0004–0.011	0.003	0.0007–0.0084	0.0041	0.0005–0.0133	0.0035	na
4,4'-DDD	0.0002–0.0023	0.0011	0.0005–0.0021	0.0013	0.0008–0.002	0.0011	na
2,4'-DDD	nd–0.0004	0.0002	0.0001–0.0006	0.0003	0.0002–0.00042	0.0003	na
4,4'-DDE	0.0002–0.09	0.023	0.0011–0.046	0.0077	0.0008–0.0033	0.002	na
2,4'-DDE	nd–0.0011	0.0002	0.00003–0.001	0.0002	0.00004–0.0002	0.0001	na
Total DDT	0.003–0.18	0.05	0.0081–0.0742	0.036	0.0052–0.08	0.025	na
(DDE+DDD)/DDT	0.1–2.0	0.8	0.1–1.9	0.4	0.04–0.7	0.3	na

nd = not detected; na = not analyzed.

324 mg/kg dw (4,4'-DDT). The average concentrations were in the following orders:

Tengeru: 4,4'-DDT > 2,4'-DDT > 4,4'-DDD ≈ 4,4'-DDE > 2,4'-DDD ≈ 2,4'-DDE
ACU maize farm: 4,4'-DDT > 2,4'-DDT > 4,4'-DDD ≈ 4,4'-DDE > 2,4'-DDD > 2,4'-DDE
Akheri garden: 4,4'-DDT > 2,4'-DDT > 4,4'-DDD > 2,4'-DDD ≈ 4,4'-DDE > 2,4'-DDE
Mbimba: 4,4'-DDT > 2,4'-DDT > 4,4'-DDE > 4,4'-DDD >2,4'-DDD > 2,4'-DDE
MBOCU: 4,4'-DDT >4,4'-DDE > 2,4'-DDT > 4,4'-DDD >2,4'-DDD ≈ 2,4'-DDE

The compositions of total DDT, that is, the relative distribution of DDT isomers and their degradation products in total DDT in soil at different sampling sites, showed that 4,4'-DDT was the dominant contaminant in the soil samples. The concentrations of 4,4'-DDT were greater than those of 2,4'-DDT in all samples. This indicated that the main source of contamination in the samples from the studied sites was technical DDT. The ratio of 2,4'-DDT/4,4'-DDT can be used to distinguish DDT pollution caused by technical DDT from that due to dicofol. Dicofol would produce a higher 2,4'-DDT/4,4'-DDT ratio than technical DDT. Dicofol is known to contain approximately 3 to 7% DDTs as impurities (Qiu et al.,2005).

Table 1c. Distribution of the concentrations of HCH isomers in soil samples at Tengeru.

Compounds	Concentrations (mg/kg dw)						
	Burial point (2 sampling points)						
	30 cm, n = 2		**75 cm, n = 2**		**150 cm, n = 2**		**220 cm**
	Min–Max	**Mean**	**Min–Max**	**Mean**	**Min–Max**	**Mean**	**n = 1**
α-HCH	0.0041–5000	2500	0.113–2.44	1.3	0.0027–0.0113	0.007	0.034
β-HCH	0.0052–18000	9000	0.0743–74	37	0.0087–0.147	0.078	0.274
γ-HCH	0.003–13000	6500	nd–37.2	19	0.0021–0.007	0.0046	0.041
δ-HCH	0.0121–3200	1600	0.036–9.0	4.52	0.0034–0.0631	0.0333	0.076
ε-HCH	0.003–3011	1510	nd–1.24	0.622	0.0014–0.033	0.0172	0.20
Total HCH	0.0273–42200	21100	0.223–123.3	62	0.0183–0.261	0.14	0.623
α-/γ-HCH	0.4–1.4	0.9	0.1–15	8	1.3–1.6	1.5	0.8
	Down slope (10–60 m from burial point, 8 sampling points)						
	30 cm, n = 8		**75 cm, n = 8**		**150 cm, n = 8**		**220 cm**
	Min–Max	**Mean**	**Min–Max**	**Mean**	**Min–Max**	**Mean**	
α-HCH	0.0002–2.4	0.302	0.00021–0.08	0.014	0.0003–1.0	0.15	na
β-HCH	0.0001–2.42	0.472	0.0004–0.0752	0.015	0.0005–0.173	0.0443	na
γ-HCH	0.0001–0.204	0.0263	0.0001–0.0085	0.0017	0.0001–0.009	0.0015	na
δ-HCH	0.0002–0.5	0.0673	0.00032–0.10	0.015	0.0003–0.0019	0.0007	na
ε-HCH	0.00003–0.17	0.0333	nd–0.027	0.0041	0.0001–0.014	0.0047	na
Total HCH	0.001–4.5	0.9	0.0017–0.29	0.049	0.0016–1.20	0.20	na
α-/γ-HCH	1.2–12	4.0	0.7–33	8.4	1.7–110	23	na
	Up slope (5–20 m from burial point, 9 sampling points)						
	30 cm, n = 9		**75 cm, n = 9**		**150 cm, n = 9**		**220 cm**
	Min–Max	**Mean**	**Min–Max**	**Mean**	**Min–Max**	**Mean**	
α-HCH	0.0003–0.013	0.0031	0.0001–0.0077	0.0012	0.0001–0.0083	0.0015	na
β-HCH	0.0004–0.0453	0.0082	0.0003–0.0054	0.0019	0.00011–0.0049	0.001	na
γ-HCH	0.0001–0.016	0.0026	0.00005–0.0006	0.0003	0.00004–0.0003	0.0001	na
δ-HCH	0.0001–0.201	0.027	0.0002–0.0076	0.0013	0.0001–0.0007	0.0002	na
ε-HCH	0.00004–0.047	0.0066	0.00004–0.001	0.0002	0.00002–0.0004	0.0001	na
Total HCH	0.0014–0.322	0.048	0.001–0.014	0.005	0.0005–0.015	0.003	na
α-/γ-HCH	0.8–3.8	1.9	0.5–29	5.1	1.1–28	8.1	na

nd = not detected; na = not analyzed.

DDE and DDD are the two main products of DDT dechlorination. DDT is dechlorinated to DDE under aerobic conditions and reductively dechlorinated to DDD under anaerobic conditions. The ratios of (DDE + DDD)/DDT were employed to assess the significance of degradation (transformation) of DDT in the soil. A high value of (DDE + DDD)/DDT ratio (> 1) indicates aged and microbiologically significantly degraded DDT, while a small value (< 1) indicates non-significantly degraded DDT or recent input (Zhang et al., 2006).

The (DDE+DDD)/DDT ratios were generally low (<1) in most soil samples from Tengeru, ACU maize farm, and all samples from Akheri vegetable garden and Mbimba, suggesting that no significant degradation had occurred to the DDT residues at these sites. The high (DDE+DDD)/DDT ratios (>1) in some soil samples from Tengeru, MBOCU and ACU maize farm indicated that the DDT residues had been significantly transformed into their degradation products at those points.

Hexachlorocyclohexanes

The highest concentrations of total HCH (α-HCH + β-HCH + γ-HCH + δ-HCH + ε-HCH) varied from 1.4 mg/kg dw (ACU maize farm) to 42200 mg/kg dw (Tengeru). The concentrations of the HCH isomers were up to 18000 mg/kg dw (β-HCH, Tengeru).

The composition of total HCH and the average concentrations of HCH isomers revealed a heterogeneous nature in soil samples from Tengeru, ACU maize farm and Akheri vegetable garden. This may be related to the isomerization of HCH isomers during transformation process in soil as well as the differences

Table 2. Concentrations of pesticides in soil at ACU maize farm.

Compounds	Concentration (mg/kg dw), burial point and 5 –10 m away					
	10 cm, n = 13		30 cm, n = 9		75 cm, n = 5	
	Min-Max	Mean	Min-Max	Mean	Min-Max	Mean
4,4'-DDT	0.048–3.0	0.59	0.008–1.403	0.40	0.016–0.59	0.313
2,4'-DDT	0.011–0.95	0.20	0.002–1.10	0.34	0.004–0.30	0.122
4,4'-DDD	0.0069–0.72	0.15	0.0006–0.33	0.065	0.0014–0.11	0.038
2,4'-DDD	0.0025–0.212	0.044	0.0002–0.122	0.028	0.0005–0.059	0.02
4,4'-DDE	0.0124–0.301	0.10	0.0041–0.25	0.061	0.0023–0.068	0.035
2,4'-DDE	0.0004–0.031	0.01	0.0002–0.035	0.012	0.0001–0.0131	0.0072
Total DDT	0.101–5.20	1.10	0.022–3.20	0.90	0.0244–1.0	0.534
(DDE+DDD)/DDT	0.2–0.9	0.5	0.14–1.3	0.4	0.12–0.4	0.22
α-HCH	0.001–0.36	0.083	0.00031–0.033	0.01	0.0015–1.20	0.241
β-HCH	0.0003–0.083	0.027	0.0002–0.054	0.016	0.001–0.213	0.047
γ-HCH	0.0011–0.052	0.011	0.0001–0.014	0.0051	0.0007–0.011	0.0045
δ-HCH	0.0002–0.23	0.045	0.0001–0.054	0.0077	0.0003–0.0054	0.0032
ε-HCH	nd–0.0063	0.0005	nd–0.0003	0.00003	nd	nd
Total HCH	0.0034–0.64	0.17	0.001–0.13	0.039	0.0034–1.40	0.295
α-HCH/γ-HCH	0.04–91	11.1	0.3–15	4.3	1.1–110	24
Pentachlorobenzene	0.0004–0.0281	0.007	0.00012–0.003	0.001	0.00011–0.0014	0.0005
Hexachlorobenzene	0.0091–2.40	0.51	0.0017–0.3	0.054	0.0031–0.019	0.0092
Pentachloroanisole	0.0003–0.0141	0.004	nd–0.004	0.0014	nd–0.004	0.001
trans-Chlordane	nd–0.0021	0.0004	nd–0.00041	0.0001	nd–0.00024	0.0001
cis-Chlordane	nd–0.0018	0.00021	nd–0.0001	0.00002	nd–0.0001	0.00002
oxy-Chlordane	nd	nd	nd	nd	nd	nd
Heptachlor	nd–0.043	0.005	nd–0.0006	0.0001	nd–0.00012	0.00004
cis-Heptachloroepoxide	nd–0.0001	0.00001	nd–0.0103	0.002	nd–0.00002	0.00001
trans-Heptachloroepoxide	nd–0.00051	0.0001	nd–0.061	0.012	nd	nd
Aldrin	0.0022–0.70	0.063	0.00013–0.016	0.0034	0.0002–0.0032	0.0012
Dieldrin	0.111–39.3	5.60	0.0009–2.0	0.431	0.0271–0.182	0.0763
Endrin	nd–0.830	0.086	nd–0.023	0.006	nd–0.0045	0.0015
Endosulfan-I	0.0057–17	1.62	nd–0.204	0.0392	nd–0.03	0.0086
Endosulfan-II	0.01–20	2.52	0.0003–0.45	0.12	0.0038–0.075	0.024
Methoxychlor	nd–0.00072	0.0001	nd–0.0007	0.0001	nd–0.0006	0.00012
Mirex	nd–0.0012	0.0002	nd–0.0001	0.00001	nd	nd

in physico-chemical properties and degradation rates (ATSDR, 2005). β-HCH was the dominant HCH isomer in soil samples from Tengeru. This can be explained by the fact that β-HCH is the most stable and persistent in soil. The equatorial configuration of the chlorine atoms in the molecular structure of β-HCH confers the greatest metabolic stability to this isomer. The persistence of β-HCH in soils is also due to higher K_{ow} and lower vapour pressure than other isomers, which enhance its adsorption to the soil and reduce loss by evaporation from the soils (Willett et al., 1998). β-HCH was also found to be the dominant HCH isomer in some studies (Concha-Grana et al., 2006; Wang et al., 2006; Zhang et al., 2006). The α-isomer was the predominant HCH isomer in all samples from Mbimba, indicating technical HCH as the source of contamination. The γ-isomer was the predominant HCH isomer in all samples from MBOCU, indicating lindane as the main source of contamination. The α-HCH/γ-HCH ratio can be used to identify the source of HCHs contamination. The ratios of α-HCH/γ-HCH are in the range of 4 to 15 for technical mixtures, and almost zero for lindane, or range from 0.2 to 1 due to transformation of γ-HCH into α-HCH. Technical HCHs mixture contains the isomers in the following percentages: 60-70% α-HCH, 5-12% β-HCH, 10-15% γ-HCH, 6-10% δ-HCH and 3-4% ε-HCH, while lindane contains >99% γ-HCH (ATSDR, 2005; Willett et al., 1998; Walker et al., 1999; Gong et al., 2004). The α-HCH/γ-HCH ratios were generally low in most samples from Tengeru, Akheri vegetable garden and all samples from MBOCU, indicating lindane as the main source of HCH contamination. High ratios may be related to

Table 3. Concentrations of pesticides and degradation products in soil at Akheri vegetable garden.

Compounds	Concentrations (mg/kg dw)						
	Burial point	Down slope, 5 m		Down slope, 20 m		Up slope, 5–20 m, n = 5	
Depth (cm)	50	10	50	10	50	10	30
4,4'-DDT	70	44	0.045	0.10	0.015	0.012–0.038	0.006–0.047
2,4'-DDT	54	40	0.014	0.025	0.0031	0.0025–0.012	0.0022–0.0102
4,4'-DDD	18	15	0.015	0.013	0.0014	0.001–0.0021	0.0007–0.0014
2,4'-DDD	7.2	6.3	0.0044	0.0044	0.0006	0.0003–0.0013	0.0004
4,4'-DDE	5.7	4.75	0.0052	0.0204	0.0016	0.0015–0.0112	0.0005–0.0023
2,4'-DDE	1.10	0.78	0.0005	0.002	0.0003	0.0001–0.0003	0.0001–0.0003
Total DDT	160	110	0.084	0.165	0.022	0.019–0.058	0.010–0.062
(DDE+DDD/DDT)	0.3	0.4	0.4	0.3	0.2	0.2–0.5	0.1–0.3
α-HCH	nd	0.924	0.0005	0.0022	0.00024	0.0001–0.0016	0.0004–0.0055
β-HCH	nd	nd	0.0007	0.0023	0.00024	0.00013–0.0012	0.00022–0.0034
γ-HCH	4.10	1.51	0.02	0.015	0.0016	0.0002–0.0011	0.0001–0.0015
δ-HCH	1.70	20	0.0012	0.0021	0.0001	0.0001–0.00013	0.0001–0.0006
ε-HCH	nd	nd	nd	nd	nd	nd	nd–0.0002
Total HCH	5.80	23	0.023	0.022	0.0026	0.0007–0.0025	0.0008–0.01
α-HCH/γ-HCH	0	0.6	0.03	0.1	0.2	0.2–10	0.5–11.3
Pentachlorobenzene	56.2	5.50	0.0083	0.08	0.0008	0.0005–0.0014	0.0002–0.0003
Hexachlorobenzene	2300	310	0.83	6.30	0.0333	0.021–0.12	0.0024–0.014
Pentachloroanisole	1.0	0.61	0.0023	0.0734	0.0011	0.0004–0.0022	0.0003–0.001
trans-Chlordane	nd	nd	0.0001	0.0003	nd	nd–0.0001	nd–0.0001
cis-Chlordane	1.5	nd	0.00011	0.0001	nd	nd–0.00003	nd–0.00004
oxy-Chlordane	nd	nd	nd	nd	nd	nd	nd
Heptachlor	nd	1.6	nd	0.0004	0.0001	nd–0.00012	nd–0.0002
cis-Heptachloroepoxide	0.36	0.35	0.00003	0.00002	nd	0.00001–0.00003	0.00001–0.00003
trans-Heptachloroepoxide	1.90	2.0	nd	nd	nd	nd–0.0002	nd–0.00014
Aldrin	6.20	0.42	0.0045	0.01	0.0003	0.0013–0.011	0.0004–0.001
Dieldrin	920	170	3.0	4.314	0.12	0.143–3.51	0.0048–0.096
Endrin	35	3.23	0.049	0.093	0.0013	0.0005–0.029	0.00002–0.0004
Endosulfan-I	4340	1100	3.0	2.50	0.042	0.0065–0.252	0.0002–0.01
Endosulfan-II	2820	800	2.50	3.10	0.049	0.022–0.25	0.0007–0.011
Methoxychlor	0.067	0.039	nd	nd	nd	nd	nd
Mirex	0.162	nd	0.0003	0.0005	nd	nd–0.00011	nd

transformation of γ-HCH into α-HCH. Both low and high α-HCH/γ-HCH ratios were found in samples from ACU maize farm and Mbimba, suggesting both lindane and technical HCH as the sources of HCH contamination. The high α-HCH/γ-HCH ratios found in this study are comparable to the findings from other studies in which the contamination was related to technical HCH. For example, α-HCH/γ-HCH ratios of up to 52 have been reported (Concha-Grana et al., 2006; Zhang et al., 2006).

Endosulfans

The highest concentrations of endosulfan-I and endosulfan-II ranged 0.31–4340 mg/kg dw and 0.21–2820 mg/kg dw, respectively and the total concentrations of endosulfans were up to 7200 mg/kg dw (Akheri vegetable garden). Endosulfan was mainly represented as endosulfan-I in most soil samples from Tengeru, MBOCU (10 cm) and some samples from Akheri vegetable garden (burial point and nearby points), which was similar to the technical formulation, suggesting lack of significant degradation or recent inputs. The concentrations of endosulfan-II were greater than those of endosulfan-I in all samples from ACU maize farm, Mbimba, most samples from Akheri vegetable garden, and some samples from MBOCU (50 cm, 100 cm and 10 m from burial point), which differs from the technical formulation and can be related to old input and the faster degradation of endosulfan-I than endosulfan-II, in soils. The half-lives for endosulfan-I and endosulfan-II are about 60 and 800 days, respectively. Another possible

Table 4. Concentrations of pesticides and degradation products in soil at TaCRI-Mbimba and MBOCU (mg/kg dw).

Sample site	TaCRI-Mbimba			MBOCU			
Sample point	Burial point		10 m from burial point	Burial point			10 m from burial point
Depth (cm)	50	100	50	10	50	100	10
4,4'-DDT	17	45	0.023	324	2.20	12	1.30
2,4'-DDT	7.10	24	0.01	76	0.41	2.30	0.30
4,4'-DDD	10.3	10	0.001	19	0.22	2.70	0.20
2,4'-DDD	4.40	5.30	0.001	6.10	0.1	0.1	0.1
4,4'-DDE	6.0	21	0.004	123	3.10	2.90	2.60
2,4'-DDE	0.80	4.0	0.001	6.20	0.10	0.10	0.03
Total DDT	45.6	110	0.04	554	6.10	20	4.50
(DDE+DDD)/DDT	0.9	0.6	0.2	0.4	1.3	0.4	1.9
α-HCH	0.90	9.0	0.001	nd	0.08	0.01	0.004
β-HCH	0.40	0.80	0.0001	nd	0.01	0.01	0.004
γ-HCH	0.60	1.0	0.00012	11	0.05	0.11	0.012
δ-HCH	0.50	0.70	0.0001	nd	0.01	0.01	0.01
ε-HCH	nd	nd	nd	nd	nd	nd	nd
Total HCH	2.40	12	0.0013	11	0.15	0.14	0.03
α-HCH/γ-HCH	1.5	8.9	8.3	0	1.6	0.1	0.3
Pentachlorobenzene	11	1.30	0.0001	6.0	0.05	0.10	0.01
Hexachlorobenzene	16	4.20	0.001	1003	0.50	0.50	0.10
Pentachloroanisole	1.0	0.70	0.00004	1.10	0.05	0.01	0.02
trans-Chlordane	nd	29	0.0004	nd	nd	nd	nd
cis-Chlordane	nd	25	0.001	nd	nd	nd	nd
oxy-Chlordane	nd	nd	nd	nd	nd	nd	nd
Heptachlor	nd	0.50	nd	nd	nd	nd	nd
cis-Heptachloroepoxide	nd	1.10	0.0001	nd	0.003	nd	nd
trans-Heptachloroepoxide	nd	nd	nd	nd	nd	nd	nd
Aldrin	1.40	2.0	0.0002	1.50	0.10	0.01	0.04
Dieldrin	18	480	0.02	602	9.50	3.0	7.10
Endrin	0.70	8.0	0.0003	18	0.03	0.02	0.01
Endosulfan-I	1.20	3.50	0.0002	2100	0.62	0.04	0.04
Endosulfan-II	1.30	6.0	0.0003	1400	1.0	0.30	0.20
Methoxychlor	nd	0.10	nd	nd	nd	nd	nd
Mirex	nd	nd	nd	nd	nd	nd	nd

reason is that endosulfan-I undergoes isomerization to endosulfan-II especially at high temperature or when exposed to light (ATSDR, 2000).

Aldrin, dieldrin and endrin

The highest concentrations of aldrin, dieldrin and endrin were as follows, respectively: Tengeru 2.2, 20.2 and 0.09 mg/kg dw; Akheri 6.2, 920 and 35 mg/kg dw; ACU maize farm 0.7, 39.3 and 0.83 mg/kg dw; Mbimba 2.0, 480 and 8 mg/kg dw, and MBOCU 1.5, 602 and 18 mg/kg dw. The concentrations of dieldrin were greater than those of aldrin and endrin in all samples from all the studied sites. This can partly be explained by the fact that, apart from being a pesticide, dieldrin is also formed as a metabolite of aldrin (FAO, 2000; ATSDR, 2002).

Hexachlorobenzene, pentachlorobenzene and pentachloroanisole

Hexachlorobenzene, pentachlorobenzene and pentachloroanisole were detected in 74 to 100% of the soil samples from the studied sites. The highest concentrations for these compounds varied from 2 to 2300, 0.028 to 56.2 and 0.014 to 1.1 mg/kg dw, respectively. The highest concentrations were generally found in Akheri samples and the lowest were found in Tengeru and ACU samples. Pentachloroanisole is the main degradation product of pentachlorophenol and pentachloronitrobenzene (UNECE, 2009); therefore, its detection indicated contamination due to these compounds.

Chlordanes and heptachlors

Chlordane isomers, heptachlor and its degradation products were detected in soil samples mainly from Akheri and Mbimba. The highest concentrations of the compounds in Akheri samples were: *cis*-chlordane 1.5 mg/kg dw, *trans*-chlordane 0.0003 mg/kg dw, heptachlor 1.6 mg/kg dw, *cis*-heptachloroepoxide 0.36 mg/kg dw and *trans*-heptachloroepoxide 2 mg/kg dw. The highest concentrations of *trans*-chlordane, *cis*-chlordane, heptachlor and *cis*-heptachloroepoxide in samples from Mbimba were 29, 25, 0.5 and 1.1 mg/kg dw, respectively. The proportions of the pesticides and their degradation products suggested lack of effective degradation.

Methoxychlor and mirex

Methoxychlor and mirex were detected in some of the samples. Their highest concentrations varied from 0.0003 mg/kg dw (Tengeru) to 0.162 mg/kg dw (Akheri). Their concentrations were generally low suggesting that the contamination status of these compounds was attributed to environmental sources, that is, there were no significant sources at the studied sites.

Distribution of the pesticides and degradation products

There are strong positive correlations in the concentrations of the compounds at each studied site, indicating common sources. There were significant differences in the concentrations of the main pesticides and degradation products among the sampling depths in soil at Tengeru burial points, whereby the distribution for DDT, DDE and DDD was 75 cm > 30 cm > 150 cm ≈ 220 cm, while for HCH isomers, aldrin, dieldrin, pentachlorobenzene and hexachlorobenzene was 30 cm > 75 cm > 150 cm ≈ 220 cm ($p < 0.05$). The distribution pattern for DDT and DDE was slightly different from that observed in a study which compared the concentrations of DDT and DDE in soil samples collected at 30-40, 60-70 and 90-100 cm depths from an undisturbed contaminated site in Mato Grosso, Brazil, in which the highest concentrations at the most contaminated point were found at the 30 to 40 cm depth (Villa et al., 2006). This fact is explained because some years ago the affected area was filled up with foreign soil and turned over and then the original deposit was at different depth depending on the sampling point. The concentrations of pesticides and metabolites showed significant differences for some of the points and depths at Akheri (10 cm > 50 cm), ACU maize farm (10 cm > 30 cm ≈ 75 cm-except fo DDTs and HCHs) and MBOCU (10 cm > 50 cm). These findings suggested that the vertical movement of the compounds was very slow maybe due to their low wate solubility and their strong interactions with so components (FAO, 2000). There were no significan differences in the concentrations of the mai contaminants among the sampling depths at the followin sites/points: Tengeru up and down slope (30 cm ≈ 75 cm ≈ 150 cm), ACU maize farm (10 cm ≈ 30 cm ≈ 75 cm DDTs and HCHs) and Akheri (10 cm ≈ 30 cm) ($p > 0.05$ indicating even distribution of the compounds among the sampling depths. The concentrations of the pesticides and degradation products increased with depth in soi samples from Mbimba and MBOCU (100 cm > 50 cm and some sampling points at Tengeru (150 cm > 75 cm and 30 cm), suggesting accumulation at lower depths despite that the differences were not quite significant.

The highest concentrations of the pesticides and degradation products in soil were found at the sampling points which were located within the burial areas and the concentrations showed a tendency of decreasing with increase in distance from those points. The concentrations in samples collected from points that were located down slope were significantly greater than those from up slope points, indicating that the distribution of the compounds was influenced by run-off, although this was only limited to short distances.

The distribution of the pesticides and degradation products in soil at the studied sites had been influenced by the management operations (e.g. turning over of soil, mixing with foreign soil and burying), agricultura operations (e.g. cultivation and irrigation activities) carried out especially at Tengeru, ACU maize farm and Akheri vegetable garden, as well as the environmenta conditions (e.g. soil properties) and the physico-chemica properties of the compounds.

The results of the present study are comparable to the data from different locations, although the sampling depths are slightly different. The findings from other studies have generally shown that the concentrations of most of the organochlorine compounds decrease with depth, while fluctuations or lack of direct correlation between concentrations and depths are observed in disturbed soils such as cultivated soils. Wang et al. (2006) reported that the concentrations of most organochlorine compounds (HCHs and DDTs) decreased with depth, while fluctuations were observed in the plow layers of some cultivated soils due to some cultivation activities and batch irrigation. The sampling depths were 0-20, 20-25, 25-30, 30-40, 40-50, 50-60, 60-80 and 80-100 cm. Another study reported that at several points the highest concentration of HCH isomers was found in the superficial samples, whereas the more contaminated sample was at 10 cm depth and in others between 36 and 48 cm, despite that no significant difference was found among the sampling depths (0-20 to 85–95 cm)

(Concha-Grana et al., 2006).

Conclusions

High concentrations of organochlorine pesticides and major degradation products were detected in samples from all the studied sites. The results indicated aged contaminants but which had not undergone significant degradation. The concentrations of the compounds generally decreased with increase in depth and distance from the burial points. Therefore, the mobility of these compounds in soil was generally low certainly due to their low water solubility and their strong interactions with soil components, mainly clay and organic matter. However, there were some variations in the distribution of the compounds among the sampling depths as some showed even distribution or accumulation downwards indicating some risks for groundwater contamination. It is recommended that remediation should be carried out at the burial points.

ACKNOWLEDGEMENTS

The German Academic Exchange Service (DAAD) and the University of Dar es Salaam are gratefully acknowledged for funding this work. The late Prof. Michael A. Kishimba is greatly acknowledged for guidance during this study. Sincere thanks to Prof. Dr. Dr. Karl-Werner Schramm, Mr. Bernhard Henkelmann, and the German Research Center for Environmental Health, Munich-Germany for facilitation of the laboratory analyses. Thanks are also due to Mr. Norbert Fischer and Mrs. Silke Bernhoft (Germany) for their technical contributions during the laboratory analyses.

REFERENCES

Åkerblom M (1995). Guidelines for environmental monitoring of pesticide residues in the SADC region, SADC/ ELMS monitoring techniques series, Sweden.

ATSDR (2002). Toxicological profile for aldrin and dieldrin, Atlanta, GA: U.S. Department of Health and Human Services, Public Health Service.

ATSDR (2005). Toxicological profile for hexachlorocyclohexane, Atlanta, GA: U.S. Department of Health and Human Services, Public Health Service.

ATSDR (Agency for Toxic Substances and Disease Registry) (2000). Toxicological profile for endosulfan, Atlanta, GA: U.S. Department of Health and Human Services, Public Health Service.

Buyuksonmez F, Rynk R, Hess TF, Bechinski E (1999). Occurrence, Degradation and Fate of Pesticides During Composting, Compost Sci. Util. 7:4.

Concha-Grana E, Turnes-Carou MI, Muniategui-Lorenzo S, Lopez Mahia P, Prada-Rodriguez D, Fernandez-Fernandez E (2006). Evaluation of HCH isomers and metabolites in soils, leachates, river water and sediments of a highly contaminated area. Chemosphere 64:588-595.

Extension Toxicological Network (EXTOXNET) (1993). Movement of Pesticides in the Environment, Oregon State University.

FAO (1998). Prevention and disposal of obsolete and unwanted pesticides stocks in Africa and the Near East, FAO, Rome.

FAO (2000). Assessing soil contamination – A reference manual, FAO, Rome.

FAO (2001). Baseline study on the problem of obsolete pesticide stocks, FAO, Rome.

FAO (Food and Agriculture Organization of the United Nations) (1996). Disposal of bulk quantities of obsolete pesticides in developing countries, FAO, Rome.

Gałuszka A, Migaszewski ZM, Manecki P (2011). Pesticide burial grounds in Poland: a review. Environ Int. 37(7):1265-1272.

Gong ZM, Xu FL, Dawson R, Cao J, Liu WX, Li BG, Shen WR, Zhang WJ, Qin BP, Sun R, Tao S (2004). Residues of hexachlorocyclohexane isomers and their distribution characteristics in soils in the Tianjin area, China, Arch. Environ. Contam. Toxicol. 46: 432-437.

Harrison SA (1990). The Fate of Pesticides in the Environment, Agrichemical Fact Sheet No.8, Pesticide Education Program, Penn State Cooperative Extension.

Mansour M (1993). Fate and Prediction of Environmental Chemicals in Soils, Plants and Aquatic Systems, Boca Raton: Lewis Publishers, P.291.

Manvelyan E, Simonyan L, Anakhasyan E (2006). Hazards of burial site in Armenia, A report, J. Pestici. Action Netw. UK.

McEwen FL, Stephenson GR (1979). The Use and Significance of Pesticides in the Environment, A Wiley Interscience Publication, New York, P.538.

Pesticide Action Network (PAN UK) (2005). Buried pesticide threat in Belarus, Press release. P.29.

Qiu X, Zhu T, Yao B, Hu J, Hu S (2005). Contribution of dicofol to the current DDT pollution in China, Environ. Sci. Technol. 39:4385-4390.

Rauf VG, Bashkin, VN, Galiulina, RA (2002). Behaviour of persistent organic pollutants in the air-plant-soil system: a review. Water Air Soil Pollut. 137:179-191.

Ritter L, Solomon KR, Forget J, Stemeroff M, C.O'Leary D (1995). Persistent Organic Pollutants (POPs): An Assessment Report. United Nations Environ. Program.

Schramm KW, Henkelmann B, Kotalik J (2008). Standard operating procedures for determination of pesticides, PAHs and PCBs, Accreditated Testing Laboratory According to DIN EN ISO/IEC 17025, Helmholtz Research Centre for Environmental Health. Inst. Ecol. Chem. Munich-Germany.

Sun WJ, Cai JM, Huang B (1997). Study on the residues of pesticides in tea garden soil and tea. J. Fujian Agric. Univ. 26:9-43.

UNECE (United Nations Economic Commission for Europe) (2009) Task Force on Persistent Organic Pollutants (POPs), Summary of expert reviews of pentachlorophenol (PCP), UNECE.

Villa RD, Dores EFG, Carbo L, Cunha MLF (2006). Dissipation of DDT in a heavily contaminated soil in Mato Grosso, Brazil, Chemosphere 64:549-554.

Waldron AC (1992). Pesticides and Ground Water Contamination, Bulletin 820, Ohio Cooperative Extension Service, The Ohio State University.

Walker K, Vallero DA, Lewis RG (1999). Factors influencing the distribution of lindane and other hexachlorocyclohexanes in the environment. Environ. Sci. Technol. 33:4373-4378.

Wang X, Piao X, Chen J, Hu J, Xu F, Tao S (2006). Organochlorine pesticides in soil profiles from Tianjin, China. Chemosphere 64:1514-1520.

Willett KL, Ulrich EM, Hites RA (1998). Differential toxicity and environmental fates of hexachlorocyclohexane isomers: Critical review. Environ. Sci. Technol. 32:2197-2207.

Zhang HB, Luo YM, Zhao QG, Wong MH, Zhang GL (2006). Residues of organochlorine pesticides in Hong Kong soils, Chemosphere 63:633-641.

An assessment of surface water pollution status around Gboko abattoir

S. T. Ubwa[1] , G. H. Atoo[1], J. O. Offem[2], J. Abah[1] and K. Asemave[1]

[1]Department of Chemistry, Benue State University, Makurdi, Nigeria.
[2]Department of Pure and Applied Chemistry, University of Calabar, Cross River State, Nigeria.

An assessment of surface water around Gboko abattoir was conducted to ascertain the pollution status of water around the area. The results showed mean values as: total dissolved solids (TDS), 1026.78 mg/L; TSS, 565.22 mg/L; dissolved oxygen (DO), 5.0 mg/L; phosphate, 8.89 mg/L and biochemical oxygen demand (BOD), 484.64 mg/L. These values were above regulatory standards. The values of; pH (6.6), sulphate (86.91 mg/L), nitrate (41.45 mg/L) and chemical oxygen demand (COD) (903.30 mg/L) fell within WHO maximum permissible limits for drinking water. The mean concentrations of: Pb, 0.2892 ppm; Cr, 0.0598 ppm; Ni, 0.1031 ppm and Cd, 0.0110 ppm were above the control and standard safe limits for: Pb, 0.01 mg/L; Cr, 0.05 mg/L; Ni, 0.02 and 0.07 mg/L and Cd, 0.003 mg/L in water. Zn had mean concentration value of 0.7635 ppm which was below standards but higher than the control water sample. Correlation coefficients revealed positive and significant correlations between the pairs of metals in water. The results of this study showed that the activities at the abattoir were contributing to the pollution load of water in the area. It is therefore recommended that the activities of the abattoir should be monitored closely by relevant agencies in order to prevent full-blown environmental problems and attendance health hazards in the near future.

Key words: Abattoir, surface water, heavy metals, physico-chemical parameters, correlation coefficient.

INTRODUCTION

Environmental pollution has generally become a threat to the existence of mankind and the ecosystem. Some pollution effects may lead to metabolic disorders and undesirable changes which in many cases cause severe injuries and health hazards (Yahaya et al., 2009). Abattoir activities may be another source of pollution since human activities such as animal production and meat processing have been reported to impact negatively on soil and natural water composition leading to pollution of the soil, natural water resources and the entire environment (Adesemoye et al., 2006). It has also been reported that animals which graze on contaminated plants and drink from polluted waters, as well as marine lives that breed in heavy metal polluted waters also accumulate such metals in their tissues and milk if lactating (Yahaya et al., 2009). When such animals are killed, these metals are released in the soil as natural sink but subsequently leached out into nearby streams or water bodies. Abattoir can be defined as a premise approved and registered by controlling authorities for hygienic slaughtering, inspection, processing, effective preservation and storage of meat products for human consumption (Alorge, 1992). However, meat processing activities in Nigeria are generally carried out in unsuitable buildings and by untrained personnel or butchers who are most of the time unaware of sanitary principles (Olanike, 2002). The major activities involved in the operations of an abattoir are: receiving and holding of livestock; slaughter and carcass dressing of animals; chilling of carcass products; carcass boning and packaging; freezing of finished carcass and cartooned product; rendering processes; drying of skins; treatment of wastes and transport of processed material.

Abattoir activities are aimed at optimizing the recovery of edible portions of the meat processing cycle for human consumption. However, significant quantities of secondary waste materials are also generated during this process. For example, blood, fat, organic and inorganic solids, salts and chemicals added during processing operations are produced as wastes (Red Meat Abbattoir Association, 2010; Steffen and Kirsten Inc., 1989).

Various parts of cattle such as muscle, blood, liver, kidney, viscera and hair have been found to contain heavy metals (Kruslin et al., 1999; Jukna et al., 2006). In ruminants, the first stomach or paunch contains undigested materials called paunch manure, which can contain long hairs, whole grain sand large plant fragments. The faeces of livestock (animal manure) consist of undigested food, most of which are: cellulose-fibre; undigested protein; excess nitrogen from digested protein; residue from digested fluids; waste mineral matter; worn-out cells from intestinal linings; mucus and bacteria. Other components of undigested food include; foreign matter such as dirt consumed, calcium, magnesium, iron, phosphorous, sodium, etc (Ezeoha and Ugwuishiwu, 2011). Abattoir effluent wastewater has a complex composition and can be very harmful to the environment. For example, discharge of animal blood into streams would deplete the dissolved oxygen (DO) of the aquatic environment. Improper disposal of paunch manure may exert oxygen demand on the receiving environment or breed large population of decomposers (micro-organisms), some of which may be pathogenic. Also improper disposal of animal faeces may cause oxygen-depletion in the receiving environment. It could also lead to eutrophication of the receiving system and increase rate of toxins accumulation in biological systems (Nwachukwu et al., 2011).

Mohammed and Musa (2012) reported that the improper disposal of abattoir effluent could lead to transmission of pathogens to human which may cause an outbreak of water borne diseases e.g.diarrhoea, pneumonia, typhoid fever, asthma, wool sorter diseases, respiratory and chest diseases, etc. Studies have shown that *Escherichia coli* infection source was reported to be undercooked beef which has been contaminated in abattoirs with faeces containing the bacterium (Bello and Oyedemi, 2009; Patra, 2007). It had also been reported that abattoir activities are responsible for the pollution of surface and underground waters, reduction of air quality as well as quality of health of residents within the surrounding environment (Katarzyna et al., 2009; Odoemelan and Ajunwa, 2008).

The main purpose of this work therefore, is to study the pollution status of surface water around Gboko abattoir and assess whether the pollution load is sufficient to affect the health of the inhabitants of the area who depend on this stream as their source of domestic and irrigation water. The results of the study will assist the regulatory bodies monitor more closely the activities at the abattoir as well as create public awareness about the health implications of abattoir activities on the environment and also establish a data bank for future reference.

MATERIALS AND METHODS

Study area

Gboko is one of the largest and most populous Local Government Areas in Benue State. It has a land mass of: Area, 1,835 km^2; Density, 196.9 inh./km^2 with a population of 361, 325 people according to National Population Commission census of (2006). It is bounded by Tarka Local Government on the North, Ushongo Local Government to the south, Buruku Local Government on the east, Gwer on the west. It lies between latitude 7°05′–7°31′N and longitude 9°13′–9°35′E in the savannah region of Nigeria with typical savannah vegetation and climate.

The integrated Gboko abattoir is located on km 8 Gboko-Aliade road and close to a stream. Several animals (cows, goats, sheep and pigs) are slaughtered in this abattoir. Normal abattoir operations are carried out every week from Monday to Saturday during morning hours (6 am -11 am) (Figure 1).

Sample collection

Six sampling stations were mapped out along the course of the stream in the abattoir area at a distance of 50 m from each other. The sampling stations were coded AW, BW, CW, DW, EW and FW. Six replicate samples were collected from each of these stations and pooled together to obtain a representative sample for that station. A water sample (coded GWctrl) was collected at a point 60 m upstream, and served as a control. Water samples were collected in plastic containers previously cleaned by washing in non-ionic (Omo) detergent. They were rinsed with tap water and thereafter soaked in 10% HNO_3 for 24 h and finally rinsed with de-ionized water and ready for use (Akan, 2010). During sampling, sample bottles were first rinsed with the sampled water three times and then filled to the brim. The samples were labeled and transported to the laboratory, stored in a refrigerator at about 4°C prior to analysis. A total of forty two samples were collected for investigation. The field investigation was carried out between the months of July and August (peak of rainy season) during which the waste from abattoir was expected to leach or drain down to the stream.

Sample preparation and analysis

Each sample (100 ml) was transferred into a beaker and 5 ml of concentrated HNO_3 was added. The beaker with the content was placed on a hot plate and evaporated down to about 20 ml. The beaker was cooled and another 5 ml of concentrated HNO_3 added. Each beaker was then covered with a watch glass and returned to the hot plate for more heating with the addition of few drops of HNO_3 until the solution appeared light coloured and clear. The walls of the beaker and the watch glass were washed down with distilled water and the sample filtered to remove insoluble materials that could clog the atomizer. The volumes of the samples were made up to the mark (100 ml) with distilled water (Radojevic and Bashkin, 1999). A blank sample was similarly treated so as to give room for blank correction. This was done by transferring 100 ml of distilled water into a beaker and digested as described above. Calibration standards were prepared from stock solutions by dilution and were matrix matched with the acid concentration of the digested samples. The digested samples were then analyzed for heavy

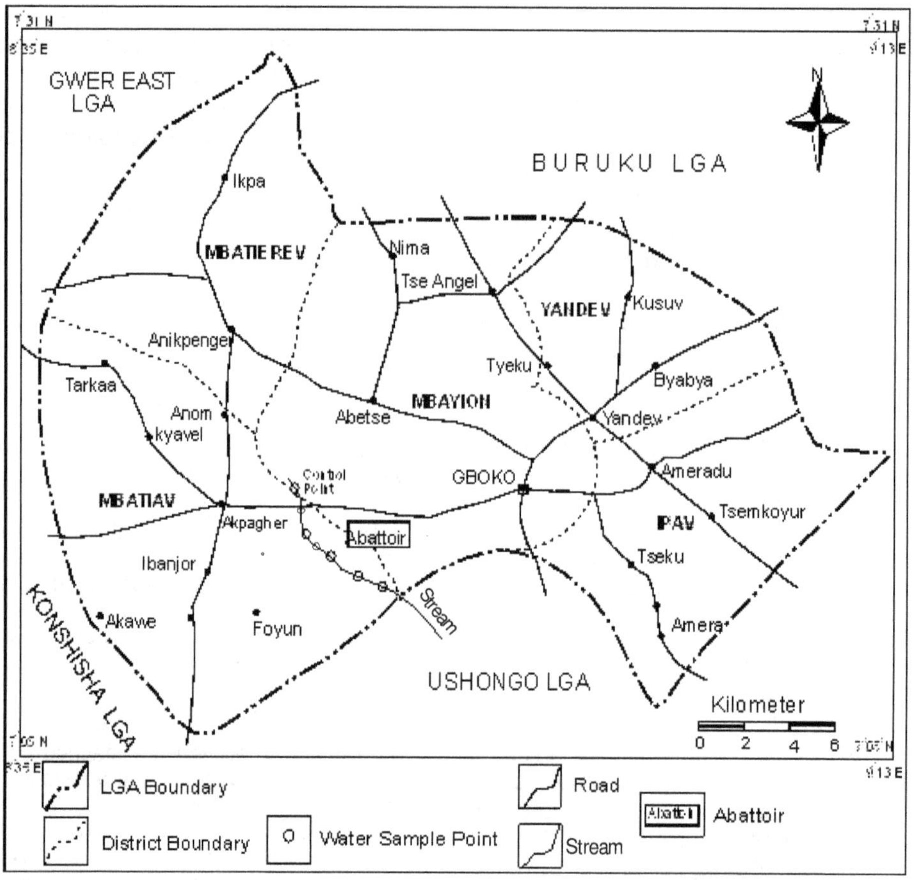

Figure 1. Map of Gboko local government area showing the location of abattoir and water sample points. Source: Ministry of lands and survey Makurdi.

metals using Atomic Absorption Spectrophotometer (model AA6800 Shimadzu-automated) at National Research Institute for Chemical Technology (NARICT), Zaria.

pH determination

The pH meter (Hanna HI9024 Microcomputer pH) was calibrated with two buffer standard solutions (pH 4 buffer and pH 9 buffer). After that, the pH reading of the water sample was taken on the spot.

Temperature

The temperature of the water sample was taken on the spot using a mercury thermometer.

Total suspended solids (TSS)

100 ml of each water sample was filtered through a pre-weighed filter paper. The filter paper was dried at 103- 105°C. TSS was determined by using the following formula (Anon, 1992).

$$TSS(mg/L) = \frac{final\ mass - initial\ mass}{Amount\ of\ sample\ taken} \times 1000$$

Total dissolved solids (TDS)

This was by the evaporation method. Evaporation dish was weighed and later 100 ml of the water sample introduced into the weighed dish and dried in an oven operated at 103°C for one hour to a constant weight. After drying, it was transferred to a desiccator and left to cool for one hour. The dish was finally weighed with its content. The difference in mass gives a measure of the total dissolved solids of the sample (HachWater Analysis Hand book 1983).

Dissolved oxygen (DO)

Dissolved oxygen was determined using Winkler Titration Method as described by (Ademoroti, 1996).

$$DO(mg/L) = \frac{16000 \times M \times V}{V_1 - 2.0}$$

Where M = Molarity of the thiosulphate solution, V = Volume of thiosulphate used for titration, V_1 = Volume of the bottle with the stopper in place.

Biochemical oxygen demand (BOD)

The BOD was determined using Winkler titration method. The water

Table 1. Mean concentration (ppm) of heavy metal in abattoir adjacent stream.

Sampling point	Metal				
	Cd	Zn	Ni	Cr	Pb
AW	0.0115 ± 0.0043	0.1860 ± 0.0303	0.0569 ± 0.0243	0.0617 ± 0.0161	0.2076 ± 0.0209
BW	0.0056 ± 0.0049	0.2788 ± 0.2046	0.1827 ± 0.2172	0.0868 ± 0.0552	0.3040 ± 0.1993
CW	0.0131 ± 0.0010	3.7445 ± 1.7111	0.0967 ± 0.0730	0.0453 ± 0.0035	0.2892 ± 0.0944
DW	0.0135 ± 0.0005	0.1124 ± 0.0092	0.0847 ± 0.1011	0.0365 ± 0.0124	0.1780 ± 0.1258
EW	0.0135 ± 0.0005	0.1135 ± 0.0530	0.1086 ± 0.0299	0.0503 ± 0.0000	0.3782 ± 0.0735
FW	0.0090 ± 0.0019	0.1458 ± 0.0218	0.0887 ± 0.0580	0.0780 ± 0.0214	0.3782 ± 0.0105
GW (ctrl)	0.0012 ± 0.0006	0.0164 ± 0.0082	0.0038 ± 0.0017	0.0012 ± 0.0006	0.0002 ± 0.0001
MEAN	0.0110 ± 0.0020	0.7635 ± 0.0260	0.1031 ± 0.0430	0.0598 ± 0.0037	0.2892 ± 0.072
RANGE	0.0056 - 0.0135	0.1124 - 3.7445	0.0569 - 0.1827	0.0365 - 0.0868	0.1780 - 0.3782

sample was collected in the BOD bottle and incubated at 20°C in the dark for 5 days. The BOD on day five was determined using the same procedure for DO above. The mass of oxygen obtained in day 5 was subtracted from the mass of oxygen on day 1 to determine the BOD (mg/L) using the formula (Ademoroti, 1996).

BOD_5 (mg/L) = (DO_1 – DO_5)

Chemical oxygen demand (COD)

Titrimetric method was employed in the determination of COD. A 10 ml of 0.125 M $K_2Cr_2O_7$ was added to 20 ml of the water sample using a pipette in a refluxing flask. Glass beads or anti -bumping chips were added. Then 30 ml of concentrated H_2SO_4 was added slowly and with gentle swirling. The flask was connected to the condenser and refluxed for 2 h. After that, the flask was cooled and the condenser washed with distilled water into the flask and diluted to about 150 ml. The excess dichromate was titrated with 0.05 M ferrous ammonium sulphate (FAS) using 2 drops of ferroin as indicator. A blank mixture was prepared and treated using the same procedure (Ademoroti, 1996).

$$mg/L\ COD = \frac{(V_b - V_s) \times M \times 16000}{mL\ sample}$$

Where: V_b = mL FAS used for blank, V_s = mL FAS used for sample, M = molarity of FAS.

Total phosphate (PO_4^{3-})

One milliliter of concentrated H_2SO_4 was added to 10 ml of the water sample drop wise and mixed. The mixture was boiled for about five minutes, filtered and allowed to cool. The filtrate was titrated with 5 M NaOH solution to phenolphthalein end point. This mixture was transferred to a 50 ml volumetric flask and diluted to 40 ml. A few drops of H_2SO_4 were added until the solution turned clear again. Thereafter, a solution of ammonium molybdenum was added followed by solid ascorbic acid and made to mark. An intense blue complex of molybdenum blue was formed which was measured using a UV-Visible spectrophotometer (model Jenway 6305) at a wavelength of 882 nm.

Sulphate (SO_4^{2-})

50 ml of the water sample was measured into a beaker and the pH adjusted to 5.0 using methyl red indicator with HCl solution. The sample was heated to boiling and $BaCl_2$ solution was added slowly until complete precipitation was achieved. The precipitate was digested at 80°C for 2 h and filtered through a weighed Whatman No.42 filter paper and then washed with small portions of warm distilled water followed by $AgNO_3$ solution. The filter paper was dried at 100°C in an oven for 1 h, cooled and weighed. Amount of SO_4^{2-} was calculated using the formula (Emmanuel et al., 2012).

$$SO_4^{2-}(mg/L) = \frac{mg of BaSO_4}{Volume of sample} + 411.5$$

Nitrate (NO_3^-)

Nitrate was determined using UV-Vis. spectrophotometer (HACH DR 2000). The stored programme number for nitrate (355) was entered and the wavelength dialed to 500 nm. A sample cell was filled with distilled water and placed in the sample holder of the spectrophotometer. This serves as the blank. Another sample cell was filled with the water sample to which one nitrate powder pillow was added and swirled and allowed to react for two minutes. The sample cell was placed in the sample holder and closed. The result was read directly (Emmanuel et al., 2012).

RESULTS AND DISCUSION

Table 1, Figures 2 and 3 present the mean concentrations (ppm) of heavy metal in water of the stream adjacent to the abattoir. The results show differences in metal concentrations at various sampling stations and their controls. Zinc was the most abundant metal (Figure 3) recorded ranging from 0.1124-3.7449 ppm and a mean value of 0.7635 ppm. The maximum value of 3.7445 ppm is slightly higher than maximum permissible limit of 3 mg/L for zinc in water (WHO, 2006, 2004) but lower than Nigeria's permissible limits of 5 mg/L (SON, 2003). The concentration of lead ranged from 0.1780-0.3782 ppm and a mean of 0.2892 ppm. This value is higher than the maximum permissible limit of 0.01 mg/L for the concentration of lead in water (WHO, 2006; EU, 1998, SON, 2003). The high concentration of Pb in water could be due to anthropogenic activities taking place at the abattoir. Cd, Ni and Cr levels ranged

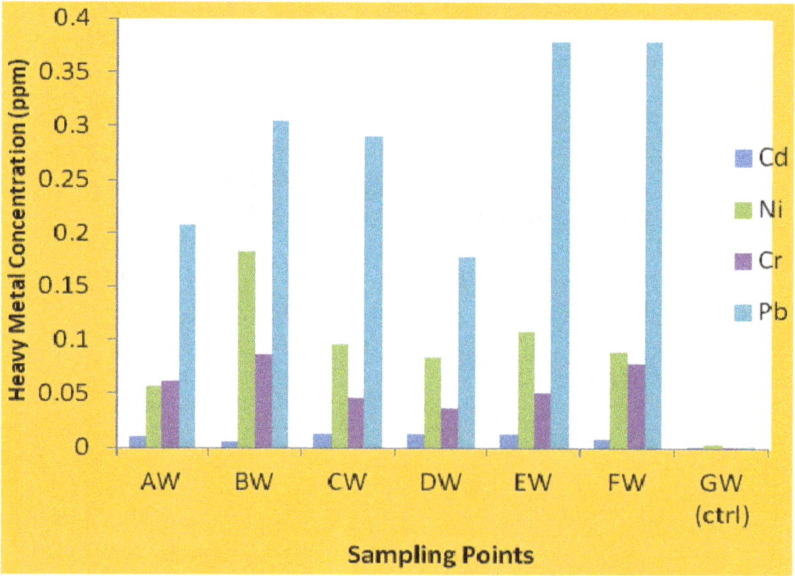

Figure 2. Mean concentration (ppm) of heavy metal in abattoir adjacent stream.

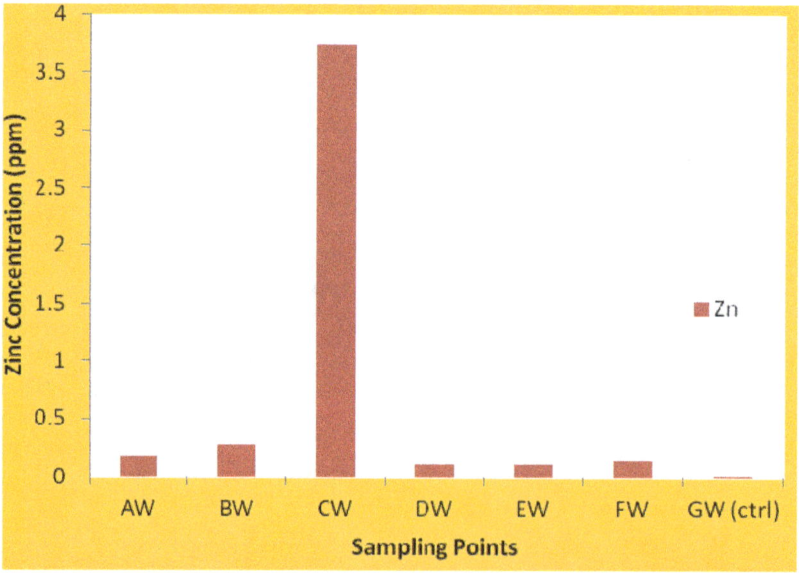

Figure 3. Mean concentration (ppm) of Zinc in abattoir adjacent stream.

from 0.0056-0.0135, 0.0569-0.1827 and 0.0365-0.0868 ppm, respectively which are also higher than set standards. Generally, the concentration of metals in water in the study area is higher than the control and standards set by Nigeria and international organizations that monitor quality of drinking water. It has been reported that heavy metals, reaching excessive levels, can exert serious impact on humans, animals and plants because they are not biodegradable as they are retained indefinitely in the ecological systems and in the food chain (OmPrakash et al., 2011). Table 2 and Figure 4 are results of the physicochemical parameters of the abattoir's adjacent stream. The temperature ranges from 30.20 to 30.80°C and is lower than 32 to 34°C reported by Osibanjo and Adie (2007). Temperature influences the amount of dissolved oxygen in water which in turn influences the survival of aquatic organisms. The pH ranged from 6.50-6.70 with a mean value of 6.60 and falls within WHO standards and compares well with 4.9 to 7.2 reported by Masse and Masse (2002). TSS and TDS values ranged from 190.00 to 710.14 and 250.00 to 2700.00 mg/L, respectively. Their mean values are

Table 2. Mean physicochemical properties of water.

Parameter	AW	BW	CW	DW	EW	FW	GW(ctrl)	Range
T (°C)	30.20 ± 1.23	30.30 ± 0.10	30.70 ± 2.10	30.80 ± 1.00	30.50 ± 0.09	30.40 ± 1.15	30.40 ± 0.11	30.2 - 30.8
pH	6.70 ± 0.09	6.55 ± 2.11	6.50 ± 1.70	6.50 ± 0.73	6.56 ± 0.91	6.51 ± 0.27	6.70 ± 1.42	6.50 - 6.70
TDS (mg/L)	650.0 ± 3.17	700.0 ± 3.10	2700.0 ± 1.08	1320.0 ± 2.10	250.0 ± 1.00	540.7 ± 2.40	110.0 ± 5.10	250 - 700
TSS (mg/L)	350.6 ± 2.01	670.4 ± 1.70	1070.0 ± 0.97	710.14 ± 2.14	400.23 ± 0.9	190.0 ± 1.01	12.0 ± 0.09	190 - 710
DO (mg/L)	3.20 ± 0.00	2.60 ± 0.10	9.00 ± 0.07	8.00 ± 0.00	2.00 ± 0.01	5.00 ± 2.01	5.90 ± 0.02	2.00 - 9.00
Sulphate (mg/L)	38.46 ± 0.67	40.00 ± 0.89	170.00 ± 0.90	120.00 ± 1.00	93.00 ± 2.00	60.00 ± 5.01	12.00 ± 0.03	38.46 - 170
Phosphate (mg/L)	8.00 ± 1.00	6.50 ± 0.03	10.1 ± 3.01	16.00 ± 1.09	8.00 ± 0.90	4.70 ± 0.10	2.10 ± 0.03	4.70 - 16.0
Nitrate (mg/L)	9.40 ± 0.25	34.00 ± 3.00	68.00 ± 1.08	62.00 ± 0.04	9.30 ± 0.10	66.00 ± 4.02	5.00 ± 0.00	9.30 - 68.0
BOD (mg/L)	356.50 ± 5.01	516.00 ± 0.05	861.70 ± 4.07	812.60 ± 0.90	342.00 ± 0.30	19.00 ± 1.04	12.00 ± 0.08	19.00 - 861
COD (mg/L)	536.00 ± 1.06	1456.0 ± 5.01	1508.0 ± 9.03	928.0 ± 0.98	444.0 ± 3.04	548.0 ± 1.90	156.0 ± 0.06	444 - 1508

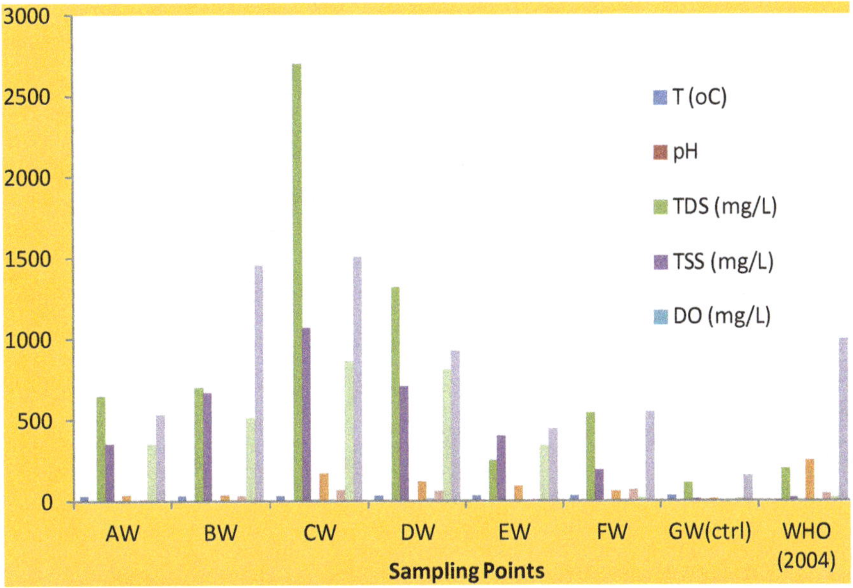

Figure 4. Mean physicochemical properties of water.

565.22 and 1,026.78 mg/L TSS and TDS respectively which is above the WHO maximum permissible limit for TSS (20 mg/L) and TDS (200 mg/L) and also higher than the control. TSS relatively measures the physical or visual observable dirtiness of a water resource. TDS are an indication of the degree of dissolved substances such as metal ions in the water (Efe, 2005). DO have a range of 2.0-9.0 mg/L with a mean value of 50 mg/L. This value is higher than WHO permissible limit of 4 mg/L and also, higher than the control. Low DO may result in anaerobic conditions that cause bad odour. BOD and COD ranges between 19.0-861.7 and 444-1508 mg/L with mean values of 484.63 and 903.3 mg/L, respectively. The mean value for BOD is higher than the control and WHO allowable limit of 20 mg/L (WHO, 2006). COD has a mean value lower than the permissible limit of 1000 mg/L (WHO, 2006). Though, the mean value of COD is lower, some points like BW and CW have values of 1,456 and 1,508 mg/L respectively which are higher than the WHO standards. BOD and COD are indices of organic pollution. BOD is not a specific pollutant indicator, but rather a measure of the amount of oxygen required by bacteria and other microorganisms engaged in stabilizing decomposable organic matter over a specified period of time. A high oxygen demand indicates the potential for developing a DO sag as the microbiota oxidizes the organic matter in the water. Since nearly all organic compounds are oxidized in the COD test, COD results are always higher than BOD results. This was confirmed in this study with some samples (BW and CW) exceeding the value of 1000 mg/L set by WHO.

Nitrate concentration ranges from 9.4-68 mg/L with a mean value of 41.45 mg/L. Even though the mean concentration value is low, some points analyzed have values higher than the set standards. It is reported that nitrate concentration above the permissible value by

Table 3. Correlation matrix for pairs of the analyzed elements in water.

Element	Cd	Zn	Ni	Cr	Pb
Cd	1.000				
Zn	0.334	1.000			
Ni	0.240	0.113	1.000		
Cr	0.249	-0.042	0.787*	1.000	
Pb	0.552	0.169	0.712	0.797*	1.000

*Correlation is significant at the 0.05 level (2-tailed).

45 mg/L is dangerous to pregnant women and poses a serious health threat to infants less than three to six months of age because of its ability to cause methaemoglobinaemia (Gelperim et al., 1975). Nitrates have a high potential to migrate to ground water since they are very soluble and do not bind to soil (Punmia and Jain, 1998). Phosphates were at relatively high concentration. All the water samples were above 5 mg/L maximum permissible limit except for FW (4.7 mg/L). Phosphate enter water ways from human, animal waste and other sources like phosphorus rich bedrock, industrial effluents, fertilizer run-off, laundry and cleaning. Phosphates in water increase the tendency of troublesome algae to grow in the water (Esry, 1991). This causes eutrophication or over fertilization as it chokes up the water ways and uses up large amounts of oxygen.

Sulphate concentration ranges from 40-170 mg/L with a mean value of 86.91 mg/L. This is lower than the maximum permissible limit of 250 mg/L set by WHO but higher than the control value of 12 mg/L. This implies that the activities in the abattoir are contributing to the pollution load of the stream and long term effect may subsequently lead to contamination of the surrounding water body. The result of the correlation analysis using Pearson's correlation coefficient (Table 3) showed that all the metals were positively correlated except for Cr and Zn which were negatively correlated. Pb, Cr, and Ni were significantly correlated. The positive and significant correlations between metals in stream water samples suggest similar source.

Conclusion

The results of this study revealed that the physicochemical parameters of the surface water around Gboko abattoir exceeded WHO, EU, SON/NAFDAC recommended safe limits (2004). The mean concentrations of Pb, Cr, Ni and Cd were also higher than the regulatory permissible limits. The positive correlations between the pairs of metals in the surface water suggest common anthropogenic source. Generally, the values of the physicochemical parameters and heavy metals were higher in the surface water around the abattoir than the control samples. This implies that the activities at the Gboko abattoir were contributing to th pollution load of the surface water in the area and thi has potential for full-blown environmental problems in th near future if no controlled.

ACKNOWLEDGEMENT

The researchers are thankful to the National Researc Institute for Chemical Technology, Zaria, Nigeri (NARICT) where this analysis was done.

REFERENCES

Ademoroti CMA (1996). Standard Methods for Water and Effluent Analysis. pp. 20-49.

Adesemoye AO, Opere BO, Makinde SCO (2006). Microbial Content o abattoir waste water and its contaminated soil in Lagos, Nigeria. Afr J. Biotechnol. 5(20):1963-1968.

Akan JC, Abdulrahman FI, Yusuf E (2010). "Physical and chemica parameters in abattoir waste water sample".Maiduguri metropolis Nigeria. Pac. J. Sci. Technol. 11(1):640-647.

Alorge DO (1992). Abattoir Design, Management and Effluent Disposa in Nigeria, University of Ibadan press: In; Nwankwoala, H.O.;Pabon D.; Amadi, P.A. (2009). Seasonal Distribution of Nitrate and Nitrit Levels in Eleme Abattoir Environment, Rivers State, Nigeria. J. Appl Sci. Environ. Manage. 13(4):35-38.

Anon R (1992). Standard methods of Water and Wastewate Examination. 18thedition American Public Health Associatior Washington, D.C. 2:172.

Bello YO, Oyedemi DTA (2009). The Impact of Abattoir Activities and Management in Residential Neighbourhoods: A Case Study o Ogbomoso, Niger. J. Soc. Sci. 19(2):121-127.

Efe SI, Ogban FE, Horsfall M, Akporhornor EE (2005). Variation o physicochemical characteristics in water resources quality in wester Niger delta region, Nigeria. J. Appl. Sci. Environ. Manag. 9:191-193.

Emmanuel OO, Solomon A (2012). Quality of Sachet Water and Bottle Water in Bolgatanga Municipality of Ghana. Res. J. Appl. Sci. Eng Technol. 4(9):1094-1098. ISSN: 2040-7467.

Esry SA, Potash JB, Shiff C (1991). Effects of improved water suppl and sanitation on ascariosis, diarrhea, dracunculiasis, hookworm infection, schistosomiasis and tachorna. Bull. WHO (5):610-614.

EU (1998). European Union Drinking Water Standards http://www.lenntech.com/EU's-drinking-water-standards.htm.

Ezeoha SL, Ugwuishiwu BO (2011). Status of Abattoir Waste Research in Nigeria. Niger. J. Technol. 30(2).

Gelperim A, Moses UK, Bridge C (1975). "Relationship of high nitrat community water supply to infants and fetal mortality". Ill. Med. J. 147:155-156.

HachWater Analysis Hand book (1983). Loveland Colorado. pp. 181-278. http://dx.doi.org/10.1016/j.ecoenv.2006.01.005: PMid:16551477. http://dx.doi.org/10.1016/S0378-4274(96)80302-X

http://www.docstoc.com/docs/103302144/Waste-Management-%EE%9F% A6-Red-Meat-Abattoirs.

Jukna C, Jukna V, Suigzdaite (2006). Determination of Heavy metals in Viscera and Muscles of cattle. Bulg. J. Vet. Med. 9(1):35-41.

Katarzyna RA, Monkiewicz J, Andrzej G (2009). Lead, cadmium, arsenic, copper and zinc contents in hair of cattle living in the area contaminated by a copper smelter in 2006-2008. Bull. Vet. InstPulawy. 53:703-706.

Kruslin E, Hodel CM, Schurgast H (1999). Progress in diagnosis of Chronic Toxic metal Poisoning by Hair analysis. Toxicol. Lett. 88:84. http://dx.doi.org/10.1016/S0378-4274(96)80302-X

Masse DI, Masse L (2002). Characterization of effluents from six hog slaughter houses in Eastern Canada and evaluation of their impact influent treatment systems. Can. J. Agric. Eng. 42:139-146.

Mohammed S, Musa JJ (2012). Impact of Abattoir Effluent on River Landzu, Bida, Nigeria. J. Chem, Biol. Phys. Sci. 2(1):132-136.

NAFDAC (National Agency for Food and Drug Administration and Control) (2004). Guidelines for Production and Registration of Packaged Water.New guidelines and regulations. NAFDAC, Abuja, Nigeria 1-10.

Nwachukwu MI, Akinde SB, Udujih OS, Nwachukwu IO (2011). Effect of abattoir wastes on the population of proteolytic and lipolytic bacteria in a Recipient Water Body (Otamiri River). Glob. Res. J. Sci. 1:40-422.

Odoemelan SA, Ajunwa O (2008). Heavy Metal Status and Physico-chemical Properties of Agricultural Soil amended by short term application of animal manure. J. Chem. Soc. Niger. 30:60-63.

Olanike KA (2002). Unhygienic operation of a city abattoir in south western Nigeria. Environmental implication AJEAM/RAGEE. 4(1):23-28.

OmPrakash M, Ashish G, Mahipatsingh, Rajayashree P (2011). Determination of Toxic Trace Metals Pb, Cd, Ni, and Zn in Soil by Polarographic Method. Int. J. Chem. Tech. Res. 3(2):599-604. CODEN (USA): IJCRGG ISSN: 0974-4290.

Osibanjo O, Adie GU (2007). Impact of effluent from Bodija abattoir on the physicochemical parameters of Oshunkaye stream in Ibadan. Afr. J. Biotechnol. 6(15):1806-1811.

Patra RC, Swarup D, Naresh R, Kumar P, Nandi D, Shekhar P, Roy S, Ali SL (2007). Tail hair as an indicator of environmental exposure of cows to lead and cadmium in different industrial areas. Ecotoxicol. Environ. Saf. 66:127-131. http://dx.doi.org/10.1016/j.ecoenv.2006.01.005: PMid:16551477.

Punmia BC, Jain AK (1998). Wastewater Engineering. Laxmi publications, New Delhi. pp. 1-660.

Radojevic M, Bashkin VN (1999). Practical Environmental Analysis. The Royal Society of Chemistry: Cambridge, UK. P. 466.

Red Meat Abbattoir Association (2010). Waste Management-Red Meat Abattoir. Retrieved from SON (2003). Standard Organization of Nigeria Safe Drinking Water Regulation. http://www.docstoc.com/docs/103302144/Waste-Management-%EE%9F% A6-Red-Meat-Abattoirs.

Steffen, Roberts, Kirsten Inc. (1989). Water and waste-water management in the red meat industry (P. 36).WRC Report No. 145 TT41/89. WRC, Pretoria.

WHO (2004). Guidelines for Drinking Water Quality. 3rd Edition Vol.1. Recommendation WHO: Geneva, Switzerland.

WHO (2006). World Health Organisation Drinking Water Standards, LENNTECH.http://www.lenntech.com/WHO's-drinking-water-standards06.htm.

Yahaya MI, Mohammed S, Abdullahi BK (2009). Seasonal Variations of Heavy Metals Concentration in abattoir Dumping Site Soil in Nigeria. J. Appl. Sci. Environ. Manag. 13(4):9-13.

24

Study of the seasonal variations in Turag river water quality parameters

A. K. M. Lutfor Rahman[1*], M. Islam[1], M. Z. Hossain[1] and M. A. Ahsan[2]

[1]Department of Chemistry, Jagannath University, Dhaka 1100, Bangladesh.
[2]Bangladesh Council of Scientific and Industrial Research (BCSIR), Dhaka, Bangladesh.

The purpose of this study was to assess the degree of pollution of Turag river water by determining various physico-chemical parameters. Water samples were collected six times per year during wet and dry season at the following three locations: Tongi Railway Bridge, Bishwa Ijtema field and Ashulia. Most of the measured physicochemical parameters exceeded permissible limit of drinking water. The recorded pH ranged from 6.6 to 7.98 and Electrical Conductivity (EC) from 160 to 1107 µs/cm. The recorded dissolve oxygen (DO) varied from 0.11 to 6.8 mg/L and biological oxygen demand (BOD) ranged from 10 to 180 mg/L while chemical oxygen demand ranged from 21 to 220 mg/L and free CO_2 value from 5 to 22 mg/L. The concentration ranges of heavy metals and arsenic in ppb were as follows: Zinc (Zn) (0.04 to 0.4), cadmium (Cd) (0.043 to 2), arsenic (As) (1.15 to 4.8), (lead) Pb (2.29 to 18.62) and mercury (Hg) (0.12 to 1.45). Due to the increased values of the parameters pH, DO, BOD, COD and free CO_2 water from these locations was not suitable for human consumption without appropriate treatment.

Key words: Tongi railway bridge, Ijtema field, Ashulia, industrial effluent, aquatic ecosystem.

INTRODUCTION

Dhaka, the capital and most populated city in Bangladesh is now a member of mega city family in the world. Rapid and unplanned urbanization, commercial development along with population pressure have made an environmentally polluted city in the world (Haigh, 2004; Karn and Harada, 2001). Urbanization is the major demographic development which is occurring very fast and with larger magnitude in the developing countries. Its transforms the existing landscape without considering the possible consequences and requirements for environmental sustainability (Brookfield, 1988). Urbanization and Industrialization near the river bank has created pollution problem. Modern civilization is dependent on large-scale use of a wide range of metals and most of them are naturally present only at trace levels in the hydrosphere (biosphere) (Chow, 1968). The major route by which heavy metals are dispersed in the biosphere is associated with the disposal of industrial effluents. The Turag river is the upper tributary of the Buriganga, a major river in Bangladesh. The most pollution sources of Turag river water are various consumer goods industries (soap and detergent), garments industries, pharmaceuticals industries, lots of tanneries, dyeing industries, aluminum industries, battery manufacturing, match industries, ink manufacturing industries, textile, paint, iron industries, pulp and paper factories, chemical factories, frozen food factories and Steel workshop etc. Most of the industries discharge their effluents directly or indirectly into the Turag river without any treatment causing pollution of the surface water. Moreover, many sewerage and municipal sewage drainage system have become a dumping ground of all kinds of solid, liquid and chemical waste that polluted the river bank.

Consequently, complex mixture of hazardous chemicals, both organic and inorganic are released into Turag river water resulting in different chemical and biochemical interactions in the river system and thus deteriorate the water quality. For this reason water causes the adverse effect of surrounding land and aquatic ecosystem as well as subsequent impact on the livelihood of the local community. In this study, the water quality parameters: pH, DO, BOD, COD, free CO_2, and

*Corresponding author. E-mail: lrahman1973@gmail.com.

Figure 1. Google view of entire study area of the present work.

heavy metals of As, Zn, Cd, Pb and Hg were estimated at different locations of Turag river.

MATERIALS AND METHODS

Study area

The study areas were located at different places of the Turag River (Figure 1). In Turag, samples were collected near Tongi Railway Bridge, Bishwa Ijtema field and Ashulia near Masimpur, Bhatulia. It is located at latitude 23°52'53.69"N and longitude 90°24'11.96"E. The Bishwa Ijtema, is situated at 23°53'19.25"N and 90°23'31.41"E by the Turag River in Tongi. Tongi Railway Bridge of Turag was the first sampling site which receives various kinds of industrial and domestic effluents. Presences of pilgrims staying at the riverside for 3 to 7 days without proper sanitation system contribute to the pollution of the Bishwa Ijtema. Garbage disposal into the river resulted to heavy pollution. The pollution and encroachment is excessively high at the Ashulia point which is situated at N23°53'07.13" and E 90°20'37.49".

Sample collection and analysis

Three water samples were collected from three major station of Turag River like Tongi Railway Bridge, Ijtema field and Ashulia. Plastic bottles of 500 ml were used for collecting samples. Prior to collection, the bottles were cleaned by detergent solution and then it was treated with 5% HNO_3 acid over night and finally washed with de-ionized water followed by repeated washing with sample water so as to avoid contamination. All the samples were taken with grab sampling. After sampling, the bottles were kept in air tight and labeled properly for identification. The parameters, pH was measured using 826 pH mobile (metrohm), DO, BOD, TDS, salinity, conductivity were measured by using portable multiparameter meter sensation ^{TM} 156 (HACH). Hardness and free CO_2 were measured by titrimetric method. Chemical oxygen demand is measured as a standardized laboratory assay in which a closed water sample is incubated with a strong chemical oxidant under specific conditions of temperature and for a particular period of time (Statistical year book, 2007). Concentration of heavy metals such as As, Cd, Zn, Pb and Hg in Turag river were estimated by using atomic absorption spectrophotometer (AAS) using standard analytical method.

RESULTS AND DISCUSSION

The result obtained on some physicochemical parameters of the Turag river water samples are presented in Tables 1 and 2.

Table 1 reveals that the pH value varied from 6.6 to 7.85 which is within the permissible limit for diverse uses like irrigation, domestic and recreational, according to standard value of DoE (pH 6 to 9). The normal range for pH in surface water systems is 6.5 to 8.5 and for groundwater systems 6 to 8.5 (Gob, Environment Conservation Rules, 1997). pH greatly affects biological activity. It also affects some properties of water body, activity of organism and effectiveness of toxic substances present in the aquatic environment. The measurement of alkalinity and pH is needed to determine the corrosiveness of the water. The electrical conductivity (EC) is usually used for indicating the total concentration of charged ionic species in water. The total study area reveals the low condition for EC except the station of Railway Bridge (1107 µs/cm). High EC shows that a large amount of ionic substances like sodium, iron, potassium etc which are present in industrial effluent are found in textile effluent (Kabir et al., 2002). The values in all measuring samples with the exception of the Railway Bridge station, were in accordance with FAO drinking water standard (1000 µs/cm). The total dissolved solids (TDS) mainly indicate the presence of various kinds of minerals like ammonia, nitrite, nitrate, phosphate, alkalis, some acids, sulphates and metallic ions etc which are comprised both colloidal and dissolved solids in water. It is also an important chemical parameter of water (Kabir, 2002). The TDS values of the study area lies between 100 and 580 mg/L. The salinity of water indicates the presence of ionic substances that may come from the reaction of metals and acids containing in water. The highest salinity value of the surface water was 0.52 mg/L and the lowest value observed was 0.08 mg/L. EC has

Table 1. Physicochemical parameters of Turag river water samples collected at six month over the year.

Parameter	Collection period and station																	
	August-10			October-10			December-10			February-10			April-10			June-10		
	RB	IF	Ashulia	RB	IF	Ashulia	RB	IF	Ashulia	RB	IF	Ashulia	RB	IF	Ashulia	RB	IF	Ashulia
pH	7.45	7.85	7.96	7.25	7.63	7.42	7.1	6.6	6.95	6.95	6.9	6.65	7.98	7.2	7.19	7.4	7.53	7.48
EC (µS/cm)	321	160	215	456	324	354	924	467	658	1107	584	766	887	570	715	589	309	458
Salinity (ppt)	0.25	0.1	0.1	0.34	0.1	0.19	0.55	0.12	0.4	0.54	0.35	0.48	0.4	0.21	0.38	0.32	0.1	0.14
TDS (mg/L)	271	100	125	320	125	219	545	220	400	580	270	420	545	200	353	339	100	210
DO (mg/L)	4.8	6.8	6.5	4.1	5.9	5.1	2.1	4.5	4.5	0.11	4.2	3.5	0.19	5.1	4.8	4.8	5.4	5.4
BOD (mg/L)	80	10	30	76	15	24	180	25	50	174	30	60	160	27	46	73	13	28
COD (mg/L)	45	21	25	60	25	31	220	50	84	210	55	91	205	49	68	55	21	25
FreeCO_2 (mg/L)	12	6	8	10	5	10	22	10	12.8	21	11.5	13	20	12.7	11	13	6	8
Hardness (mg/L)	60	58.17	45.8	68	65.1	58.5	179	145	133.8	204	148.9	139.8	178	138.8	129.4	64	60	49.5

RB = Railway Bridge, IF = Ijtema field.

Table 2. Concentration of heavy metals of Turag river in dry season.

Metals (ppb)	Name of stations								
	Railway bridge			Ijtema field			Ashulia		
	Point 1	Point 2	Point 3	Point 1	Point 2	Point 3	Point 1	Point 2	Point 3
As	4.5	3.45	4	2.5	4.8	2.85	3.15	2.15	2
Cd	1	1	1	1	1	1	2	2	2
Pb	8.24	16.71	18.62	5.87	4.74	5.53	2.06	5.89	4.25
Zn	0.05	0.04	0.4	0.3	0.2	0.058	0.45	0.05	0.06
Hg	0.25	1.25	0.3	0.56	0.65	0.15	0.45	1.45	1.25

good relation with TDS and salinity. From the result it was observed that the EC value increased with increasing TDS and salinity.

An adequate supply of dissolve oxygen is essential for the survival of aquatic organism (Dara, 2002). Dissolved oxygen (DO) is needed for waste degradation and decomposition by microorganism. Fish in water containing excessive dissolved gases may suffer from "gas bubble disease"; however, this is a very rare occurrence. The bubbles or emboli block the flow of blood through blood vessels cause death. On the other hand, the decrease of dissolve oxygen concentration is dangerous for aquatic life. Dissolve oxygen (DO) values were found in the range from 0.11 to 6.5 mg/L during dry season followed by wet season. The concentrations were lower from December to April and higher from June to October. The lowest value of 0.11 was observed at Railway Bridge point in February. The highest value 6 mg/L for DO was found in 2006 decreased to 0.1 to 2.1 mg/L in the present study. This reduction in value is due to high discharge of organic material, e.g. from sewage treatment works, storm overflows, agricultural slurry, silage liquor. Such low value do not supports the survival of aquatic life. Enrichment by nutrients results in lower oxygen levels leading to eutrophication. According to the environmental quality standard (EQS), the following requirements for DO are prescribed: 6 mg/L for drinking, 4 to 5 mg/L for recreation, 4 to 6 mg/L for fish and livestock and 5 mg/L for industrial application.

When BOD levels are high, dissolved oxygen (DO) levels decrease because the oxygen that is available in the water is being consumed by the bacteria (Sawyer et al., 2003). Since less dissolved oxygen is available in the water, fish and other aquatic organisms may not survive. If there is no organic waste present in the water, there would not be as many bacteria present to decompose it and thus the BOD will tend to be lower and the DO level will tend to be higher. Higher BOD values were found in Turag River at Tongi Railway Bridge station in December to April (dry period). According to BOD values ranging from 73 to 180 mg/L the most polluted area is Tongi Railway Bridge. The higher concentrations were found during dry season compared to the wet period of the year. The lowest value was found in Turag River at Ijtema field station in wet period. The values at Ijtema field ranged from 10 to 30 mg/L and at Ashulia site in the range of 24 to 60 mg/L. The permissible limit for BOD for drinking water is 0.2 mg/L, for recreation 3 mg/L, for fish 6 mg/L and 10 mg/L for irrigation. (Bangladesh standard) (DoE, Department of Environment, 1991, 1997).

The biological oxygen demand (BOD) was higher in Tongi Railway Bridge because this station flows-through the densest urbanized and industrialized area. Most of the industries are situated near the bank of this station. They discharge organic material, e.g. from sewage treatment works, storm overflows, domestic waste water (human waste and food waste) and industrial waste water (from tannery, textile and food processing industries), agricultural slurry, silage liquor. Moreover, municipal waste materials directly or within the sewerages are dumped into this station (Subramanian, 2004).

COD can be related empirically to BOD. The higher values of COD were found in Railway Bridge because most of the industries are situated near the bank of this station. Their waste materials (organic/inorganic substances) are discharged directly into this river. The higher concentrations were observed during the dry season of the year and lower concentration during the wet period. The higher value was found in December to April compared with other month. The highest COD value was found in December at Railway Bridge and lowest value was found at Ijtema field in June and August. Permissible limit for drinking water is 4 mg/L (DoE/BEMP, 2003).

Almost all natural waters contain some dissolved carbon dioxide which they gain in several ways (Water Quality Criteria, Environmental Studies Board, 1972). Carbon dioxide gas (CO_2,) is present in the air to the extent of 0.03% by volume and 0.05% by weight. As rain falls through the air, it absorbs some of this gas. Carbon dioxide quickly combines with water to form carbonic acid, a weak acid. The presence of carbonic acid in waterways may be good or bad depending on the water's pH and alkalinity. If the water is alkaline (high pH), the carbonic acid will act to neutralize it. But if the water is already quite acidic (low pH), the carbonic acid will only make things worse by making it even more acidic. The standard value limit of free carbon dioxide is 8 to 12 mg/L for fresh water. It is lucky for fish that "free" carbon dioxide (by "free" we mean it is not combined with anything) levels rarely exceeded 20 mg/L (milligrams per liter), because most fish are able to tolerate this carbon dioxide level without bad effects. Here the highest value of free CO_2 is 22 mg/L and lowest is 5 mg/L. When carbon dioxide levels are high and oxygen levels are low, fish have trouble respiring (taking up oxygen) and their problems become worse as water temperatures rise.

Hardness of water is due to the presence of chloride, sulfate, carbonate, bicarbonate etc. salt of Ca^{2+} and Mg^{2+}. According to the DoE (Department of Environment), (EQS: Environmental Quality Standard for Bangladesh, PRB, 1991) standard, the permissible limit of Hardness of drinking water is 200 to 500 ppm. Table 1 reveals the experimental stations did not exceed the permissible limit. From Table 1, the hardness increases in dry season and it decreases in wet season (June to November – Wet season; December to May – Dry season). Presence of heavy metals in the samples experimented were illustrated in Table 2 and 3.

At the Tongi Railway Bridge station, the value of As and heavy metals Cd, Pb, Zn and Hg are as follows: (1.15 to 4.5, 0.043 to 1, 2.29 to 18.62, 0 to 0.4 and 0.17 to 1.25 ppb). At the station in Ijtema field, the values of heavy metals (As, Cd, Pb, Zn and Hg) are (1.36 to 4.8, 0.05 to 1, 2.79 to 5.87, 0 to 0.3 and 0.18 to 0.65 ppb). At the station in Ashulia the values of heavy metals (As, Cd, Pb, Zn and Hg) are (1.25 to 3.15, 0.092 to 2, 2.69 to 5.89, 0 to 0.45 and 0.12 to 1.45 ppb). Drinking water standards for Pb is 50 ppb, for Zn is 5000 ppb, for Cd is 5 ppb, for Hg is 1 ppb and for As is 50 ppb (Bangladesh standard). At all studied points heavy metals values are lower compared to permissible limit representing at present no significant threat to the ecosystem.

Conclusions

Concerning all measured parameters (especially DO, BOD and COD), it could be concluded that pollution of Turag water reached critical point with increasing tendency day by day. Consequently, in order to decrease pollution from various sources appropriate steps must be taken immediately. If the necessary steps are not taken, very soon it would be a source danger point for water pollution. Compared to other two stations, the pollution level of Tongi Railway Bridge is significantly higher because, it passes through industrialized and more unplanned residential area. On the basis of monitoring of selected physico-chemical parameters in the last five years there is no doubt of worsening the Turag water quality day by day.

Table 3. Concentration of heavy metals of Turag river in wet season.

Metals (ppb)	Name of stations								
	Railway bridge			Ijtema field			Ashulia		
	Point 1	Point 2	Point 3	Point 1	Point 2	Point 3	Point 1	Point 2	Point 3
As	1.15	1.43	2.58	1.36	2.3	1.85	1.25	1.6	1.25
Cd	0.06	0.043	0.272	0.166	0.168	0.05	0.153	0.092	0.143
Pb	7.47	11.04	2.29	4.55	3.25	2.79	2.69	2.85	3.25
Zn	nf	nf	nf	nf	nf	nf	nf	nf	nf
Hg	0.17	0.19	0.18	0.18	0.21	0.18	0.23	0.24	0.12

nf = Not found.

It is alarming that the pollution concentration is speedily escalating day by day as different types of industries and land uses are developing along the banks of the Turag River which lead to more pollution generation and more encroachment on river bank. At the same time, it is a matter of fortune that the pollution level of the Turag has not yet gone beyond treatability and the river has not experienced the massive grasp of encroachment like the Buriganga. Therefore, recovery process of the river must start immediately. The pollution level of the river is increasing sharply and can cause serious problem in near future. From this study of the surface water quality of the major river around Dhaka City, it can be concluded that the water of this river may represent serious threat to the ecosystem. Although, some parameters may not at critical pollution level, the condition of the river side urbanization and industrialization with no proper waste treatment may cause severe water pollution.

ACKNOWLEDGEMENT

We gratefully acknowledge for the laboratory and other support from the Bangladesh Council of Scientific and Industrial Research (BCSIR) and Jagannath University.

REFERENCES

Brookfield HC (1988). "The new great age of clearance and beyond: What sustainable development is possible?" University of California Press, Berkeley.

Chow TJ (1968). "Water Pollution Control Federation, 40: 399-411.

Dara SS (2002). A text Book of environmental Chemistry and pollution Control. S. Chand and Company Limited, New Delhi, pp. 216.

DoE, Department of Environment (1991, 1997), Environ. Qual. Stand. Bangladesh. July.

DoE/BEMP (2003). Fourth Dhaka Water Supply Project, Dhaka Water Resour. Manage. Programme.

GoB, Environment Conservation Rules (1997). Bangladesh Gazette, The Ministry of Environment and Forest, The Government of the People's Republic of Bangladesh, 27th August.

Haigh MJ (2004). Sustainable Management of headwater Resources: The Nairobi 'Headwater' Declaration (2002) and Beyond, Asian J. Water, Environ. Pollut., 1(1-2): 17-28.

Kabir ES, Kabir M, Islam SM, Mia CM, Begum N, Chowdhury DA, Sultana SM, Rahman SM (2002). "Assessment of effluent quality of Dhaka export processing zone with special emphasis to the textile and dying industries". Jahangirnagar Uni. J. Sci,. pp. 137-138.

Karn SK, Harada H (2001). Surface water pollution in three Urban Territories of Nepal, India, and Bangladesh. Environ. Manage., 28(4): 438-496.

Sawyer CN, McCarty PL, Parkin GF (2003). "Chemistry for Environmental Engineering and Science. (5th ed.). New York: McGraw-Hill.

Statistical year book (SYB) (2007). Statistical year book of Bangladesh, Annual report.

Subramanian B (2004). Water quality in South Asia. Asian J. Water Environ. Pollut., 1(1-2): 41-54.

Water Quality Criteria, Environmental Studies Board (1972). National Acad. Sci.

Separation of lead(II) paraffin-embedded tissues from liver loggerhead turtles specimens by organic-solution-processable functionalized-nano graphene prior to determination by flame atomic absorption spectrometry (FAAS)

Ali Moghimi

Department of Chemistry, Varamin (Pishva) Branch Islamic Azad University, Varamin, Iran.
E-mail: alimoghimi@iauvaramin.ac.ir, kamran9537@yahoo.com.

A novel and selective method for the fast determination of trace amounts of Pb(II) ions in liver loggerhead turtles specimens has been developed. A reliable and rapid method for preconcentration and modification of organic-solution-processable functionalized-nano graphene with Isopropyl 2-[(isopropoxy carbothioyl) disulfanyl] ethanethioate (IIDE) in order to prepare an effective sorbent for the preconcentration and determination of lead. The sorption capacity of modified organic-solution-processable functionalized-nano graphene (IIDE MS) was 82.34 mg.g^{-1} and the optimum pH for the quantitative recovery of lead was found as 5.3. The optimum flow rate, sorbent amount and sample volume were 8 ml.min^{-1}, 300 mg and 50 ml, respectively. 10 ml of 0.1 mol.L^{-1} HCl was the most suitable eluent. The recommended method is simple and reliable for the determination of lead without any notable matrix effect and successfully applied to environmental water samples. The limit of detection of the proposed method is 7.5 ng/ml. The method was applied to the extraction and recovery of Pb(II) in different water samples. In the present study, we report the application of preconcentration techniques still continues increasingly for trace metal determinations by flame atomic absorption spectrometry (FAAS) for quantification of lead in Formalin-fixed paraffin-embedded (FFPE) tissues from liver loggerhead turtles. This method exhibits the superiority in compared to the other adsorption reagents because of the fact that there is no necessity of any complexing reagent and optimum pH of solution presents in acidic media. In this method, the relative standard deviation (RSD) of 2.7%.

Key words: Organic-solution-processable functionalized-nano grapheme, lead, Isopropyl 2-[(isopropoxy carbothioyl)disulfanyl] ethanethioate (IIDE) -modified silica-gel, formalin-fixed paraffin-embedded (FFPE), tissues from liver loggerhead turtles.

INTRODUCTIONS

Lead is one of the most ubiquitous elements in the environment and recognized as a major health risk to humans and animals (Released on World Environment Day, 2001). Direct determination of trace metals especially toxic metal ions such as mercury, tin, lead and metalloids arsenic, antimony and selenium from various samples requires mostly an initial and efficient pre-concentration step (Leyden et al., 1976). This pre-concentration is required to meet the detection limits as well as to determine the lower concentration levels of the analyte of interest (Jones et al., 1983). This can be performed simply in many ways including liquid and solid phase extraction techniques (Nambiar et al., 1998; Caroli et al., 1991). The application of solid phase extraction technique for pre-concentration of trace metals from different samples results in several advantages such as the minimal waste generation, reduction of sample matrix effects as well as adsorption of the target species on the solid surface in a more stable chemical form (Alexandrova and Arpadjan, 1993).

The normal and selective solid phase extractors are those derived from the immobilization of the organic compounds on the surface of solid supports which are mainly nano polyurethane forms (Arpadjan et al., 1997), filter paper (Leyden and Luttrell, 1975), cellulose (Gennaro et al., 1983) and ion exchange resins (Shamsipur et al., 2005). Silica gel, alumina, magnesia and zirconia are the major inorganic solid matrices used to immobilize the target organic modifiers on their surfaces (Unger et al., 1979) of which silica gel is the most widely used solid support due to the well documented thermal, chemical and mechanical stability properties compared to other organic and inorganic solid supports (Boudreau and Cooper, 1989). The surface of silica gel is characterized by the presence of silanol groups, which are known as weak ion exchangers, causing low interaction, binding and extraction of the target analytes (Kvitek et al., 1982). For this reason, modification of the silica gel surface with certain functional groups has successfully been employed to produce the solid phase with certain selectivity characters (Bruening et al., 1991). Two approaches are known for loading the surface of solid phases with certain organic compounds and these are defined as the chemical immobilization which is based on chemical bond formation between the silica gel surface groups and those of the organic modifier, and the other approach is known as the physical adsorption in which direct adsorption of the organic modifier with the active silanol groups takes place (Unger et al., 1979).

Selective solid phase extractors and pre-concentrators are mainly based on impregnation of the solid surface with certain donor atoms such as oxygen, nitrogen and sulfur containing compounds (Mahmoud, 1997, 1999; Mahmoud and Soliman, 1997; Tong et al., 1990; Dadler et al., 1987). The most successful selective solid phases for soft metal ions are sulfur-containing compounds, which are widely used in different analytical fields. Amongst these sulfur-containing compounds are dithiocarbamate derivatives for selective extraction of lead(II) (Mahmoud, 1998, 1999) and pre-concentration of various cations (Leyden et al., 1976; Moghimi et al., 2009; Tehrani et al., 2005) and 2- mercaptobenzothiazol-modified silica gel for on-line pre-concentration and separation of silver for atomic absorption spectrometric determinations (Moghimi et al., 2009). Ammonium hexa-hydroazepin-1-dithiocarboxylate (HMDC)-loaded on silica gel as solid phase pre-concentration column for atomic absorption spectrometry (AAS) and inductively coupled plasma atomic emission spectrometry (ICP-AES) was reported (Alexandrova and Arpadjan, 1993). Mercapto-modified silica gel phase was used in pre-concentration of some trace metals from seawater (Moghimi et al., 2009). Sorption of lead (II) by some sulfur containing complexing agents loaded on various solid supports (Moghimi et al., 2011) was also reported. 2-Amino-1 cyclopentene-1-dithiocaboxylic acid (ACDA) for the extraction of silver(I), mercury(II) and lead(II) (Moghimi 2006), 2-[2-triethoxysilyl-ethylthio] aniline for the selective extraction and separation of palladium from other interfering metal ions (Tehrani et al., 2005) as well as thiosemicarbazide for sorption of different metal ions (Moghimi et al., 2011) and thioanilide loaded on silica gel for pre-concentration of palladium(II) from water (Tehrani et al., 2005) are also sulfur containing silica gel phases.

Ion adsorption onto solid chelating nano polymer materials is now considered as one of the most promising techniques for selective concentration, removal and recovery of metal ions from a wide variety of sources. Among different types of polymer adsorbent, polymer fibers have attracted great interest in recent years (Tahaei et al., 2008). This can be related to their structure and characteristics, like high specific surface, small cross-section, uniformity in diameter (in macroscopic scale) and long length of fiber to diameter (Abdouss and Mousavi, 2012, Moghimi, 2006).

Raw acrylic fibers (RAF), due to their chemical and thermal stability, are a good substrate for the modification. The mentioned advantages are mainly attributed to the high adsorption capacities, fast adsorption equilibrium, high recycling rate and low cost of these polymeric fibers. The use of commercial fiber and introducing functional groups on its structure are of the important methods for producing ion adsorbent fibers. The properties of the fiber can be maintained in this method (Abdouss and Mousavi, 2012).

In our previous attempts, we modified solid phase extraction (SPE) membrane disks with suitable compounds for selective determination of lead (Tuzen et al., 2009). Meanwhile, other investigators have successfully utilized these sorbents for quantitative extraction and monitoring trace amounts of lead (Tahaei et al., 2008).

This study focuses on the utilization of IIDEMS as an efficient organic-solution-processable functionalized-nano graphene in the preconcentration step of lead(II) ions prior to flame atomic absorption spectrometry (FAAS) determination. The influences of some analytical conditions on the preconcentration procedure, such as initial pH, sample volume, eluent type and volume, sorbent amount, flow rate, etc. were investigated.

The second aim of this study was the selection of an appropriate method for the analysis of Formalin-fixed paraffin-embedded (FFPE) tissue were based on present work with atomic absorption spectrophotometric determination of lead(II).

EXPERIMENTAL

Apparatus

Determination of Pb^{2+} contents in working samples were carried out

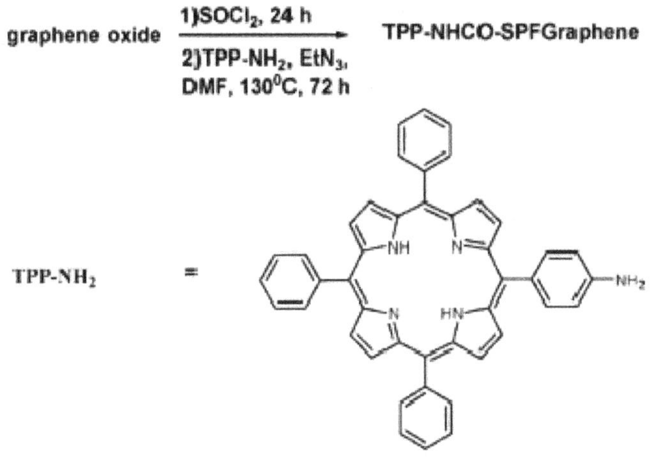

Scheme 1. Synthesis scheme of TPP-NHCO-SPFGraphene.

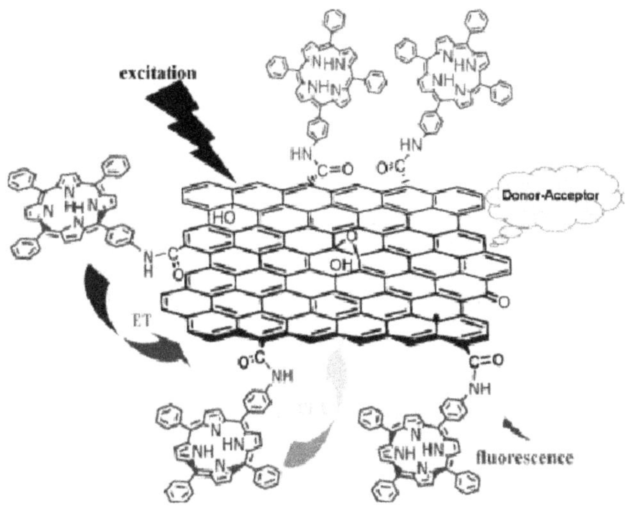

Scheme 2. Schematic representation of part of the structure of the covalent TPP-NHCO-SPFGraphene.

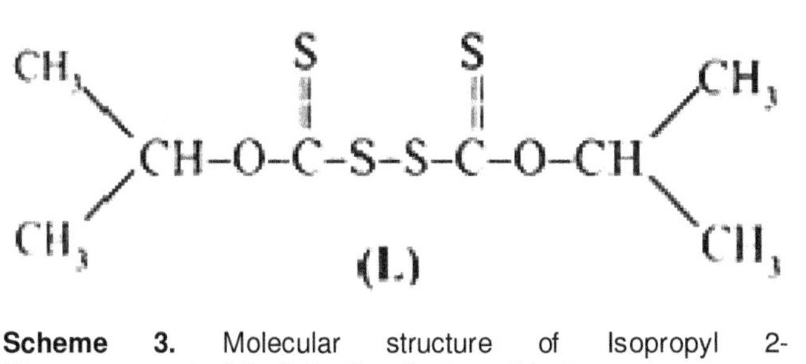

Scheme 3. Molecular structure of Isopropyl 2-[(isopropoxycarbothioyl)disulfanyl] ethanethioate.

by a Varian spectra A.200 model atomic absorption spectrometer equipped with a high intensity hallow cathode lamp (HI-HCl) according to the recommendations of the manufacturers. Instrumental parameters were selected according to the manufacturer's suggestion. The infrared spectra of the materials were recorded on a Perkin Elmer spectrum 100-IR spectrometer (Waltham, MA, USA) using KBr disk in the range of 4000 to 400 cm^{-1}. Heidolph PD 5201 (Schwabach, Germany) model peristaltic pump with eight heads was used for controlling the flow of the liquid into column. The pH measurements were carried out by an ATC pH meter (EDT instruments, GP 353).

Reagents

Organic-solution-processable functionalized-nano graphene was used as the support material. All the chemicals used in this study were of analytical grade. Doubly distilled deionized water was used in all experiments. The standard solution of lead(II) (1000 mgL^{-1}) for the calibration of AAS was purchased from Merck (Darmstadt, Germany). The other concentrations of the standard solutions were prepared by diluting this solution. A stock solution of lead(II) was prepared by dissolving appropriate amount of $Pb(NO_3)_2$ Merck (Darmstadt, Germany) in doubly distilled deionized water and the other concentrations of the working solutions were obtained by diluting this stock metal solution. The pH of the solutions was adjusted to desired values with 0.1 M HCl and/or 0.1 M NaOH solutions. Working solutions were prepared by appropriate dilution of the stock solution.

Synthesis of L

Iodine (1 mmol) in CH_2Cl_2 (10 ml) was added to a stirred solution of potassium *o*-isopropyl(dithiocarbomate) (1 mmol) in CH_2Cl_2 (10 ml) and stirred for 1 h. The reaction mixture was washed with 10% aqueous $Na_2S_2O_3$ (2×10 ml) and H_2O (2×10 ml). The organic layer was dried over $MgSO_4$ and evaporated under reduced pressure. More purification carried out with the re-crystallization in hexane so that pale yellow crystals of L were obtained in 90% yield (0.24 g). The structure and purity of L was confirmed by elements analysis, NMR and IR Spectroscopy. ^1H NMR (CCl_4). δ (ppm): 1.43 (t, 12H, CH_3), 5.63 (m, 2H, CH). IR (KBr). ν_{max} (cm^{-1}): 2979.8 (s), 2869.9 (w), 1463.9 (s), 1442.7 (s), 1373.0 (s), 1271.1 (s, b), 1145.6 (s), 1082.2 (s), 1048.0 (s, b), 898.8 (s), 796.5 (s), 690.5 (m) (Scheme 3).

Synthesis of TPP-NHCO-SPFGraphene

The first organic-solution-processable functionalized-graphene (SPFGraphene) hybrid material with porphyrins. The synthesis of the porphyrin–Graphene nanohybrid, 5-4 (aminophenyl)-10, 15, 20-triphenyl porphyrin (TPP) and grapheme oxide molecules covalently bonded together via an amide bond (TPP-NHCO-SPF Graphene, Schemes 1 and 2) was carried out using an amine-functionalized prophyrin ($TPP-NH_2$) and Graphene oxide in N,N-dimethylformamide (DMF), following standard chemistry. Large-scale and water-soluble Graphene oxide was prepared by the modified Hummers method (Becerril et al., 2008; Liu et al., 2008; Hummers and Offeman, 1958). Results of atomic force microscopy characterization have confirmed that this graphene material can be easily dispersed at the state of complete exfoliation, which consists of almost entire single-layered Graphene sheets in H_2O (Becerril et al., 2008; Liu et al., 2008; Hummers and Offeman, 1958).

$TPP-NH_2$ and Graphene oxide molecules are covalently bonded together by an amide bond. Much care has been taken to make sure all the unreacted TPP-NH2 has been removed using extensive solvent washing, sonication, and membrane filtration. Details are given in the experimental part. The attachment of organic molecules to Graphene oxide has made TPP-NHCO-SPFGraphene soluble in DMF and other polar solvents

Figure 1. SEM micrograph of IIDEMS.

(Becerril et al., 2008; Liu et al., 2008; Hummers and Offeman, 1958).

Preparation of Isopropyl 2-[(isopropoxy carbothioyl)disulfanyl]ethanethioate modified organic-solution-processable functionalized-nano graphene

All acids were of the highest purity available from Merck and were used as received. Methanol and Chlorofom were of High-performance liquid chromatography (HPLC) grade from Merck. Analytical grade nitrate salts of litium, sodium, potassium, magnesium, calcium, strontium, barium, zinc, cadmium, copper(II) nickel, cobalt(II), and Pb(II) were of the highest purity. Ultra pure organic solvents were obtained from E. Merck, Darmstat, Germany. The stock standard solution of Pb (II) was prepared by dissolving 0.1000 g of the Pb(II) powder in 10 ml concentrated nitric acid and diluted to 1000 ml with water in a calibrated flask. Working solutions were prepared by appropriate dilution of the stock solution. In order to prepare a 0.1% Isopropyl 2-[(isopropoxy carbothioyl)disulfanyl] ethanethioate (IIDE) solution, 0.1 g of the reagent was dissolved in 10 ml of acetone and 8 ml of concentrated ammonia solution was added. The final volume of this solution was diluted to 100 ml with water. 100 ml of reagent solution was added into four grams of organic-solution-processable functionalized-nano graphene suspended in 100 ml water and then mixed on a magnetic stirrer for 24 h. The final product was filtered, washed with doubly distilled deionized water and then dried at 100°C in an oven overnight.

Column preparation

A glass column (5.5 cm height × 9 mm i.d.) was packed with a known amount of IIDEMS between two layers of glass wool into the column. A definite volume of lead(II) solution (1 µg ml^{-1}) was passed through the column. All the column studies were performed at room temperature of 25°C.

Preconcentration procedure

The preconcentration method was tested with synthetic lead(II) solutions prior to its application to the real samples. For this purpose, an aliquot of 50 ml of standard lead(II) solution (1 µgml^{-1}) was taken and the pH of the solution was adjusted to desired value with HCl and/or NH_3. The model solution was pumped through the column at a flow rate of 8 ml min^{-1}controlled with a peristaltic pump. The bound metal ions were eluted from IIDEMS with 10 ml of 0.1 mol L^{-1}HCl solution. The lead(II) concentration in the eluate was determined by FAAS. The recoveries of lead(II) were calculated from the ratio of concentration found by FAAS to that calculated theoretically. The general preconcentration procedure described above was carried out to optimize the experimental conditions such as pH, amount of adsorbent, flow rate, type, concentration and volume of the elution solutions, etc.

Analysis of sample paraffin-embedded tissues from liver loggerhead turtles specimens

Selected areas from fresh frozen tissues from liver loggerhead turtles specimens were sliced in three pieces (numbered as 1, 2 and 3) of approximately 10 × 5 × 2 mm each. Sets of pieces of set 1 (controls), were placed into a vacuum chamber at 50°C overnight to dry (until a constant weight was obtained) and the sets 2 and 3 were subjected to the standard 10% buffered formalin fixation and paraffin embedding31 histological process using a tissue processor (Tissue-Tek VIP, Sakura Finetek USA Inc., Torrance, CA). After the paraffin embedding process, tissues were subsequently excised from the blocks with a titanium knife and deparaffinized in xylene at 55°C for 1 h in the tissue processor (the set 2), or with hexane at 20°C for 1 week with frequent changes of the solvent in handling-based procedure (the set 3). Xylene was of a grade routinely used for the FFPE process and hexane was of "Optima" grade (Fisher Scientific). Upon deparaffinization, the tissue samples were dried in a vacuum chamber until constant weight was obtained. Each dried sample (of the sets 1-3) was divided into three portions (5 to 10 mg each) to be further analyzed as triplicates.

RESULTS AND DISCUSSION

Characteristics of the material

The sorbent surface was examined by scanning electron microscopy (SEM) and SEM micrograph of IIDEMS was given in Figure 1. The surface of the organic-solution-processable functionalized-nano grapheme changed after the modification with Isopropyl 2-[(isopropoxy carbothioyl)disulfanyl]ethanethioate (IIDEMS). The surface of organic-solution-processable functionalized-nano graphene was smoother while the modified sorbent surface had a tendency to form agglomerates.

The modification of the sorbent was confirmed by interpreting the infrared spectra of organic-solution-processable functionalized-nano graphene and IIDEMS in Figure 2. The main characteristic absorption band between 3420 and 3646 cm^{-1}is due to O-H stretching mode of silanol groups of organic-solution-processable functionalized-nano graphene and also the remaining adsorbed water (Ekinci-Dogan and Akcin, 2007). The absorption band at 1633 cm^{-1} in the FTIR spectra of organic-solution-processable functionalized-nano graphene (Figure 2a) assigned to water bending mode (Hatay et al., 2008). The broad and intense band at 1047 cm^{-1} is related to the

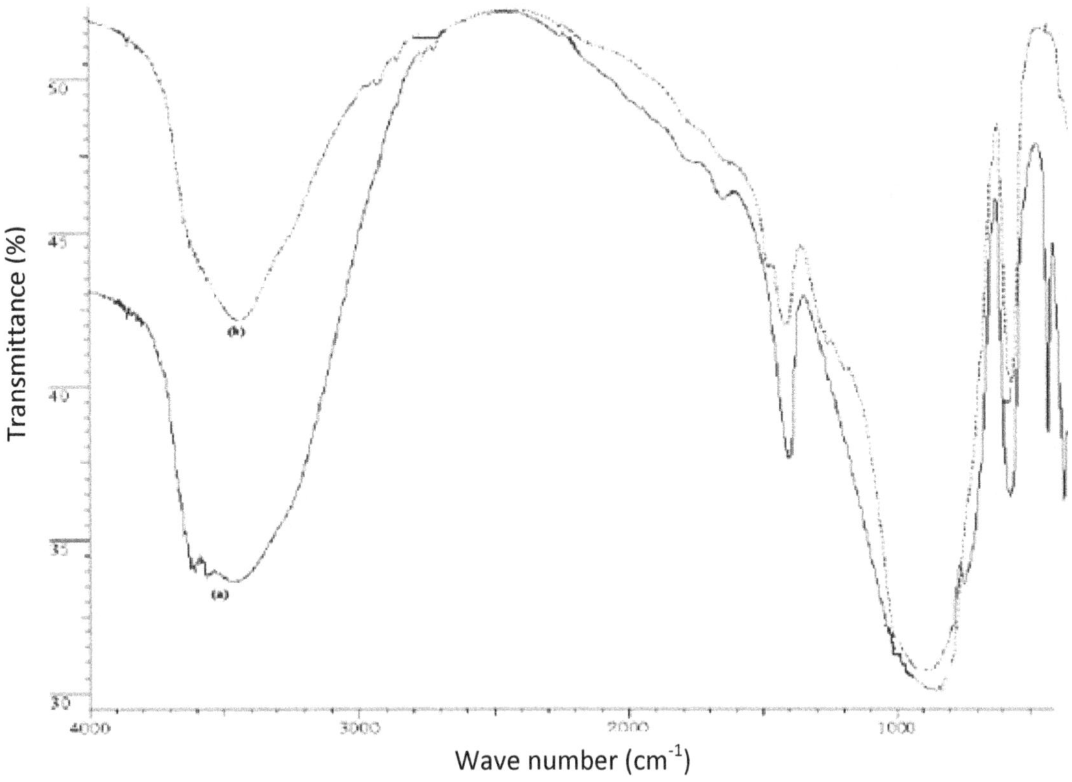

Figure 2. SEM micrograph of IIDEMS.

anti-symmetric stretching vibration of Si-O-Si groups (Roldan et al., 2005). The modified sorbent shows the characteristic stretching vibrations of C-H groups at 2924 and 2856 cm^{-1} in Figure 2b. Also an absorption band at 1285 cm^{-1} in the FTIR spectra of IIDEMS indicates C=S stretching vibration in IIDE. These evidences confirm the modification of organic-solution-processable functionalized-nano graphene with Isopropyl 2-[(isopropoxy carbothioyl)disulfanyl]ethanethioate (IIDE).

Effect of pH on the recovery of lead

The pH of the sample solutions were adjusted to different values between 2-8 by addition of hydrochloric acid or a suitable buffer such as sodium aceate-acetic acid or sodium dihydrogen phosphate- disodium hydrogen phosphate, and then solutions passed through the column. Eventually, the metal ions were stripped by 5 ml of HNO_3, 1 M followed by flame atomic absorption determination of the eluted Pb(II). Then, percentage recovery at various pH values was determined (Figure 4). According to the results shown in Figure 4 up to pH 4.5-5.5, complete recoveries are obtained. However, at higher pH values, percentage recovery decreases. This is due to fact that in an acidic solution the protonation of IIDE occurs and there is a weak tendency for retention between Pb(II)and 1-nitroso-2-naphthol-3,6-disulfonic acid, whereas at higher values (pH>5.7), Pb(II)reacts with hydroxide ions to produce $Pb(OH)_2$. Therefore, sodium aceate-acetic acid buffer with pH=5.3 was used for the preconcentration step. Other solvents used for dissolving IIDE were 5 ml of HNO_3, 1 M. The influences of these solvents on the recoveries as a function of pH are compared and shown in Figure 4.

Evaluation of the role of the ligand

Some preliminary experiments were performed for investigation of absence or presence of IIDE on the quantitative extraction of Pb(II). It was concluded that the surface itself does not show any tendency for the retention of Pb(II), but introduction of 100 ml portions of aqueous Pb(II) samples containing 10 µg of Pb(II)and 10 mg of IIDE leads to satisfactory its retention (Table 1). The latter case is most probably attributed to the existence of a considerable interaction between Pb(II) and the IIDE. It should be mentioned that formation of stable complexes between Pb(II) and IIDE at pH 5.3 is probably due to an ion pair formation mechanism. However, at pH higher than 5 the retention and percentage recovery of Pb(II)are negligible.

Table 1. The effect of presence of IIDE on extraction percent of lead (II)[a].

IIDE	pH	Extraction percent of lead (II)
Absence	2-6	0.03(6.6)[b]
Presence	2-6	98.6(2.4) to 65(2.5)

[a] Initial samples contained 10 µg of lead (II) in 100 ml of water, [b] Values in parentheses are RSDs based on five individual replicate analyses.

Choice of eluent

In order to select the most appropriate eluent for the quantitative stripping of the retained Pb(II) on the column, 5 ml of various non organic solvents were tested. The results are shown in Table 6. As can be seen, the best eluting solvents were found to be 5 ml of 0.1 $molL^{-1}$ HCl, resulting in quantitative elution of Pb(II) from the column.

Effect amount of counter anion

In order to investigate the effect of counter ion on the recovery Pb(II) ions by the modified column, different counter anions were tested (Table 2), it is immediately obvious that the nature of the counter anion strongly influences the retention of Pb(II) ions by the column. The results revealed that the IIDE behaves as a neutral ionophore in the pH range 4.5-5.3 (Liu et al., 2008) so that the Pb(II) ions are retained as ion pair complexes by the column. As seen, acetate ion is the most efficient counter anion for the SPE of Pb(II)ions. The influence of the concentration of sodium acetate ion on Pb(II) recovery was investigated, and the results are shown in Table 2. As seen, the percent recovery of Pb(II) increased with the acetate concentration until a reagent concentration of about 0.1 M is reached, beyond which the recovery remained quantitative.

Moreover, acetate ion acts as a suitable buffering agent, while it effectively contributes to the ions- pair formation; thus, in the SPE experiments, there was no need for the addition of any buffer solution.

The influence of flow-rate

One of the most important parameters affecting solid phase extraction is the speed of the process. Hence, the effect of flow-rates on extraction efficiencies was investigated. It was found that in the range of 1 to 15 $ml.min^{-1}$, the retention of Pb(II)was not considerably affected by the sample solutions flow-rates and leads to reproducible and satisfactory results (Figure 3). Thus, the flow-rate was maintained at 8.0 $ml.min^{-1}$ throughout the experiment.

Quantity of the IIDE

The optimum amount of IIDE for the quantitative extraction of Pb(II) was also investigated by adding various amounts of it to solution (between 2-20 mg). The results are listed in Table 3. The experimental results revealed that the extraction of Pb(II) was quantitative using a sample solution containing more than 10 mg IIDE. Hence, subsequent extractions were performed with 15 mg of IIDE.

Analytical performance

When solutions of 10 µg Pb(II) in 10, 50, 100, 500, and 1000 ml solutions under optimal experimental conditions were passed through the column, the Pb(II) was quantitatively retained in all cases. Thus, the break-through volume for the method must be greater than 1000 ml, providing a concentration factor of >200. The limit of detection (LOD) of the method for the determination of Pb(II) was studied under the optimal experimental conditions. The LOD based on 3σ of the blank is 7.5 ng/ml.

In order to investigate the selective separation and determination of Pb(II) ions from its binary mixtures with various metal ions, an aliquot of aqueous solutions (50 ml) containing 10 µg Pb(II) and mg amounts of other cations was taken and the recommended procedure was followed. The results are summarized in Table 4. The results show that the Pb(II)ions in binary mixtures are retained almost completely by the modified column, even in the presence of up to about 100 mg of various ions. Meanwhile, retention of other cations by the column is very low and they can separated effectively from the Pb(II)ion. It is interesting to note that, in other experiments, we found that in the presence of high enough concentrations $NH_2OH.HCl$ as a suitable reducing agent (> 0.5 M) (Liu et al., 2008).

Analysis of water samples

To assess the applicability of the method to real samples, it was applied to the extraction and determination of Pb(II) from different water samples. Tap water (Tehran, taken after 10 min operation of the tap), rain water (Tehran, 25 January, 2012), Snow water (Saveh, 30 February, 2012) and Sea water (taken from Caspian sea, near the Mahmoud-Abad shore) samples were analyzed. Development of a methodology for the determination of

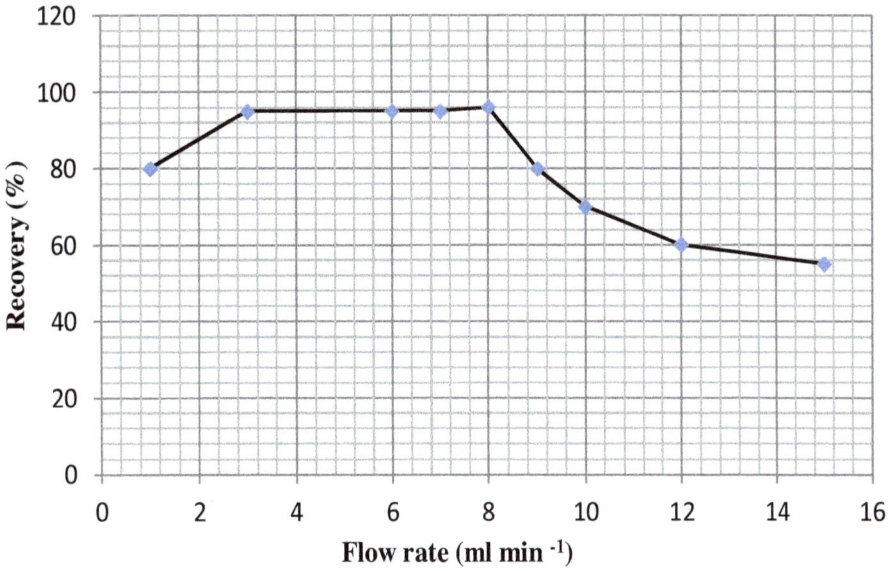

Figure 3. Effect of flow rate of sample solutions on the recovery of lead (II) by IIDEMS.

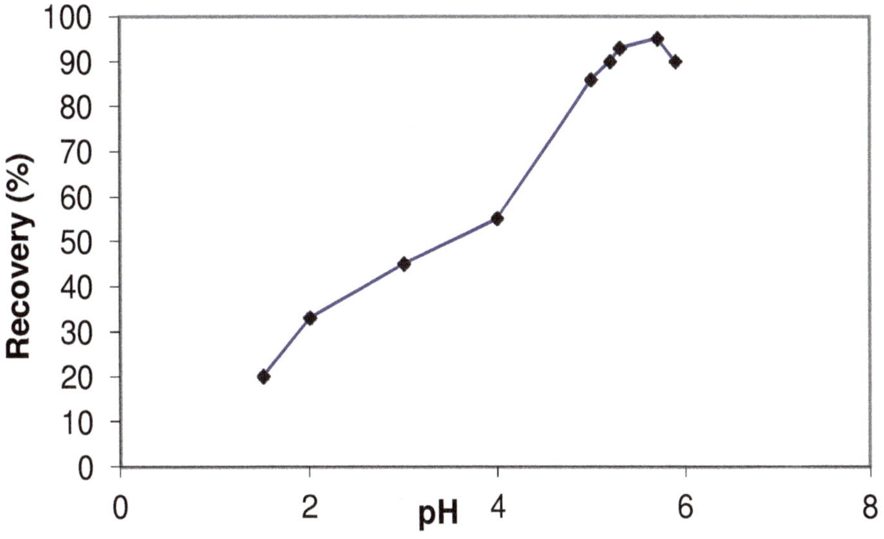

Figure 4. Effect of pH on the recovery of lead (II) by IIDEMS.

Table 2. Percent recovery of lead (II) from the modified of organic-solution-processable functionalized-nano graphene in the presence of 0.01 M of different counter anions[a].

Counter anion	%Recovery
Cl^-	28.5
Br^-	22.6
ClO_4^-	30.7
SCN^-	42.8
Picrate	75.7
Acetate	92.6

[a]Initial samples contained 10 µg of lead (II) in 100 ml of water.

Table 3. Influence of the IIDE amount on the recovery of lead (II)ions[a].

IIDE amount (mg)	Recovery (%) of lead(II)
2	30(2.7)[b]
5	45(2.6)
8	80(2.5)
10	95.8(2.4)
15	99.0(2.5)
20	98.1(2. 4)

[a]Initial samples contained 10 µg of each lead (II) in 100 ml water, [b]Values in parentheses are RSDs based on five individual replicate analysis.

Table 4. Separation of lead (II) from binary mixtures [a].

Diverse ion	Amounts taken(mg)	% Found	%Recovery of Pb^{2+} ion
Na^+	92.0	1.19(2.6)[b]	98.9(1.9)
K^+	92.9	1.30(2.0)	98.9(2.1)
Mg^{2+}	14.2	0.68(1.8)	99.2(2.0)
Ca^{2+}	20.3	2.20(2.0)	98.5(2.7)
Sr^{2+}	2.80	2.87(2.2)	98.2(2.0)
Ba^{2+}	2.90	3.15(2.3)	98.3(2.8)
Mn^{2+}	2.26	1.73(2.5)	97.3(2.8)
Co^{2+}	2.33	1.20(2.7)	98.8(2.9)
Ni^{2+}	1.90	2.17(2.4)	98.0(2.4)
Zn^{2+}	2.10	1.77(2.0)	98.4(2.2)
Cd^{2+}	2.35	1.90(2.3)	98.2(2.6)
Cu^{2+}	1.90	2.77(1.4)	97.6 (2.5)
Hg^{2+}	0.60	2.81(2.9)	97.7(2.4)
Ag^+	2.45	3.45(2.9)	96.6(2.5)
Cr^{3+}	1.70	2.92(2.1)	97.3(2.6)
UO^{2+}	2.60	2.85(2.2)	98.3(2.0)

[a]Initial samples contained 10 μg Pb^{2+} and different amounts of various ions in 100 ml water (0.1 M acetate ion), [b] Values in parentheses are RSDs based on five individual replicate analysis.

Table 5. Recovery of lead (II) added to 1000 ml of different water samples (contaning 0.1 Macetate at pH= 5.3).

Sample	Pb^{2+} added (μg)	Pb^{2+}determined(ng.mL^{-1})	ICP-AES
Tap water	0.0	1.74(2.0)[a]	ND[b]
	10.0	11.98(2.2)	11.7
Snow water	0.0	4.86(2.1)	ND
	10.0	14.94(2.2)	14.7
Rain water	0.0	2.65(2.3)	ND
	10.0	12.76(2.9)	12.3
Sea Water	0.0	12.67(2.3)	12.5
	10.0	22.95(2.0)	23.1
Development of a methodology for the determination of Pb^{2+} in FFPE tissue	0.0	N.D[b]	N.D[b]
	10.0	9.95(2.2)	10.05(2.5)

[a] Values in parentheses are %RSDs based on five individual replicate analysis, [b] Not detected.

Pb(II) in FFPE tissue was performed in a number of steps to optimize the major factors affecting the precision of the analysis (Table 5). As can be seen from Table 5 the added Pb(II) ions can be quantitatively recovered from the water samples used.

Effect of the type, concentration and volume of the elution solutions

In order to determine the most suitable solution for the elution of lead, three different types of eluting agents (HCl, HNO_3 and EDTA) were evaluated. HCl was found to be most effective eluent (Table 6). 0.1 molL^{-1} and 0.2 molL^{-1} of HCl solutions give quantitative recoveries (≥99%). Various volumes of 0.1 molL^{-1} HCl were also examined as eluent and the results were represented in Figure 5. Since quantitative recovery (≥99%) was obtained with 10 ml of 0.1 mol L^{-1}HCl solution, it was selected as an eluent for the recovery of lead by IIDE-modified silica-gel (IIDEMS).

Adsorption capacity

Adsorption capacity of IIDEMS for lead(II) ions was

Table 6. Effect of the type and concentration of the elution solutions.

Type of the elution solution	Concentration (mol L^{-1})	Recovery [a] (%)
	0.05	17
HCl	0.1	99
	0.2	97
	0.05	28
HNO_3	0.1	57
	0.2	95
	0.05	90
EDTA	0.1	91
	0.2	87

[a] Mean of three replicates.

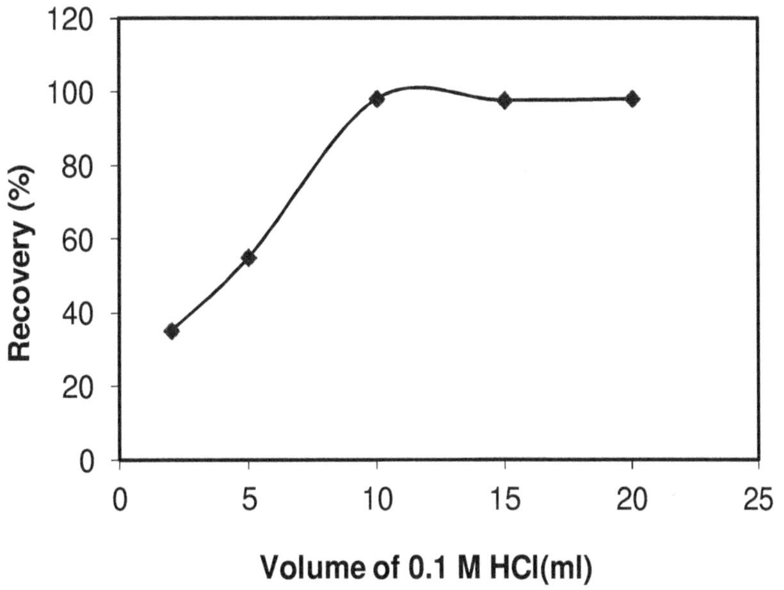

Figure 5. Effect of the volume of 0.1 mol L-1 HCl solution on the recovery of lead (II) by IIDEMS.

determined by a batch method. The initial lead(II) ion concentration was changed from 100 to 1000 mgL^{-1} for the investigation of adsorption capacity of modified adsorbent. The pH of 50 ml of Pb(II) solution was adjusted to optimum value and then 300 mg of modified sorbent was added to lead(II) solutions and mixed on a digitally controlled magnetic stirrer at a rate of 200 rpm. Suspended solids were separated from the adsorption medium by centrifugation at 4500 rpm for 3 min and lead(II) ion concentrations were then measured using FAAS. The equilibrium data were analyzed by Langmuir and Freundlich (Hummers and Offeman, 1958) isotherm models.

$$\text{Langmuir equation} \quad \frac{1}{q_e} = \frac{1}{q_{max}} + \left(\frac{1}{q_{max}K_L}\right)\frac{1}{C_e} \tag{1}$$

$$\text{Freundlich equation} \quad \ln q_e = \ln K_F + 1/n \ln C_e \tag{2}$$

Where q_e and q_{max} are the equilibrium and monolayer sorption capacities of the sorbent (mg g^{-1}), respectively, C_e is the equilibrium metal ion concentration in the solution (mgL^{-1}) and KL is the equilibrium constant (L mg^{-1}) related to the free energy of biosorption. K_F (L.g^{-1}) and n (dimensionless) are Freundlich sorption isotherm constants. The Langmuir and Freundlich isotherm plots

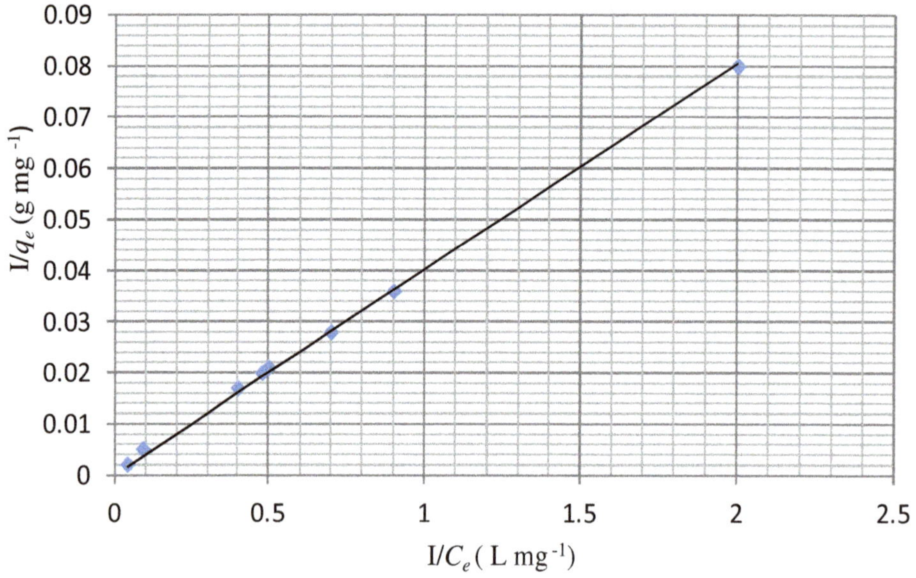

Figure 6. Langmuir isotherm plot for the sorption of lead (II) by IIDEMS.

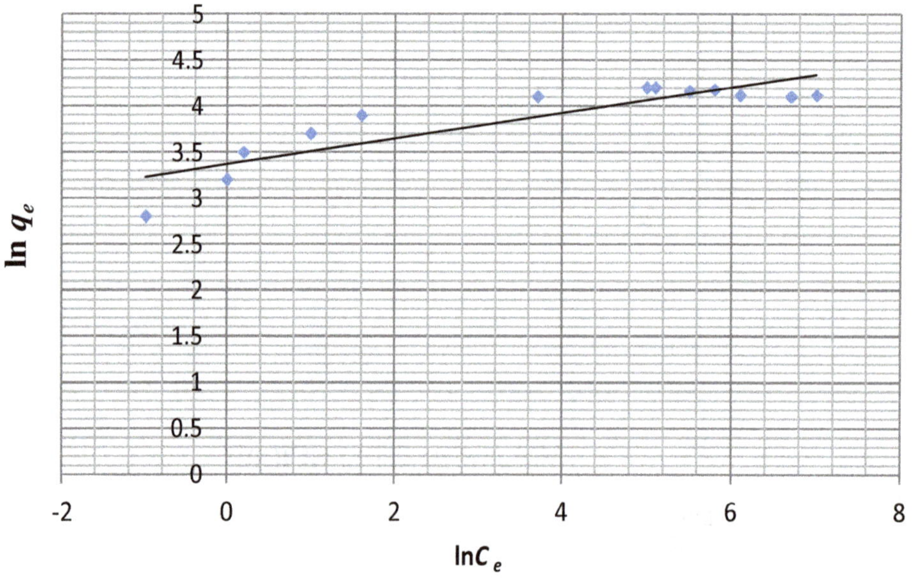

Figure 7. Freundlich isotherm plot for the sorption of lead (II) by IIDEMS.

are shown in Figures 6 and 7, respectively and the model constants are presented in Table 7. The Langmuir isotherm model provided a better correlation than the Freundlich isotherm model. It could be concluded that the sorption of lead(II) onto IIDEMS was monolayer and the maximum monolayer sorption capacity of the sorbent was found to be 82.34 mg. g^{-1}.

The adsorption capacity of IIDEMS is comparable to and moderately higher than that of many corresponding sorbents reported in the literature (Cui et al., 2007; Liu et al., 2008). The dimensionless separation factor, *R*L, was also evaluated and calculated according to Equation 3.

$$R_L = \frac{1}{1 + K_L C_o} \quad (3)$$

*R*L values can be used for the interpretation of the sorption type and it was reported that, when 0<*R*L<1, the sorption process is favorable (Liu et al., 2008). Further, the *RL* value for this study was 1.60×10^{-3}, therefore,

Table 7. Isotherm model parameters for the adsorption of Pb^{2+} onto IIDEMS.

Langmuir			Freundlich			
q_{max} (mg.g^{-1})	K_L(L mg$^{\prime}$)	r_L^2	R_L	n	K_F(L.g^{-1})	r_F^2
74.45	0.68	0.9989	1.60×10^{-3}	6.30	29.80	0.850

adsorption of lead(II) IIDEMS was favorable.

Analytical features

Precision

The precision of the proposed method for the determination of lead was investigated at the optimum experimental conditions (sample volume: 50 ml; pH: 5.3; flow rate: 8 ml min-1; eluent: 10 ml 0.1 M HCl). 10 independent sorption and elution cycles were carried out by following the recommended procedure. The precision of the preconcentration method was evaluated by using the relative standard deviation (RSD) and was found to be 2.7%. The mean recovery of ten replicates was 98±3% at a confidence level of 95%. The precision of the proposed method was good and the recovery of lead was quantitative.

Calibration graph

A linear calibration curve was obtained in the concentration range of 0.21.0 µg ml^{-1}. The calibration equation was A = 0.0025 + 0.0398C, where A is the absorbance and C is the lead concentration in µg ml^{-1}. Correlation coefficient was 0.9989 and the average values of triplicate readings for each standard solution were used for the calculations.

Detection limit

The value of detection limit based on three times the standard deviation of blank signal (N = 20) was 7.5 ng ml^{-1}. The detection limit of the proposed method is comparable to those obtained by other methods described in the literature (Becerril et al., 2008; Tuzen et al., 2009). The corresponding limit of quantification was calculated from ten times the standard deviation of blank signal and found as 20.2 ng ml^{-1}. As seen, the recovered lead ion reveals that the results are quite reliable and are in satisfactory agreement with those obtained by ICPAES.

Conclusions

The lead ions were quantitatively recovered (≥95%) and determined by using IIDE modified organic-solution-processable functionalized-nano graphene sorbent. The optimum recoveries for lead were obtained with 300 mg sorbent, at pH 5.3 and 8 ml. min^{-1} of flow rate. Lead ions were quantitatively recovered (>95%) with 10 ml of 0.1 mol L^{-1}HCl and the preconcentration factor was 5 at optimum conditions. The modified organic-solution-processable functionalized-nano graphene has high sorption capacity (82.34 mg.g^{-1}) and the equilibrium data followed by the Langmuir isotherm model. The precision of the proposed method evaluated as the relative standard deviation obtained from ten replicates, was 3.5%. In comparison to other solid phases (Becerril et al., 2008; Tuzen et al., 2009), high flow rates and large preconcentration factor was achieved using IIDE modified organic-solution-processable functionalized-nano graphene sorbent. While other advantages over reported methods are the high tolerances for matrix components Mazlum et al., 2003), superior sorption capacity and good reusability (Choi and Choi, 2003; Gode and Pehlivan, 2005; Tuzen et al., 2009; Soylak et al., 2001; (Narin et al., 2001).

ACKNOWLEDGEMENT

We gratefully acknowledge financial support of Department of Chemistry of Varamin branch Islamic Azad University for financial support.

REFERENCES

Abdouss M, Mousavi AA (2012). Fabrication Of Chelating Diethylenetriaminated Pan Micro- And Nano-Fibers For Heavy Metal Removal. J. Appl. Polym. Sci. 18(1):27-34.

Alexandrova A, Arpadjan S (1993). Column solid phase extraction as preconcentration method for trace element determination in oxalic acid by atomic absorption spectrometry and inductively coupled plasma atomic emission spectrometry. Analyst 118:1309-1314.

Arpadjan S, Vuchkova L, Kostadinova E (1997). Sorption of Arsenic, Bismuth, Mercury, Antimony, Selenium and Tinon Dithiocarbamate Loaded Polyurethane Foam as a Preconcentration Methodfor Their Determination in Water Samples by Simultaneous Inductively Coupled Plasma Atomic Emission Spectrometry and Electrothermal Atomic Absorption Spectrometry. Analyst 122:243-249.

Becerril HA, Mao J, Liu ZF, Stoltenberg RM, Bao ZN, Chen YS (2008). Evaluation of solution-processed reduced graphene oxide films as transparent conductors ACS Nano 2:463-469.

Boudreau SP, Cooper WT (1989). Analysis of thermally and chemically modified silica gels by heterogeneous gas-solid chromatography and infrared spectroscopy. Anal. Chem. 61:41-47.

Bruening ML, Mitchell DM, Bradshaw JS, Izatt RM, Bruening RL (1991). Effect of organic solvent and anion type on cation binding constants with silica gel bound acrocycles and their use in designing selective concentrator columns. Anal. Chem. 63:21-27.

Caroli C, Alimanti A, Petrucci F, Horvath Zs (1991). Determination of trace elements in analytical-reagent grade sodium salts by atomic absorption spectrometry and inductively coupled plasma atomic emission spectrometry after preconcentration by column solid phase extraction. Anal. Chim. Acta. 248:241-245.

Choi YS, Choi HS (2003). High Functional Inorganic Polymers Containing Main Group 13, 16 Elements in the Polymer Backbone Chain Bull. Korean Chem. Soc. 24:222-228.

Cui Y, Chang X, Zhu X, Luo H, Hu Z, Zou X, He Q (2007). Chemically modified silica gel with p-dimethylaminobenzaldehyde for selective solid-phase extraction and preconcentration of Cr(III), Cu(II), Ni(II), Pb(II) and Zn(II) by ICP-OES Microchem. J. 87:20-28.

Dadler V, Lindoy LF, Sallin D, Schlaepfer CW (1987). Selective pre-concentration and solid phase extraction of mercury(II) from natural water by silica gel-loaded dithizone phases. Aust. J. Chem. 40:1557-1561.

Ekinci-Dogan C, Akcin G (2007). Solid Phase Extraction of Lead(II) by Sorption on Grinded Eucalyptus Stem and Determination with Flame Atomic Absorption Spectrometry. Anal. Lett. 40:2524-2528.

Gennaro MC, Baiocchi C, Campi E, Mentasti E, Aruga R (1983). Preparation and characterization of iminodiacetic acid-cellulose filters for concentration of trace metal cations. Anal. Chim. Acta. 151:339-344.

Gode F, Pehlivan E (2005). Adsorption of Cr(III) ions by Turkish brown coals. Fuel Process. Technol. 86:875-884.

Hatay I, Gup R, Ersöz M (2008). ICP-OES Determination of Trace Metal Ions after Preconcentration Using Silica Gel Modified with 1,2-Dihydroxyanthraquinone J. Hazard. Mater. 150:546-551.

Hummers WS, Jr. Offeman RE (1958). Preparation of graphitic oxide. J. Am. Chem. Soc. 80:1339-1339.

Jones JS, Harrington DE, Leone BA, Bramdstedt WR (1983). Application of optical emission source developments in metallurgical. Atom. Spectrosc. 4:49-54.

Kvitek RJ, Evans JF, Carr PW (1982). Diamine/Silane-Modified controlled pore glass: The covalent attachment reaction from aqueous solution and the mechanism of reaction of bound diamine with copper. Anal. Chim. Acta 144:93-97.

Leyden DE, Luttrell GH (1975). Preconcentration of certain anions using reagents immobilized via silylation. Anal. Chim. 47:1612-1619.

Leyden DE, Luttrell GH, Nonidez WK, Werho DB (1976). ESCA-studies on activated silicagel surfaces preconcentrating heavy metal ions. Anal. Chem. 48:67-74.

Liu ZF, Liu Q, Zhang XY, Huang Y, Ma YF, Yin SG, Chen YS (2008). Substrate effect on the electronic structures of CuPc/graphene interfaces. Adv. Mater. 20:3924-3928.

Mahmoud ME (1997). Silica gel-immobilized Eriochrome black-T as a potential solid phase extractor for zinc (II) and magnesium (II) from calcium (II) Talanta. 45:309-314.

Mahmoud ME (1998). Proceeding of the 25th FACSS Conference, Austin, TX, USA, 11-15 October.

Mahmoud ME (1999). Selective solid phase extraction of mercury (II) by silica gel-immobilized-dithiocarbamate derivatives. Anal. Chim. Acta 398:297-302.

Mahmoud ME, Soliman EM (1997). Study of the selective extraction of iron (III) by silica-immobilized 5-formyl-3-arylazo-salicylic acid derivatives. Talanta 44:1063-1069.

Mazlum AM, Ensafi AA, Naeimi H, Dastanpour A, Shamelli A (2003). Synthesis and analytical application of a novel tetradentate N_2O_2 Schiff base as a chromogenic reagent for determination of nickel in some natural food samples. Russ. J. Electrochem. 39:269-272.

Moghimi A (2006). Preconcentration and Determination of Fe(III) Using Octadecyl SilicaMembrane Disks and Flame Atomic Absorption Spectrometry Oriental. J. Chem. 22(3):527-535.

Moghimi A, Abedin AR, Shahriar GS, Ghiasi R (2009). Solid phase extraction of Cd (II) using mesoporous organosilicas and determination by FAAS. Afr. J. Pure Appl. Chem. 3(3):051-059.

Moghimi A, Shahriar GS, Ghiasi R (2011). A study on the solid phase extraction of CO(II)-IIDE chelate with C18 disk and its application to the determination of trace cobalt. Afr. J. Pure Appl. Chem. 5(6):149-154.

Nambiar DC, Patil NN, Shinde VM (1998). Liquid-liquid extraction of mercury (II) with triphenylphosphine sulphide: Application to medicinal and environmental samples Fresenius. J. Anal. Chem. 360:205-211.

Narin I, Soylak M, Elci L, Dogan M (2001). Separation and Enrichment of Chromium, Copper, Nickel and Lead in Surface Seawater Samples on a Column Filled with Amberlite XAD-2000. Anal. Lett. 34:1935-1947.

Released on World Environment Day (2001). Current Status of Lead in India.

Roldan PS, Alcântara IL, Padilha CCF, Padilha PM (2005). Solid phase extraction of iron and lead in environmental matrices on amberlite xad-1180/pv, Fuel 84:305-311.

Shamsipur M, Shokrollahi A, Sharghi H, Eskandari MM (2005). Solid phase extraction and determination of sub-ppb levels of hazardous Hg2+ ions. J. Hazard. Mater. pp. 117-122.

Soylak M, Elci L, Dogan M (2001). Solid Phase Extraction of Trace Metal Ions with Amberlite XAD Resins Prior to Atomic Absorption Spectrometric Analysis. J. Trace Microprobe Tech. 19:329-344.

Tahaei P, Abdouss M, Edrissi M, Shoushtari AM, Zargaran M (2008). Preparation of chelating fibrous polymer by different diamines and study on their physical and chemical properties *Mat.-wiss. u. Werkstofftech.* 39:839-844.

Tehrani MS, Moghimi A, Waqif HS (2005). Solid Phase Extraction of Cr (III) from Natural Water by Modified Nano Polyacrylonitrile Fiber. Mater. Sci. Res. India 3(2):135-142.

Tong A, Akama Y, Tanaka S (1990). Selective preconcentration of Au (III), Pt (IV) and Pd (II) on silica gel modified with γ-aminopropyltriethoxysilane. Anal. Chim. Acta 230:179-186.

Tuzen M, Karaman I, Citak D, Soylak M (2009). Mercury(II) and Methyl Mercury Speciation on Streptococcus Pyogenes Loaded Dowex Optipore SD-2. J. Hazard. Mater. 169:345-350.

Unger K (1979). Comparison of an ordered mesoporous aluminosilicate, silica, alumina, titania and zirconia in normal-phase high-performance liquid chromatography Porous Silica, Elsevier, Amsterdam. pp. 345-350.

The preparation of activated carbon from agroforestry waste for wastewater treatment

Hesham R. Lotfy*, Jane Misihairabgwi and Mary Mulela Mutwa

Chemistry and Biochemistry Department, Science Faculty, University of Namibia, P. Bag 13301, Windhoek, Namibia.

The scope of this study was to produce activated carbon from local agroforestry wastes (marula fruit stones, jackalberry seeds, eembe seeds, efukwa shells and eembu seeds) and assess the efficiency of the produced carbons in removing dyes and metal ions from wastewater. Preparation of activated carbons from efukwa shells, marula fruit stones, jackalberry seeds, eembe seeds and eembu seeds are described in this study. The carbons were chemically activated by treatment with 50% phosphoric acid. The dyes tested in this study were methylene blue and methyl orange and the metal ions tested were lead and zinc. The carbons produced in this study were compared to commercially powdered activated carbon (PAC) and granular activated carbon (GAC). In case of Lead ion and Powdered Activated Carbon (PAC) produced, the removal efficiency for Eembe, Eembu and Jackalberry was 100%, so as for the commercial powdered activated carbon, 100%. In case of Lead ion and granular activated carbon (GAC) produced, the removal efficiency for Eembu and Marula was 100% and for the commercial granular activated carbon was also 100%. In case of Zinc ion and granular activated carbon (GAC) produced, the removal efficiency for Eembu was 91% and Eembe had zero removal and for the commercial granular activated carbon was 89%. In case of Zinc ion and powdered activated carbon (PAC) produced, the removal efficiency for Eembe, Eembu and Efukwa was around 60%, while the powdered activated carbon (PAC) produced from Jackalberry and Marula had zero removal and for the commercial powdered activated carbon (PAC) the removal was 90%. In case of methyl orange dye and powdered activated carbon (PAC) produced, the removal efficiency for Eembu, Marula and Efukwa was 84% and for the commercial powdered activated carbon was 92%. In case of methyl orange dye and granular activated carbon produced, the removal efficiency for Eembu and Marula was 78% and for the commercial granular activated carbon was 87.5%. In case of methylene blue dye and powdered activated carbon (PAC) produced, the removal efficiency for Jackalberry, 84.4%; Marula, 83% and Eembe was 81.3 % and for the commercial powdered activated carbon was 84.4%. In case of methylene blue dye and granular activated carbon (GAC) produced, the removal efficiency for Jackalberry, 83%; Eembu was 81.25% and for the commercial granular activated carbon was 92.2%. The effectiveness of the produced activated carbon in most of the cases is comparable and in some cases equivalent to that of the commercial carbons.

Key words: Activated carbon (AC), powdered activated carbon (PAC), granular activated carbon (GAC), adsorption.

INTRODUCTION

Agricultural by-products represent a considerable quantity of harvested commodity crops. The use of by-products as precursors for the production of widely used adsorbents, such as activated carbons, may impart a value-added component of the overall biomass harvested.

In most developing countries, the activated carbon is imported at high cost, limiting the quantities of safe

*Corresponding author. E-mail: hrlotfy@unam.na.

drinking water available to the people (Misihairabgwi et al., 2007).

In recent years, there has been research focusing on the use of appropriate, low cost technology for the treatment of drinking water in the developing world. Research has also been focused on the indigenous production of water treatment chemicals using locally available raw materials (Warhurst et al., 1997). Generally, the raw materials for the production of Activated Carbon (AC) are those with high carbon but low inorganic contents such as wood, lignite, peat and coal (Jabit, 2007).

Activated carbons form a large and important class of porous solids, which have found a wide range of technological applications. The characteristics of activated carbon depend on the physical and chemical properties of the precursor as well as on the activation method. In addition to the starting material and the oxidizing agent, activation time and temperature affect the structural properties of the resulting activated carbon. There are two different ways of preparing activated carbon: physical and chemical activation. The physical activation method involves: carbonization of raw material and activation at high temperature in carbon dioxide or water vapor (Petrov et al., 1999). Chemical activation is a well-known method for the preparation of activated carbon, which has been the objective of numerous studies within the last few years (Karim et al., 2006; Wu et al., 2006; Yang et al., 2006; Haimour and Emeish, 2006; Shalaby et al., 2006; Fierro et al., 2006; Pérez et al., 2006) as it presents several advantages compared to the physical activation. The advantage of chemical activation over physical activation is that it is performed in one step and at relatively low temperatures. The most important and commonly used activating agents are phosphoric acid, zinc chloride and alkaline metal compounds, such as KOH (Serrano-Gomez et al., 2005; Dabrowski et al., 2005; Li et al., 2008).

Acid activation of carbons made from wood and coal under a nitrogen atmosphere was investigated by Jagtoyen et al. (1992) and Jagtoyen and Derbyshire (1993) but no attempt was made to investigate cationic metal uptake. It was determined that the oxidation of the carbon surface was important as it produced a negative surface charge, which contributed to its ability to adsorb cations, and a linear relationship between surface charge and metal adsorptive capacity has already been demonstrated by Lima et al. (2004). The process was further evaluated and the predicted cost of producing phosphoric acid activated carbons from agricultural by-products was estimated at \$2.89/kg (Ng et al., 2003).

A wide range of carbonaceous materials can be used as the carbon precursors such as coal, peat, wood and various agricultural by-products. Recently, agricultural by-products have received an increasing attention for the production of activated carbon due to their low-cost, renewability and wide prevalence (Kadirvelu et al., 2003; Yang and Lua, 2003; Ahmedna et al., 2004; Zhang et al., 2004; Cox et al., 2000; Kadirvelu et al., 2004; Gurses et al., 2006; Mohanty et al., 2005). The production of value added products such as activated carbon will enlarge its application, reduce waste materials and generate income to rural communities in Namibia.

Heavy metal contamination exists in aqueous waste streams of many industries such as metal plating facilities, mining operations and tanners (Őzçimen and Ersoy-Meriçboyu, 2009). Activated carbons were used as adsorbent materials because of their extended surface area, microporous structure, high adsorption capacity and high degree of surface reactivity. Furthermore, the presence of different surface functional groups on activated carbon, especially oxygen groups, leads to the adsorption of ions of heavy metals (Al-Asheh et al., 2003; Naseem and Tahir, 2001).

One of the fastest growing research areas is the environmental applications of activated carbon, such as wastewater treatment. In the treatment of wastewater, it is used for purification, decolorization and the removal of toxic organics and heavy metal ions (Jabit, 2007; Lotfy, 2006; Dastgheib and Rockstraw, 2001; Wartelle and Marshall, 2001).

The demand of activated carbon increased over the years and the market growth was estimated at 4.6 % per year (Jabit, 2007). This demand can be satisfied considering the large number of raw material available for the production of activated carbon (Jabit, 2007).

The aim of the study was to produce activated carbon from local agroforestry wastes (Eefukwa shells, marula fruit stones, jackalberry seeds, eembe seeds and eembu seeds) and assess the efficiency of the produced carbons in removing dyes and metal ions from wastewater.

MATERIALS AND METHODS

The shells and seeds which were used as the carbon source were obtained from the northern regions of Namibia during April 2011.

Sample preparation

The raw material was washed with deionized water and dried at 120°C for 4 h. The dried raw material was then ground to particle size of 125 to 250 µm. The particle size was determined using a laboratory test sieve.

Chemical activation and carbonization

Six grams of each ground raw material was soaked in 50 ml of 50 % phosphoric acid at 30°C for 48 h. After 48 h, the phosphoric acid was filtered out and the activated raw material was carbonized in a muffle furnace at 300°C for 2 h in nitrogen atmosphere. After cooling each of the carbonized materials was washed with 200 ml hot distilled water, then dried for 2 h at 120°C. The dried carbon was then weighed to determine percentage yield. This procedure

Table 1. Percentage yield of PAC.

Activated carbon	Mass of raw material (g)	Yield (g)	Yield (%)
Eembe	12.0455	8.6116	71.5
Marula	12.0910	8.0447	66.5
Eembu	12.0321	6.1999	51.5
Jackleberry	12.3630	7.5670	61.2
Eefukwa	12.0530	6.5983	54.7

Table 2. Percentage yield of GAC.

Activated carbon	Mass of raw material (g)	Yield (g)	Yield (%)
Eembe	12.0936	6.5608	54.3
Marula	12.0708	7.1994	59.6
Eembu	12.0755	4.6727	38.7
Jackleberry	12.0733	4.5154	37.4

was used to prepare both powdered activated carbon (PAC) and granular activated carbon (GAC).

Calibration curves

Dye solutions with concentrations of 10, 20, 30, 40, 50, 60, 70, and 80 mg/L were prepared, maximum wavelength was determined using UV-spectrophotometer and calibration curves were drawn.

The removal efficiency

Removal of dyes

A solution of 32 mg/L of each dye was prepared in a 100 ml volumetric flask. 20 ml of the dye solution was poured in 50 ml beaker and mixed with 0.01g PAC, and in another set of experiments, mixed with 0.03g GAC. The mixture was stirred for 15 minutes and allowed to settle for 10 minutes before filtration. Residual dye concentration was determined in each case.

Removal of metal ions

Solutions of 100 mg/L lead and Zinc were prepared in two different 100 ml volumetric flasks. A 20 ml of each was mixed in 50ml beakers with 0.01 g PAC, and in another set of experiments, mixed with 0.03 g GAC, the mixtures were stirred for 15 min and allowed to settle for 10 min before they were filtered. The residual metal ion concentration of the filtrate was determined.

RESULTS AND DISCUSSION

The percentage yield for the powdered activated carbon (PAC) is shown in Table 1. Table 2 shows the percentage yield of GAC.

Figure 1 shows the lead ion concentration before and after removals with PAC while the lead ion concentration before and after removal with GAC is shown in Figure 2. For Zinc ion concentration before and after removals with PAC and GAC are shown in Figures 3 and 4, respectively. Removal of Methyl orange using PAC and GAC are shown in Figures 5 and 6. Figures 7 and 8 shows the removal of Methylene blue using PAC and GAC, respectively.

Agricultural by-products are receiving an increasing attention for the production of activated carbon due to their low-cost, renewability and wide prevalence. The production of value-added products such as activated carbon will enlarge its application and help to deal with wastewater treatment challenges in Namibia.

Percentage yield

The percentage yield of the prepared powdered activated carbons were, Eembe, 71.5%; Marula, 66.5%; Eembu, 51.5%; Jackleberry, 61.2% and Eefukwa, 54.7%. In case of granular activated carbon, Eembe, 54.3%; Marula, 59.6%; Eembu, 38.6%; Jackleberry, 37.4%. Efukwa has soft shells which cannot be ground into granulars. Most of the mass loss was during the carbonization process in the muffle furnace.

Adsorption efficiency

In case of Lead ion and PAC produced, the removal efficiency for Eembe, Eembu and Jackalberry was 100% and for the commercial PAC was also 100%. In case of Lead ion and GAC produced, the removal efficiency for Eembu and Marula was 100% and for the commercial GAC was 100% as well. In case of Zinc ion and GAC produced, the removal efficiency for Eembu was 91% but for Eembe, there was zero removal and for the

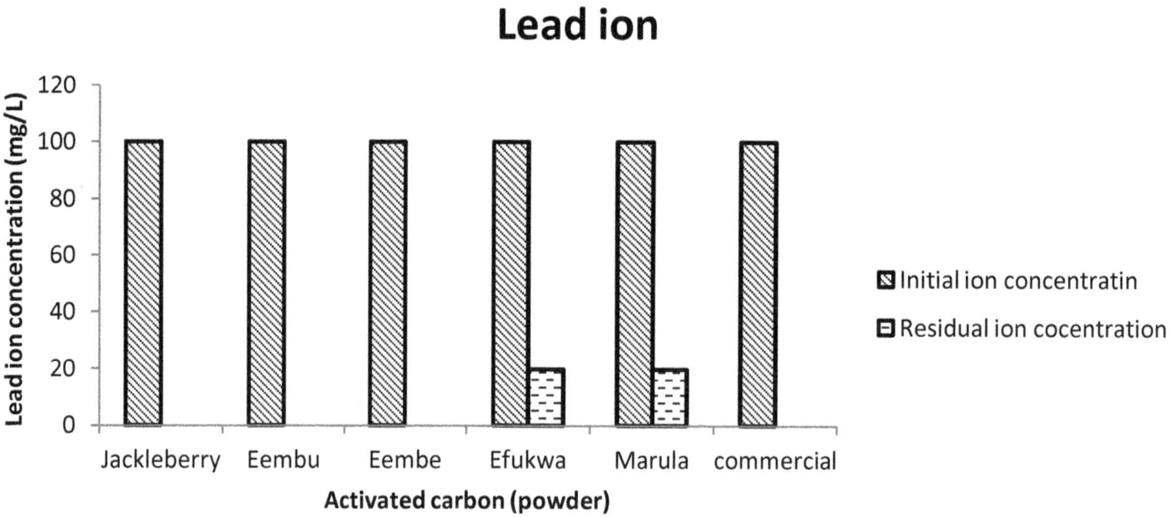

Figure 1. Lead ion concentration before and after removal with PAC.

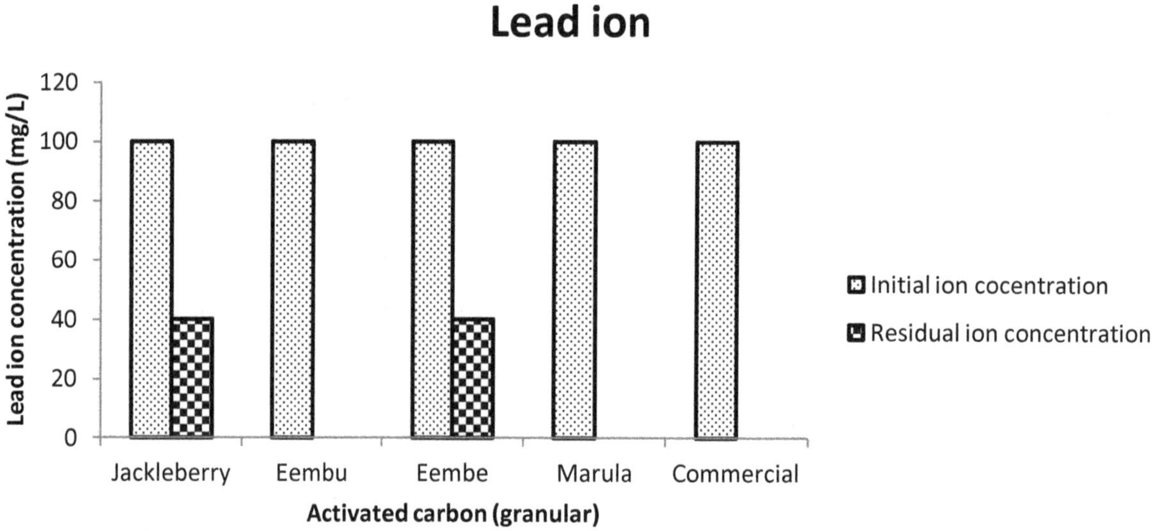

Figure 2. Lead ion concentration before and after removal with GAC.

commercial GAC was 89%, in this case the removal efficiency of the GAC produced from Eembu is higher than that of the commercial GAC. In case of Zinc ion and PAC produced, the removal efficiency for Eembe, Eembu and Efukwa was around 60%. The PAC produced from Jackalberry and Marula had zero removal and for the commercial PAC the removal was 90%. In case of methyl orange dye and PAC produced, the removal efficiency for Eembu, Eembe, Marula and Efukwa was 84% and for the commercial PAC was 92%. In case of methyl orange dye and GAC produced, the removal efficiency for Eembu, Eembe and Marula was 78% and for the commercial GAC was 87.5%. In case of methylene blue dye and PAC produced, the removal efficiency for Jackalberry, 84.4%; Marula, 83% and Eembe was 81.3% and for the commercial PAC was 84.4%, in this case the removal efficiency of the PAC produced from Jackalberry is equal to that of the commercial PAC. In case of methylene blue dye and GAC produced, the removal efficiency for Jackalberry, 83%; Eembu was 81.25% and for the commercial GAC was 92.2%.

All the adsorbents used namely; marula, jackalberry, eembe, eembu and efukwa were found to have (varying) adsorptive capacities, that can be attributed to the fact that the characteristics of activated carbon depend on the physical and chemical properties of the precursor. In general PAC was found to have higher adsorptive capacity compared to GAC, which is attributed to the

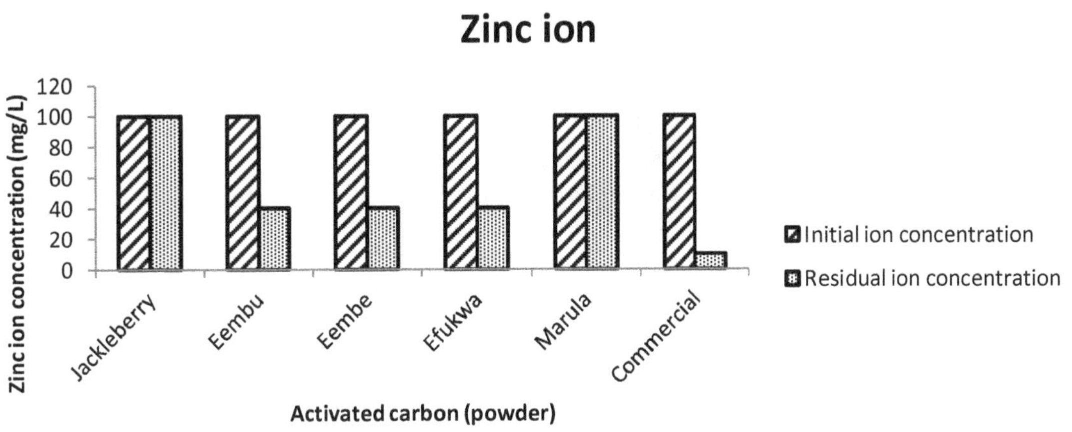

Figure 3. Zinc ion concentration before and after removal with PAC.

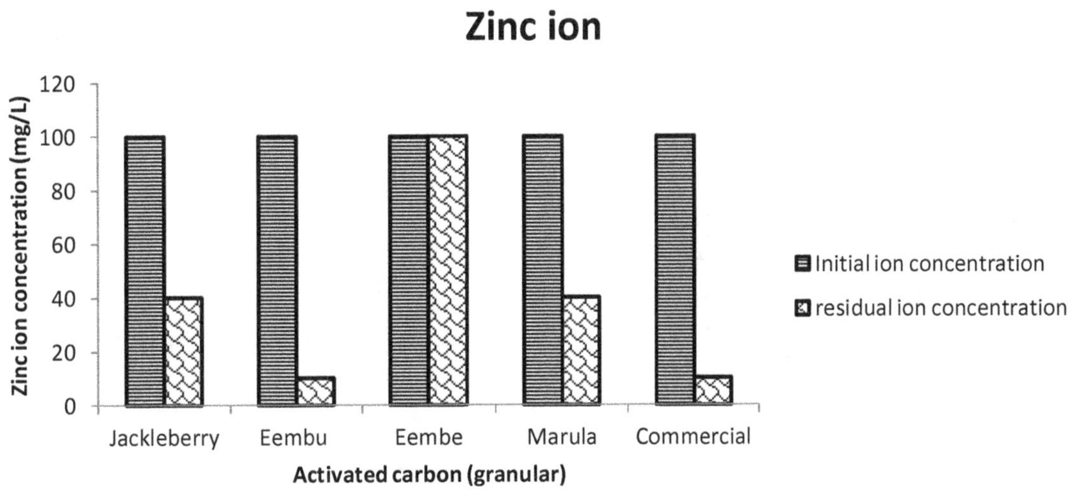

Figure 4. Zinc ion concentration before and after removal with GAC.

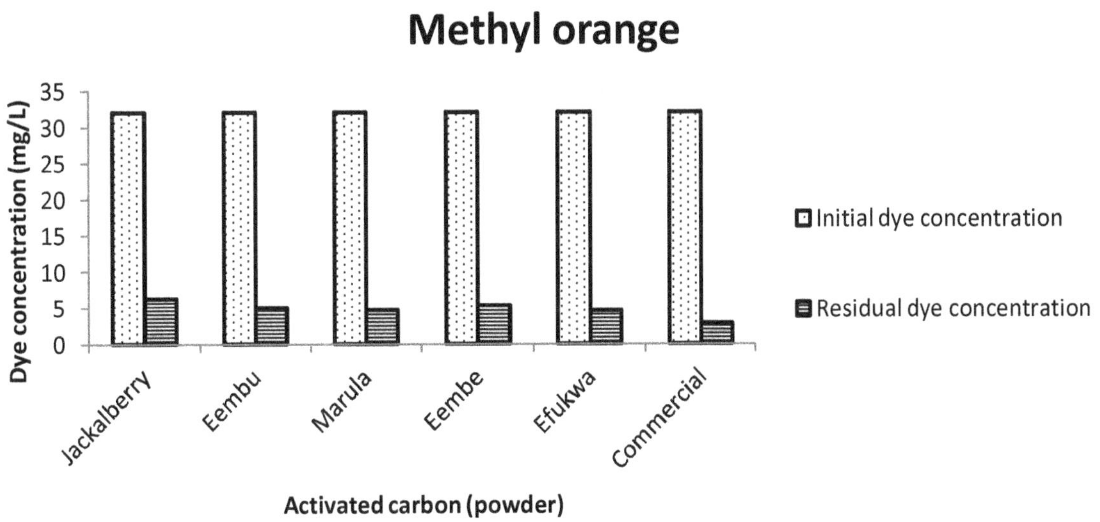

Figure 5. Concentration of Methyl orange before and after removal with PAC.

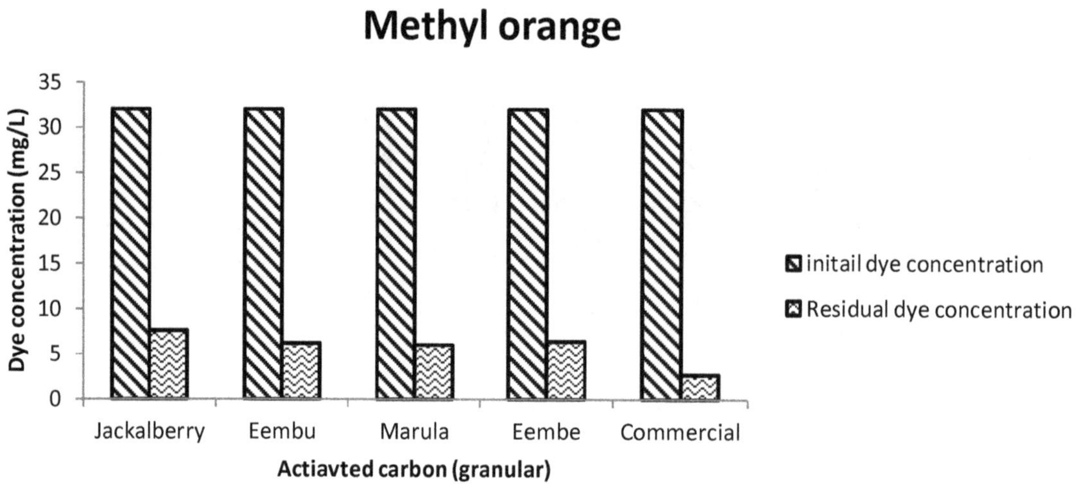

Figure 6. Concentration of methyl orange before and after removal with GAC.

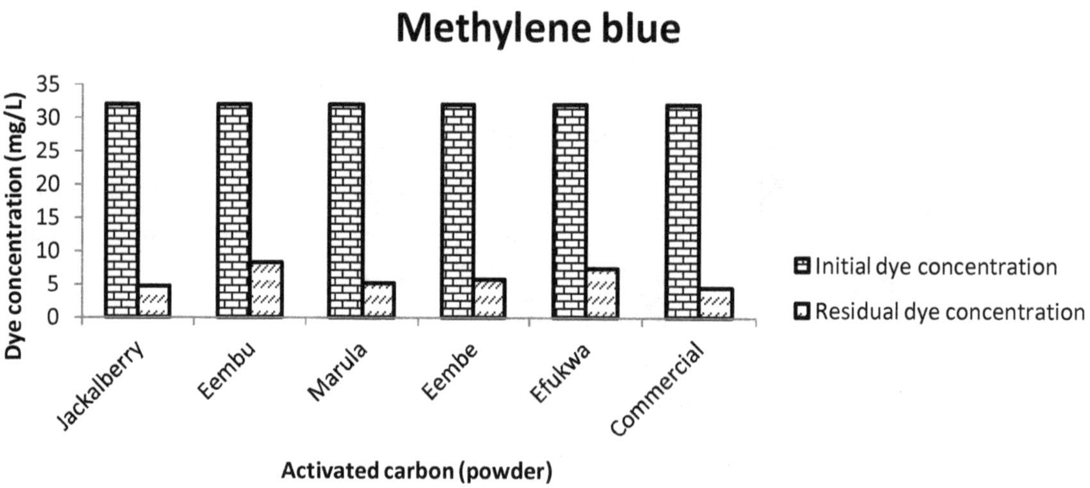

Figure 7. Concentration of Methylene blue before and after removal with PAC.

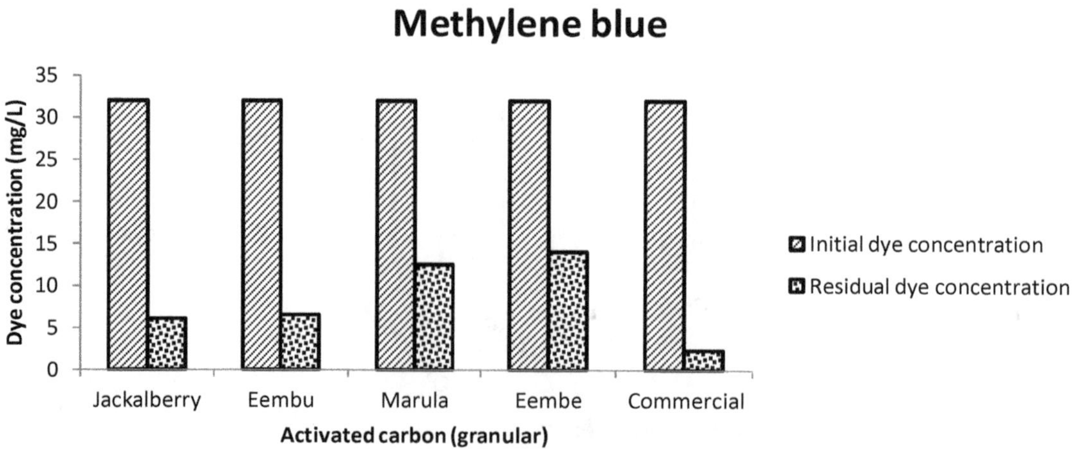

Figure 8. Concentration of Methylene blue before and after removal with GAC.

larger surface area of the PAC.

Conclusion

This study proved the possibility of producing activated carbons from the studied agriculture by-products (efukwa shells, marula fruit stones, jackalberry seeds, eembe seeds and eembu seeds). The study showed that PAC produced from Eembe, Eembu and Jackalberry and GAC produced from Eembu and Marula can completely remove lead ions from a solution containing 100 mg/L Pb^{2+}. The PAC produced from Jackalberry is equivalent to the commercial PAC in removing methylene blue dye. The GAC produced from Eembu is better than the commercial GAC in removing Zinc.

If the carbons prepared in this study are produced on a large scale, the cost of water treatment in Namibia will be greatly reduced. The predicted cost of producing phosphoric acid activated carbons from agricultural by-products was estimated at US$2.89/kg (Ng et al., 2003), which is much cheaper than the imported activated carbon. Therefore, consideration should be given to process scale-up for the local manufacturing of these carbons.

ACKNOWLEDGEMENT

Authors wish to thank Mrs. Sanette Potgieter for providing some of the seeds.

REFERENCES

Al-Asheh S, Banat F, Al-Hamed N (2003). Adsorption of pollutants from aqueous solutions using activated and non-activated oak shells: parametric and fractional factorial design study, part 1, removal of copper. Adsorption, 21: 177-188.

Ahmedna M, Marshall WE, Husseiny AA, Rao RM, Goktepe I (2004). The use of nutshell carbons in drinking water filters for removal of trace metals. Water Res., 38: 1062-1068.

Cox M, El-Shafey EI, Pichugin AA, Appleton Q (2000). Removal of mercury (II) from aqueous solution on a carbonaceous sorbent prepared from flax shive, J. Chem. Technol. Biotechnol., 75: 427-435.

Dabrowski A, Podkoscielny P, Hubicki Z, Barczak M (2005). Adsorption of phenolic compounds by activated carbon-a critical review. Chemosphere, 58: 1049-1070.

Dastgheib SA, Rockstraw DA (2001). Pecan shell activated carbon: synthesis, characterization, and application for the removal of copper from aqueous solution. Carbon, 39, 1849-1855.

Fierro V, Fernández VT, Celzard A (2006). Kraft lignin as a precursor for microporous activated carbons prepared by impregnation with ortho-phosphoric acid: synthesis and textural characterization. Microporous Mesoporous Mater., 92: 243-250.

Gurses A, Dogar C, Karaca S, Ikyildiz AC, Bayrak MR (2006). Production of granular activated carbon from waste Rosa canina sp. seeds and its adsorption characteristics for dye. J. Hazard. Mater. 131(1-3): 254-259.

Haimour NM, Emeish S (2006). Utilization of date stones for production of activated carbon using phosphoric acid, Waste Manage., 26: 651-660.

Jabit NB (2007). Production and characterization of activated carbon using local agricultural waste through chemical activation process. Industrial Crops and Products, 23: 23-28.

Jagtoyen M, Derbyshire F (1993). Some considerations of the origins of porosity in carbons from chemically activated wood. Carbon, 31: 1185-1192.

Jagtoyen M, Thwaites M, Stencel J, McEnaney B, Derbyshire F (1992). Adsorbent carbon synthesis from coals by phosphoric acid activation. Carbon, 30: 1089-1096

Karim MM, Das AK Lee SH (2006). Treatment of colored effluent of the textile industry in Bangladesh using zinc chloride treated indigenous activated carbons, J. Power Sources, 1056-1066.

Kadirvelu K, Kavipriya M, Karthika C, Vennilamani N, Pattabhi S (2004). Mercury(II) adsorption by activated carbon made from sago waste, Carbon, 42: 745-752.

Kadirvelu K, Kavipriya M, Karthika C, Radhika M, Vennilamani N, Pattabhi S (2003). Utilization of various agricultural wastes for activated carbon preparation and application for the removal of dyes and metal ions from aqueous solutions. Biores. Technol., 87: 129-132.

Li W, Zhang LB, Peng JH, Li N, Zhu XY (2008). Preparation of high surface area activated carbons from tobacco stems with K_2CO_3 activation using microwave radiation. Ind. Crops Prod., 27: 341-347.

Lima IM, Marshall WE, Wartelle LH (2004). Hardwood-based granular activated carbons for metals remediation. J. Am. Water Works Assoc., 96: 95-102.

Lotfy HR (2006). The use of different types of carbon for the removal of dyes from industrial wastewater. J. Environ. Sci., 13: 1-19.

Misihairabgwi J, Ntuli V, Kasiyamburu A, Zinyowera S, Ncube I, Chipofya V (2007). Application of locally produced activated carbons in watertreatment.http://www.bscw.ihe.nl/pub/bscw.cgi/d2607370/Misihairabgwi.

Mohanty K, Jha M, Meikap BC, Biswas MN (2005). Removal of chromium(VI) from dilute aqueous solutions by activated carbon developed from *Terminalia arjuna* nuts activated with zinc chloride. Chem. Eng. Sci., 60: 3049-3059.

Naseem R, Tahir SS (2001). Removal of Pb(II) from aqueous/acidic solutions by using Bentonite as an adsorbent. Water Res., 35: 3982-986.

Ng, C, Marshall WE, Rao RM, Bansode RR, Lasso JN, Portier RJ (2003). Granular Activated Carbons from Agricultural By-products: Process Description and Estimated Cost of Production. Bulletin Number 881. Louisiana State Univ. AgCenter, Baton Rouge, LA.

Őzçimen D Ersoy-Meriçboyu A (2009). Removal of copper from aqueous solutions by adsorption onto chestnut shell and grapeseed activated carbons. J. Hazard. Mater., 168: 1118-1125.

Petrov N (1999). Preparation of activated carbons from cherry stones, apricot stones and grape seeds for the removal of metal ions from water, in: Proceedings of the 2nd Olle Indstorm Symposium on renewable Energy-Bioenergy Stockholm, Sweden, pp. 9-11.

Pérez RU, Marín FC, Jiménez DF, Castilla CM (2006). Granular and monolithic activated carbons from KOH-activation of olive stones, D.F. Microporous Mesoporous Mater., 92: 64-70.

Shalaby CS, Mine G, Uçak-Astarlioglu, Levent A, Çigdem S (2006). Preparation and characterization of activated carbons by onestep steam pyrolysis/activation from apricot stones, Microporous Mesoporous Mater., 88: 126-134.

Serrano-Gomez V, Correa-Cuerda ME, Gonzalez-Fernandez CM, Franco-Alexandre FM, Garcia-Macias A (2005). Preparation of activated carbons from chestnut wood by phosphoric acid-chemical activation. Study of microporosity and fractal dimension. Mater. Lett., 59: 846-853.

Warhurst AM, McConnachie GL, Pollard SJT (1997). Characterisation and applications of activated carbon produced from *Moringa oleifera* seed husks by single step steam pyrolysis. Water Res., 31: 759-766.

Wartelle LH, Marshall WE (2001). Nutshells as granular activated carbons; physical, chemical and adsorptive properties. Lewis Publishers, Chelsea.

Wu FC, Tseng RL, Hu CC, Wang CC (2006). The capacitive characteristics of activated carbons comparisons of the activation methods on the pore structure and effects of the pore structure and

electrolyte on the capacitive performance, J. Power Sources, 159: 1532-1542.
Yang T, Lua AC (2006). Textural and chemical properties of zinc chloride activated carbons prepared from pistachio-nut shells. Mater. Chem. Phys., 100: 438-444.
Yang T, Lua AC (2003). Characteristics of activated carbons prepared from pistachio-nut shells by physical activation. J. Colloid Interface Sci., 267: 408-417.
Zhang T, Walawender WP, Fan LT, Fan M, Daugaard D, Brown RC (2004). Preparation of activated carbon from forest and agricultural residues through CO_2 activation. Chem. Eng. J., 105: 53-59.

Physiochemical characterisation and level of potentially toxic metals in surface water around yauri abattoir, north western Nigeria

Yahaya M. I.[1] and Agbendeh Z. M.[2]

[1]Department of Chemistry, Sokoto State University, Sokoto State, Nigeria.
[2]Department of Applied Chemistry, Federal University Dutsin-Ma, Katsina State, Nigeria.

Physicochemical and potentially toxic metals statuses of surface water samples collected around the Yauri abattoir were examined. Health risk assessment of the toxicant via ingestion was interpreted by calculating the hazard quotient (HQ). The mean concentration values for total dissolved solid (TDS), total suspended solid (TSS), dissolved oxygen (DO), phosphate, and biochemical oxygen demand (BOD) were 1026.78, 565.22, 5.0, 8.89 and 484.64 mg/L respectively. The values are above those obtained for the control water sample as well as those of international safe limits for water. The values of pH (6.6), sulphate (86.91 mg/L), nitrate (41.45 mg/L) and chemical oxygen demand (COD) (903.30 mg/L) fell within world health organization maximum permissible limits for drinking water. The mean concentrations of Co (6.93 mg/L), Cu (14.45 mg/L), Fe (64.16 mg/L), and Zn (37.14 mg/L) were above both the control and WHO, EU and EPA safe limits for metals in water. On the other hand Cd (11.47 mg/L), Ni (35.49 mg/L), and Pb (41.94 mg/L) had a mean concentration values which are higher than the international standard limits and control water sample. The calculated values of HQ show that Cd, Co, Cu and Ni were of high risk, Fe with a medium and Pb and Zn are of low risk. Correlation coefficients revealed general positive and significant correlations between the pairs of metals in water. The results of this study showed some levels of pollution of the stream water indicating that the activities at the abattoir were contributing to the pollution load of water in the area.

Key words: Yauri, abattoir, toxic metals, physicochemical.

INTRODUCTION

Pollution of the environment by toxic metals even at low levels and their resulting long term cumulative health effects are among the leading health concerns all over the world. They are non-biodegradable, thus persisting for long periods in environmental ecosystems. Environmental pollution has generally become a threat to the existence of humanity and the ecosystem. Some pollution effects may lead to metabolic disorders and undesirable changes which in many cases cause severe injuries and health hazards (Alorge, 1992).

Abattoir can be defined as a premise approved and registered by controlling authorities for hygienic slaughtering, inspection, processing, effective preservation and storage of meat products for human consumption. However, meat processing activities in Nigeria are generally carried out in unsuitable buildings and by untrained staff or butchers who are most of the time unaware of sanitary principles (Olanike, 2002).

Abattoir activities may be another source of pollution since human activities such as animal production and meat processing have been reported to influence negatively on soil and natural water composition leading to pollution of the soil, natural water resources and the entire environment (Adesemoye et al., 2006). Activities at the abattoir are aimed at optimizing the recovery of edible portions of the meat processing cycle for human consumption. However, significant quantities of secondary waste materials are also generated during this process. For example, blood, fat, organic and inorganic solids, salts and chemicals added during processing operations are produced as wastes (RMAA, 2010; Steffen and Kirsten, 1989). Various parts of cattle such as muscle, blood, liver, kidney, viscera and hair have been found to contain potentially toxic metals (Kruslin et al., 1999; Jukna et al., 2006). The faeces of livestock (animal manure) consist of undigested food, most of which are: cellulose fibre; undigested protein; excess nitrogen from digested protein; residue from digested fluids; waste mineral matter; worn-out cells from intestinal linings; mucus and bacteria. Other components of undigested food include; foreign matter such as dirt consumed, calcium, magnesium, iron, phosphorous, sodium among others (Ezeoha and Ugwuishiwu, 2011). Abattoir effluent waste water has a complex composition and can be very harmful to the environment. For example, discharge of animal blood into streams would deplete the dissolved oxygen (DO) of the aquatic environment. Improper disposal of paunch manure may exert oxygen demand on the receiving environment or breed large population of decomposers (micro-organisms), some of which may be pathogenic. Also, improper disposal of animal faeces may cause oxygen-depletion in the receiving environment. It could also lead to eutrophication of the receiving system and increase rate of toxins accumulation in biological systems (Nwachukwu et al., 2011).

Mohammed and Musa (2012) reported that the improper disposal of abattoir effluent could lead to transmission of pathogens to human which may cause an outbreak of water borne diseases like diarrhoea, pneumonia, typhoid fever, asthma, wool sorter diseases, respiratory and chest diseases. Studies have shown that *Escherichia coli* infection source was reported to be undercooked beef which has been contaminated in abattoirs with faeces containing the bacterium (Bello and Oyedemi, 2009; Patra et al., 2007). It had also been reported that abattoir activities are responsible for the pollution of surface and underground waters, reduction of air quality as well as quality of health of residents within the surrounding environment (Katarzyna et al. 2009; Odoemelan and Ajunwa, 2008).

The above situations were even more worrisome in the developing countries where research efforts towards monitoring the environment have not been given the desired attention by the stake holders (Adesemoye et al. 2006).

The main purpose of this work therefore, is to study the pollution status of surface water around Yauri abattoir and assess whether the pollution load is sufficient to affect the health of the inhabitants of the areas who depend on this stream as their source of domestic and irrigation water. The results of the study will assist the regulatory bodies monitor more closely the activities at the abattoir as well as create public awareness about the health implications of abattoir activities on the environment and also establish a data bank for future reference.

MATERIALS AND METHODS

The study area

Yauri town in Yauri Local Government Area of Kebbi state northwestern Nigeria was the study area. It is located southward on the earthen bank of River Niger and falls within latitudes 10° N and 30° N and longitudes 3° W and 6° W of the globe. The area has flat topography with a few elevated areas. It is an extension of the Sokoto plain: dotted with some doom-shaped hills and complemented by a portion of the great River Niger and its numerous tributaries, which gently meanders on the landscape. Yauri abattoir is located some meters from Yauri main market close to Yauri River. Several animals (cows, goats, sheep and cattle) are slaughtered in this abattoir. Normal abattoir operations are carried out every day of the week during morning hours (5 to 11 am) and in the afternoon and evening when the need arise.

Sample collection

Six sampling stations were mapped out along the course of the river in the abattoir area at a distance of 50m from each other. The sampling stations were coded SS_1, SS_2, SS_3, SS_4, SS_5 and SS_6. Six replicate samples were collected from each of these stations and pooled together to obtain a representative sample for that station. Water sample (coded SS_{ctrl}) was collected at a point 60m upstream and served as a control. Water samples were collected in plastic containers previously cleaned by washing in non-ionic detergent. During sampling, sample bottles were first rinsed with the sampled water three times and then filled to the brim. The samples were labeled and transported to the laboratory, stored in a refrigerator at about 4°C prior to analysis (Akan et al., 2010). A total of forty two samples were collected for the research. The field research is carried out between the months of July and December, 2012.

Sample preparation and analysis

Each sample (100 ml) was transferred into a beaker and 5ml of concentrated HNO_3 was added. The beaker with the content was placed on a hot plate and evaporated down to about 20 ml. The beaker was cooled and another 5 ml of concentrated HNO_3 added. Each beaker was then covered with a watch glass and returned to

Table 1. Mean concentration (mg/L) of potentially toxic metals in the water sample around Yauri Abattoir.

Sampling points	Cd	Co	Cu	Fe	Ni	Pb	Zn
SS_1	22.30±0.03	8.13±2.10	10.21±0.11	50.60±1.30	49.03±2.30	40.51±0.34	62.14±0.04
SS_2	13.10±0.13	3.01±0.70	8.13±2.10	77.02±0.11	18.70±1.00	66.13±0.41	41.07±0.59
SS_3	5.09±0.30	11.68±0.50	19.32±4.40	63.50±0.70	31.61±1.30	38.14±0.03	29.46±1.01
SS_4	8.14±0.21	9.40±0.47	22.31±1.50	48.17±0.50	56.06±0.40	32.10±1.50	31.47±0.09
SS_5	10.07±0.11	7.34±0.09	15.65±2.03	89.11±0.31	30.20±1.60	18.63±1.05	57.72±0.25
SS_6	9.47±0.08	2.03±1.32	11.06±1.30	56.55±2.30	27.31±0.50	56.13±0.01	60.13±0.11
SS_{ctrl}	2.07±1.04	0.63±0.02	5.53±2.30	36.20±2.60	0.59±1.33	22.01±2.01	20.52±0.10
Range	5.09- 22.30	2.03-11.68	10.21-89.11	48.17-89.11	18.70-56.06	18.63-66.13	29.46-62.14
HQ	631.00	641.00	10.85	5.94	49.29	3.88	4.35
EPA*	0.25	23.00	9.00	300.00	52.00	2.5.00	120.00
EU	0.005	N/A	2.00	0.20	0.02	0.01	N/A
WHO	0.003	N/A	2.00	N/A	0.02	0.01	3.00

N/A = Not Available, * = ug/L.

the hot plate for more heating with the addition of few drops of HNO_3 until the solution appeared light coloured and cleared. The walls of the beaker and the watch glass were washed down with distilled water and the sample filtered to remove insoluble materials that could clog the atomizer. The volumes of the samples were made up to the mark (100 ml) with distilled water (Radojevic and Bashkin, 1999). A blank sample was similarly treated so as to give room for blank correction. This was done by transferring 100ml of distilled water into a beaker and digested as described above. Calibration standards were prepared from stock solutions by dilution and were matrix matched the acid concentration of the digested samples. The digested samples were then analyzed for potentially toxic metals using atomic absorption spectrophotometer alpha star model 4 (Chem Tech Analytical) at the Centre for Energy Research and Development of the Obafemi Awolowo University, Ile–Ife, Nigeria. The instrument was operated according to the instrument handbook and data were acquired with Hewlett Packard (HP) Pavilion 3134 software.

The method used for the determination of physicochemical parameters was as described by AOAC (2005) and reported elsewhere (Anon, 1992; Lovell and Colorado, 1983; Ademorati, 1996 Emmanuel and Solomon, 2012).

RESULTS AND DISCUSSION

Potentially toxic metals concentration

Table 1 is presented the mean concentrations (mg/L) of potentially toxic metals in surface water from Yauri. The results show differences in metal concentrations at various sampling stations and their controls. Iron was the most abundant metal recorded. The higher level of Fe recorded within the study area could be related to run-off from rusted metallic roofing sheets on the houses in the area, scrap metal dump sites and the abattoir refuse dump sites. All the metals with the exception of Cd, Ni and Pb were below the international maximum permissible limit (WHO, 2006; EU, 1998).

The sources of cadmium in the urban areas are much less well defined than those of lead, but metal plating and tire enforced with metals were considered the likely common anthropogenic sources of Cd in the street dust through burning of tires and bad roads. Cadmium high mean concentration levels at all the sampling points could be attributed to the above reason and in addition to rural/urban effluents along the river course and atmospheric precipitation. Cadmium is extremely toxic that it could cause adverse health effects to end user when water with high percentage is consumed and it is also toxic to fish and other aquatic organisms.

Lead and Nickel concentrations within the study area is pointed to the fact that naturally, Pb and Ni are distributed in surface waters due to weathering of minerals and atmospheric deposition. Also, Lead and Nickel recorded high values beside the abattoir activities could be related to technical uses, most of which are: electric storage batteries, leachate from sludge containing nickel-cadmium batteries, nickel plate items and emissions from burning of fossil fuels and gasoline which contain high levels of tetraethyl lead (TEL), which is still in use despite, its ban in 2004. Generally, the concentration of metals in water in the study area is higher than the control. It has been reported that potentially toxic metals, reaching excessive levels, can exert serious impact on humans, animals and plants because they are not biodegradable as they are retained indefinitely in the ecological systems and in the food chain (Omprakash et al., 2011).

Physicochemical evaluation

The physicochemical parameters derived from surface water in Yauri around the abattoir are listed in Table 2. It could be seen from the table that the temperature range is from 30.20 to 30.80°C which is lower than 32 to 34°C reported by Osibanjo and Adie (2007). Temperature

Table 2. Mean results of the physicochemical parameters of the abattoir's surface water.

Parameter	SS_1	SS_2	SS_3	SS_4	SS_5	SS_6	SS_{CRTL}	Range
T (°C)	28.40±0.03	31.30±0.10	30.70±2.10	31.80±1.00	30.50±0.09	30.40±1.15	30.40 ±0.11	30.2-30
pH	5.30±0.09	7.55±2.11	6.50±1.70	6.50±0.73	6.56±0.91	6.51±0.27	6.70 ±1.42	6.50-6.
TDS (mg/L)	630.0±3.17	710.0±3.10	2700.0±1.08	1320.0±2.10	260.0±1.00	540.7±2.40	110.0±5.10	250-70
TSS (mg/L)	330.6±1.01	680.3±1.70	1070.0±0.97	720.14±2.14	400.23±0.9	190.0±1.01	12.0±0.09	190-71
DO (mg/L)	3.10±0.00	2.60±0.10	6.00±0.07	8.00±0.00	4.00±0.01	5.00±2.01	5.90±0.02	2.00-9.
Sulphate (mg/L)	35.064±0.25	41.60±0.89	170.00±0.90	120.00±1.00	95.11±2.05	60.00±5.01	12.00±0.03	38.46-1
Phosphate (mg/L)	8.00±1.00	7.30±0.03	9.10±3.01	16.00±1.09	8.00±0.90	4.70±0.10	2.10±0.03	4.70-16
Nitrate (mg/L)	9.40±0.25	31.09±3.20	68.00±1.08	64.06±0.04	9.30±0.10	66.00±4.02	6.00±0.00	9.30-68
BOD (mg/L)	350.50±5.01	516.00±0.05	861.70±4.07	812.60±0.90	344.00±0.30	23.00±0.04	12.00±0.08	19.00-8
COD (mg/L)	540.03±1.06	1452.0±5.01	1508.0±9.03	928.0±0.98	444.0±3.04	548.0±1.90	156.0±0.06	444-15

influences the amount of dissolved oxygen in water which in turn influences the survival of aquatic organisms. The pH ranged from 6.50 to 6.70 with a mean value of 6.60 and falls within WHO standards and compares well with 4.9 to 7.2 reported by Masse and Masse (2002). Total Suspended Solids (TSS) and Total Dissolved Solids (TDS) values ranged from 190.00-710.14 and 250.00-2700.00 mg/L respectively. Their mean values are 565.22 and 1,026.78 mg/L TSS and TDS respectively which, is above WHO maximum permissible limit for TSS (20 mg/L) and TDS (200 mg/L) and also higher than the control. Total suspended solids relatively measures the physical or visual observable dirtiness of a water resource while TDS is an indicator of the degree of dissolved substances, such as metal ions in the water (Efe et al., 2005). Dissolved oxygen (DO) has a range of 2.0 to 9.0mg/L with a mean value of 50 mg/L. This value is higher than international permissible limit of 4 mg/L and also, higher than the control. Low DO may result in anaerobic conditions that cause bad odour. Biochemical Oxygen Demand (BOD) and Chemical Oxygen Demand (COD) ranges between 19.0-861.7 and 444-1508 mg/L with mean values of 484.63 and 903.3 mg/L respectively. The mean value for BOD is higher than the control and allowable limit of 20 mg/L (WHO, 2006). COD has a mean value lower than the permissible limit of 1000 mg/L. Though, the mean value of COD is lower, some points like SS_2 and SS_3 have values of 1,456 and 1,508 mg/L respectively which are higher than the international standards. Both BOD and COD are indices of organic pollution. BOD is not a specific pollutant indicator, but rather a measure of the amount of oxygen required by bacteria and other microorganisms engaged in stabilizing decomposable organic matter over a specified period of time. A high oxygen demand indicates the potential for developing Dissolved Oxygen sag as the microbiota oxidizes the organic matter in the water. Since nearly all organic compounds are oxidized in the COD test, COD results are always higher than BOD results. This was confirmed in this study with some samples (SS_2 and SS_3) exceeding the value of 1000 mg/L set by WHO.

Nitrate concentration ranges from 9.4 to 68 mg/L with a mean value of 41.45 mg/L. Even though the mean concentration value is low, some points analyzed have values higher than the set standards. It is reported that nitrate concentration above the permissible value by 45 mg/L is dangerous to pregnant women and poses a serious health threat to infants less than three to six months of age because of its ability to cause methaemoglobinaemia (Gelperim et al., 1975). Nitrates have a high potential to migrate into ground water since they are very soluble and do not bind to soil (Punmia and Jain, 1998). Phosphates were at relatively high concentration. All the water samples were above 5mg/L maximum permissible limit except for SS_6 (4.7mg/L). Phosphate enter water ways from human, animal waste and other sources like phosphorus rich bedrock, industrial effluents, fertilizer run-off, laundry and cleaning. Phosphates in water increase the tendency of troublesome algae to grow in the water (Esry et al., 1991). This causes eutrophication or over fertilization as it chokes up the water ways and uses up large amounts of oxygen. Sulphate concentration ranges from 40 to 170 mg/L with a mean value of 86.91 mg/L. This is lower than the maximum permissible limit of 250 mg/L set by WHO but higher than the control value of 12 mg/L. This implies that the activities in the abattoir are contributing to the pollution load of the stream and long term effect may subsequently lead to contamination of the surrounding water body.

Health risk assessment

Health risk assessment is normally based on a quantification of risk level in relation to two types of adverse effects: chronic (non-carcinogenic) and carcinogenic. Chronic risk level estimated was expressed as maximum hazard quotient (HQ_{max}) calculated for a group of evaluated elements and as hazard index (HI) calculated as a sum of HQ of all evaluated elements in every sample ($HI = \Sigma HQ_i$). Characterisation of the chronic risk level consists of threshold effects (tolerance chemical

Table 3. Correlation Matrix for pairs of the analyzed potentially toxic metals in water.

Metals	Cd	Co	Cu	Fe	Ni	Pb	Zn
Cd	1.000						
Co	0.334	1.000					
Cu	0.240	0.113	1.000				
Fe	0.249	-0.042	0.787*	1.000			
Ni	0.552	0.169	0.712	0.797*	1.000		
Pb	0.025	0.073	0.039	0.161	0.391	1.000	
Zn	0.365	0.060	-0.194	0.162	0.011	0.318	1.000

* Correlation is significant at the 0.05 level (2-tailed).

level) and is based on the presumption and manifestation of adverse chronic effects until the threshold, that is, the lifetime daily exposure level tolerated by human beings the so-called reference dose (RfD), is exceeded. The characterisation of carcinogenic risk level consists of a concept of non-threshold effects – that is, no dose is safe and risk-free and each level of exposure can generate a carcinogenic response (USEPA, 1989).

In the present study, health risk from increased concentrations of HM in the surface water was evaluated in relation to its chronic as well as carcinogenic effects, based on the calculation of average daily dose estimates and defined toxicity values for toxic HM (USEPA, 1999) according to the following relationships. The chronic risk level was computed as health risk assessment using CDI and HQ indices. The CDI through water ingestion was calculated using the USEPA (1992) equation below:

$CDI = C \times DI/BW$

Where C, DI and BW represent the concentration of HM in water (microgrammes per litre), average daily intake rate (2 L/ day) and body weight (72 kg), respectively (USEPA 2005).

Conversely, the chronic risk level was calculated (HQ) for non-carcinogenic risk using the following equation by USEPA (1999):

$HQ = CDI/RfD$

Where according to USEPA, the oral toxicity RfD values are 0.0005 mg/kg-day for Cd, 0.0003 mg/kg-day for Co, 0.037 mg/kg-day for Cu, 0.3 mg/kg-day for Fe, 0.02 mg/kg-day for Ni, 0.0036 mg/kg-day for Pb and 0.3 mg/kg-day for Zn, respectively.

The scale of chronic risk level (HQ) based on average daily intake (CDI) and reference dose (milligrammes per kilogramme-day) is classified based on the ratio of CDI/RfD indicating≤1 (no risk) if>1≤5 (low risk), if>5≤10 (medium risk) and if>10 (high risk).

Correlation analysis

One-way analysis of variance with parametric Pearson's correlation between mean potentially toxic metals concentration in the water samples standard statistical methods (Table 3) showed that all the metals were positively correlated except for Fe and Co and Zn and Cu which were negatively correlated. Pb, Cd, and Ni were significantly correlated. The positive and significant correlations between metals in the surface water samples suggest common source.

Conclusions

The results of this study revealed that the physicochemical parameters of the surface water around Yauri abattoir exceeded international recommended safe limits. The mean concentrations of Pb, Ni and Cd were also higher than the regulatory permissible limits. The positive correlations between the pairs of metals in the surface water suggest common anthropogenic source. Generally, the values of the physicochemical parameters and potentially toxic metals were higher in the surface water around the abattoir than the control samples. This implies that the activities at the Yauri abattoir were contributing to the pollution load of the surface water in the area and this has potential for full-blown environmental problems in the near future if not controlled. It is therefore, recommended that the activities of the abattoir should be monitored closely by relevant agencies and constant monitoring of the river water quality is needed to record any alteration in the quality and mitigate outbreak of health disorders and the detrimental impacts on the aquatic ecosystem and through bio-magnifications may enter the food chain thereby affecting the human beings as well.

Conflict of Interest

The authors have not declared any conflict of interest.

REFERENCES

Ademorati CM (1996) . Standard methods for water and effluents

analysis. Foludex Press Ltd, Ibadan. pp. 20-49.
Adesemoye AO, Opere BO, Makinde SC (2006). Microbial Content of abattoir waste water and its contaminated soil in Lagos, Nigeria. Afr. J. Biotechnol. 5(20):1963-1968.
Akan JC, Abdulrahman FI, Yusuf E (2010). Physical and chemical parameters in Abattoir waste water sample in Maiduguri metropolis, Nigeria. The pacific J. Sci. Tech. 11(1): 640-647.
Alorge DO (1992). Abattoir Design, Management and Effluent Disposal in Nigeria, University of Ibadan press.
Anon R (1992). Standard methods of Water and Wastewater Examination. 18thedition American Public Health Association Washington, D. C. 2:172.
AOAC (2005). Official Method of Analysis Association of Analytical Chemist. Washington, DC, 15th ed. 11–14.
Bello YO, Oyedemi, DT (2009). The impact of abattoir activities and management in residential neighbourhoods: A case study of Ogbomoso, Nigeria. J. Soc. Sci. 19(2):121-127.
Efe SI, Ogban FE, Horsfall M, Akporhornor EE (2005). Variation of physicochemical characteristics in water resources quality in Western Niger delta region, Nigeria. J. Appl. Sci. Environ. Man. 9:191-193.
Emmanuel OO, Solomon A (2012). Quality of sachet water and bottled water in Bolgatanga Municipality of Ghana. Res. J. Appl. Sci. Eng. Tech. 4(9):1094-1098.
Esry SA, Potash JB, Shiff C (1991). Effects of improved water supply and sanitation on ascariosis, diarrhea, dracunculiasis, hookworm infection, schistosomiasis and tachorna. Bulletin of WHO (5):610-614.
EU (1998). European Union Drinking Water Standards http://www.lenntech.com/EU's-drinking-water-standards.htm.
Ezeoha SL, Ugwuishiwu BO (2011). Status of abattoir wastes research in Nigeria. Nig. J. Technol. 30(2).
Gelperim A, Moses UK Bridge C (1975). Relationship of high nitrate community water supply to infants and fetal mortality. Illinois's Med. J. 147:155-156.
Jukna C, Jukna V, Suigzdaite J (2006). Determination of potentially toxic metalss in Viscera and Muscles of cattle. Bulgarian J. Vet. Med. 9(1):35-41.
Katarzyna RA, Monkiewicz J, Andrzej G (2009). Lead, Cadmium, Arsenic, Copper and Zinc contents in hair of cattle living in the area contaminated by a copper smelter in 2006-2008. Bull. Vet. Inst. Pulawy. 53:703-706.
Kruslin E, Hodel CM, Schurgast H (1999). Progress in diagnosis of chronic toxic metal poisoning by hair analysis. Toxicol. Lett. 88:84.
Lovell J, Colorado W (1983). Hach Water Analysis Hand book: 181-278.
Masse DI, Masse L (2002). Characterisation of effluents from six hog slaughter houses in Eastern Canada and evaluation of their impact influent treatment systems. Canada J. Agric. Eng. 42:139-146.
Mohammed S, Musa JJ (2012). Impact of Abattoir Effluent on River Land Use, Bida, Nigeria. J. Chem. Bio. Phys. Sci. 2(1):132-136.
Nwachukwu MI, Akinde SB, Udujih OS, Nwachukwu IO (2011). Effect of abattoir wastes on the population of proteolytic and lipolytic bacteria in a Recipient Water Body (Otamiri River). Global Res. J. Sci.1(1):40 – 422.
Odoemelan SA, Ajunwa O (2008). Potentially toxic metals status an physicochemical properties of agricultural Soil amended by sho term application of animal manure. J. Chem. Soc. Nig. 30:60-63.
Olanike KA (2002). Unhygienic operation of a city abattoir in Sout Western Nigeria: Environmental implication *AJEAM/RAGEE.* 4(1) 23-28.
OmPrakash M, Ashish G, Mahipat S, Rajay P (2011). Determination c Toxic Trace Metals Pb, Cd,Ni, and Zn in Soil by Polarographi Method. Int. J. Chem. Tech. Res. 3(2):599-604.
Osibanjo O Adie GU (2007). Impact of effluent from Bodija Abattoir o the physicochemical parameters of Oshunkaye stream in Ibadan. Afr J. Biotech. 6(15):1806-1811.
Patra RC, Swarup D, Naresh R, Kumar P (2007). Tail hair as a indicator of environmental exposure of Cows to lead and cadmium i different industrial areas. Eco. Toxicol. Environ. 66:127-131.
Punmia BC Jain AK (1998). Wastewater Engineering. Laxm publications, New Delhi.
Radojevic M Bashkin VN (1999). Practical Environmental Analysis. Th Royal Society of Chemistry: Cambridge, UK. 466.
Red Meat Abattoir Association (RMAA). (2010). Waste Management Red Meat Abattoir. Retrieved fron http://www.docstoc.com/docs/103302144/Waste-Management-%EE%9F%A6-Red-Meat-Abattoirs.
Steffen, Roberts, Kirsten Inc. (1989). Water and waste-wate management in the red meat industry. (P. 36). WRC Report No. 14 TT41/89. WRC, Pretoria.
United States Environmental Protection Agency (USEPA) (1989). Ris assessment guidance for superfund (RAGS): volume I- Human healt evaluation manual (HHEM) part A: baseline risk assessment, Interin Final (EPA/540/1-89/002). Office of Emergency and Remedia Response, Washington, DC.
United States Environmental Protection Agency (USEPA) (1992) Guidelines for Exposure Assessment, EPA/600/Z-92/001. Ris Assessment Forum, Washington, DC.
United States Environmental Protection Agency (USEPA) (1999) Guidance for performing Aggregate Exposure and Risk Assessmen office of Pesticide Programs, Washington, DC.
United States Environmental Protection Agency (USEPA) (2005) Guidelines for carcinogen risk assessment. Risk Assessment Forum Washington, DC (EPA/630/P-03/001F).
World Health Organization (WHO). (2006). Drinking Water Standards .http://www.lenntech.com/WHO's-drinking-water-standards06.htm.

A physico-chemical analysis of soil and selected fruits in one rehabilitated mined out site in the Sierra Rutile environs for the presence of heavy metals: Lead, Copper, Zinc, Chromium and Arsenic

P. O. Egbenda, F. Thullah and I. Kamara

Department of Chemistry, Fourah Bay College, University of Sierra Leone, Freetown, Sierra Leone.

The accumulation of heavy metals in soils especially in mining environments is of increasing concern to researchers in the Agricultural Industry. This is because the metals are biomagnified by plants. Accumulation of heavy and trace metals in plants occur by various sources but soil is considered the major one. Consumption of vegetables and fruits containing heavy metals is one of the main ways in which these elements enter the human body. Once in the body, heavy metals are deposited in bone and fat tissues, overlapping noble minerals and cause an array of diseases. The present study investigated the concentration of heavy metals that is, Cu, Zn, Cr, As and Pb in soil as well as mango (*Mangifera indica* L.) and cashew (*Anacardium occidentale*) fruit samples collected from the Mokaba rehabilitated site in the Sierra Rutile environs, to evaluate the possible health risks to human body through food chain transfer. Atomic absorption spectrophotometry was used to estimate the levels of these metals in the fruits and soil. Results showed that the concentrations of Pb and Cu in both soil and fruits are higher than the World health average values. However, Zn and Cr were found to be below the World health average values, whereas As was not detected. Translocation factors (TF) from soil to fruits were calculated from the data on levels of metals in both soil and fruits. The sampled plants showed high translocation factor values (TF > 1in almost all cases) implying that the plants could be labeled as accumulators of pollution. Pearson's product moment correlation showed a very strong relationship between soil and fruits. It can be concluded that the crops/plants grown in the rehabilitated lands in the Sierra Rutile environs absorb significant levels of some heavy metals from the polluted soil.

Key words: Rehabilitated, heavy metals, bioaccumulation, translocation, bioavailability, biomagnified.

INTRODUCTION

Heavy metals are significant environmental pollutants, and their toxicity is a problem of increasing concern for ecological, evolutionary, nutritional and environmental reasons. In addition to being non biodegradable, heavy metals have long biological half-lives as well as the potential to accumulate in different body organs, leading to unwanted side effects. One consequence of mineral exploitations is the exposure of metals to the earth's surface. After several years of operation, the Sierra Rutile mining Company has left behind lakes, mine spoil heaps and sand tailings in many areas of land that were once viable for agricultural activities. The mine spoil heaps and sand tailings are believed to contain various heavy metals in different forms. In order to restore the mined out areas into productive agricultural resources, the Company established land and water rehabilitation programmes. The Mokaba rehabilitated land is one of the mined out sites that was rehabilitated several years before the civil war in Sierra Leone. The site is about half a mile from Mokaba town, a fairly large settlement in the Impere Chiefdom, in Bonthe District, in the Sierra Rutile environs. A variety of economic trees such as mango, guava, cashew, coconut and oil palm have been grown at the Mokaba site. Fruits from these plants are harvested and sold to communities in the Sierra Rutile environs. Notably fruits and vegetables are rich sources of vitamins, minerals and fibers. They also have beneficial anti-oxidative properties. Sadly however, plants can take up heavy metals from contaminated soils through root systems. Consumption of fruits and vegetables contaminated with heavy metals may pose a risk to human health (Sal Jasir et al., 2005). The aim of this study is to carry out a physicochemical analysis of soil and fruits (cashew, *Anacardium occidentale* and mango, *Mangifera indica* L.) from the Mokaba site to ascertain soil pollution and to provide guidance for pollution assessment and control in the rehabilitated lands in the Sierra Rutile environs.

Heavy metals in soil environment

Heavy metal contamination of soil results from anthropogenic processes such as mining, smelting procedures and agriculture as well as natural activities (Aziz et al., 2004). Heavy metals are generally present in agricultural soils at low levels. Due to their cumulative behaviour and toxicity, however, they have a potentially hazardous effect not only on crops but also on human health (Slagle et al., 2004). High concentration of heavy metals in soils is toxic for soil organisms such as bacteria, fungi and higher organisms (Elvingson and Agren, 2004). In soil Lead tightly binds itself to organic soil particles which may decrease its mobility and reduce uptake by plants (Cooper et al., 1999). It has been suggested that the mobility of lead and copper is greater in sandy soils, with apparently very little organic matter, than in organic soils. Chromium exists in two possible oxidation states in soils namely Cr(III) and Cr(VI). The Cr(VI) ions are more toxic than Cr(III) ions. Because of the anionic nature of Cr(VI), its association with soil surfaces is limited to positively charged exchange sites, the number of which decreases with increasing soil pH. Cr(VI) was found to be highly mobile in alkaline soils (Griffin and Shimp, 1978) and can be reduced to Cr(III) under normal soil pH and redox conditions. Soil organic matter has been identified as the electron donor in this reaction (Bartlett and Kimble, 1976). The presence of sulfate can enhance Cr(VI) adsorption to kaolinite (Zarchara et al., 1988). The parameters that correlated with Cr(VI) immobilization in the soils were free iron oxides, total manganese, and soil pH. On the other hand, soil properties, cation exchange capacity, surface area, and percent clay had no significant influence on Cr(VI) mobility (Rai et al., 1987). Zinc is readily adsorbed by clay minerals, carbonates, or hydrous oxides. The greatest percent of the total Zn in polluted soils and sediments is associated with Fe and Mn oxides. Zinc hydrolyses at pH>7.7 and the hydrolyzed species are strongly adsorbed to soil surfaces. It also forms complexes with inorganic and organic ligands that will affect its adsorption reactions with the soil surface (Hickey and Kittrick, 1984). Acidic and sandy soils with low organic content have a reduced capacity for zinc absorption. Copper may exist in soils in the following forms: water soluble, exchangeable, organically bound, associated with carbonates and hydrous oxides of Fe, Mn and Al, and residual. Copper is adsorbed on the soil, forming an association with organic matter, Fe and Mn oxides, soil minerals, etc., thus making it one of the least mobile of the trace metals (Ioannou et al., 2003). The metal is retained in soils through exchange and specific adsorption mechanisms. Clay mineral exchange phase may serve as a sink for Cu in noncalcareous soils. In calcareous soils, specific adsorption of Cu onto $CaCO_3$ surfaces may control Cu concentration in solution (Cavallaro and McBride, 1978). Arsenic exists as either arsenate, (AsO_4^{3-}), or arsenite, (AsO_3^{3-}) in the soil environment. Arsenite is the more toxic form of arsenic. Arsenite compounds are reported to be 4 to 10 times more soluble than arsenate compounds. In the adsorption by kaolinite and montmorillonite, maximum adsorption of As(V) occurs at pH 5. Adsorption of arsenate by aluminum and iron oxides is a maximum at pH 3 to 4 and gradually decrease with increasing pH (Anderson et al., 1976). As(III), is also strongly pH-dependent. It was observed that an increase in sorption of As(III) by kaolinite and montmorillonite occur over a pH range of 3 to 9 (Griffin and Shimp, 1978). The maximum adsorption of As(III) by iron oxide occurred at pH 7. Adsorption of As(III) is rapid and irreversible on some soils. Formation of As(III) also may lead to the volatilization of arsine (AsH_3) and methyl- arsines from soils (Woolson, 1977). The loss of organic arsenic compounds from the soil was far greater than for the inorganic source of arsenic. As(III), can be oxidized to As(V) in which manganese oxides act as primary electron acceptor (Oscarson et al., 1983).

Uptake and translocation of heavy metals in fruits

Migration of metals in the soil is influenced by physical and chemical characteristics of each specific metal and by several environmental factors; the most significant appear to be: soil type, total organic content, redox potential, and pH (Murray et al., 1999). The fate of heavy metals in polluted soils is a subject of study because of the direct potential toxicity to biota and the indirect threat to human health via the contamination of groundwater and accumulation in food crops (Martinez and Motto, 2000). Heavy metal pollution of soil enhances plant uptake causing accumulation in plant tissues and eventual phytotoxicity and change of plant community (Gimmler et al., 2002). Compounds accumulate in living organisms any time they are taken up faster than they are broken down (metabolized) or excreted (O'Brien, 2008).The soil to plant transfer factor known as Translocation Factor (TF) is one of the important parameters used to estimate the possible accumulation of toxic elements, especially radionuclides through food ingestion (El-Ghawi et al., 2005). Translocation Factor (TF) is the transfer capability of heavy metals from soil to various parts of the plant. TF of heavy metals depends upon bioavailability of metals, which in turn depends upon its concentration in the soil, their chemical forms, difference in uptake capability and growth rate of different plant species (FAO/WHO, 2011). TF > 1, indicates that the plant translocate metals effectively from root to the shoot (Baker and Brooks, 1989). Most plants translocate inorganic and nutrient constituents from roots to leaves (Roselli et. al., 2003). Higher values of TF also suggest poor retention of metals in soil and/or more translocation in plants (indicating stronger accumulation of the respective metal by that fruit). The higher uptake of heavy metals in fruits may be due to higher transpiration rate to maintain the growth and moisture content of plants (Gildon and Tinker, 1981). A related study reported highest translocation factor for heavy metals through leafy vegetables. The TF does not present the risk associated with the metal in any form. The degree of toxicity of heavy metal to human beings depends upon their daily intake (Sridhara et al., 2008). Several studies have indicated that crops grown on soils contaminated with heavy metals have higher concentrations of heavy metals than those grown on uncontaminated soil (Nabulo, 2006). The translocation factor (TF) of heavy metals can be calculated as follows:

$$TF = \frac{\text{Metal concentration in fruit (shoot)}}{\text{Metal concentration in soil in which fruit was grown (root)}}$$

MATERIALS AND METHODS

Sampling

Soil samples were collected from the Mokaba site at different locations (A, B, C, D and E) and at variable depths ranging from 0 to 10 cm from the surface. Fresh ripe cashew fruits (*A. occidentale*) and mango fruits (*M. indica* L.) were collected from the trees at locations indicated in Figure 1 (not to scale).

Sample preparation

The soil samples were air–dried in an oven at 50°C temperature until constant weight. The cashew (*A. occidentale*) and mango (*M. indica* L.) fruits were washed thoroughly with distilled water, peeled, sliced and then dried in oven at 60°C. The dried fruit and soil samples were powdered in an agate mortar, homogenized and sieved in a 60 micron sieve. The powdered samples were stored in clean stoppered sterile bottles (Djingova et al., 1993; Keane et al., 2001).

Heavy metal analysis

Five grams of soil sample was placed in a 250 ml beaker followed by addition of 50 ml distilled water and 60 ml aqua regia (HNO_3: HCl;3:1). Boiling chips were added and the mixture digested on a hot plate at 100°C for one hour and at125°C for further 15 min to concentrate to 5 ml volume. The 5 ml volume concentrate wascooled, 1 ml 30% H_2O_2 added and then heated for further 10 min. The hot solution was again cooled and then treated with 3 ml 30% H_2O_2 before heating for another 10 min. 50 ml distilled water and 25 ml HCl was added to the solution and then heated to boiling. The resulting hot solution was cooled, filtered and then diluted to 250 ml with distilled water.

10 ml HNO_3 was added to 2 g of the powdered fruit sample in a 100 ml beaker and the mixture heated at 40°C for 15 min. To the cooled digest 5 ml concentrated HNO_3 was added and then heated for another 30 min at 40°C to concentrate to 5 ml volume. 5 ml H_2O was added to the solution followed by 3 ml 30% H_2O_2. The beaker was covered and the solution gently heated until vigorous effervescence occurred. 1 ml 30% H_2O_2 was added to the solution followed by gentle heating until effervescence subsided. 5 ml concentrated HCl was added followed by 10 ml distilled water and the resulting solution heated for 15 min, cooled, filtered using ash less filter paper and then diluted to 60 ml with distilled water.

The concentrations of Pb, Cr, Cu, Zn and As were determined in soil, cashew and mango fruit samples using the PG-800 Atomic Absorption Spectrophotometer. The working standards were prepared by diluting concentrated stock solution of 1000 mg/L for Cu, Cr and Zn and 1000 µg/L for As, and Pb in deionized water. The matrix modifiers $NH_3H_2PO_4$ and Mg $(NO_3)_2$ were used.

RESULTS AND DISCUSSION

The results for pH and conductivity measurements of the soil samples in five different locations are given in Table 1. The WHO and E.U recommended limits for pH and conductivity in soil are given in Table 2. The pH of the soil at the Mokaba site has a range of 4.90 to 5.24 and mean value of 5.11 (Table 1).This implies that the soil is acidic compared to the pH values of WHO (6-5) and EU (6-8). The pH has a great impact on bioavailability in the soil. At high pH metals tend to form insoluble metal mineral phosphates and carbonates, whereas at low pH they tend to be found as free ionic species or as soluble organometals and are more bio available (Sandrin and Hoffman, 2007; Twiss et al., 2001; Rensing and Maier,

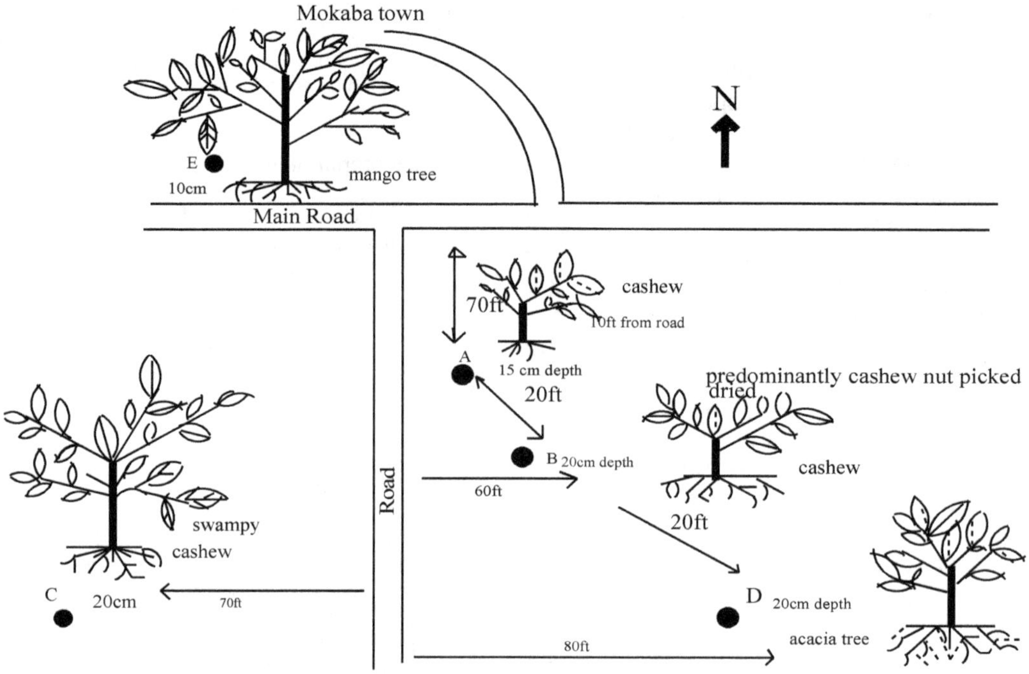

Figure 1. Sketched map of the samples collection sites.

Table 1. pH and conductivity in the soil.

Location	pH	Conductivity (S/cm)
A	4.90	0.02
B	5.24	0.01
C	5.12	0.01
D	5.14	0.01
E	5.15	0.01
Mean value	5.11	0.01

Table 2. Recommended values for pH and Conductivity by WHO and EU in soil.

Parameter	WHO	EU
pH	6-5	6-8
Conductivity (µS/cm)	250	250

2003; Naidu et al., 1997). The measured conductivity values of the soil samples (0.01-0.02 S/cm) testify the presence of trace metal ions or ionizable materials in the soil.

Table 3 gives the concentrations of the metals in the soil and fruit samples (in mg/L). The WHO recommended limits of the investigated metals Pb, Zn, Cr, Cu and Asin mg/L are given in Table 4. Both mango and cashew fruit samples showed higher average metal content for the elements Cu, Pb and Zn (Table 3) and relatively lower values for Cr. This is because Cr accumulates mainly in roots followed by stems and leaves and that only small amount of the metal is translocated to leaves (Tiwar et al., 2009; Huffmann and All away, 1973). Also the mango fruit tend to bio accumulate Pb, Zn and Cu more whereas cashew fruit tend to bio accumulate Zn and Cu more. The mango fruit bio accumulates Pb metal more than the cashew fruit whereas the cashew fruit bio accumulates

Table 3. Average concentrations of metals in soil and fruits (in mg/L).

Element	Soil (mg/L)	Mango (mg/L)	Cashew (mg/L)
Pb	0.053	0.355	0.055
Zn	0.222	0.396	0.685
Cr	0.027	0.019	0.025
Cu	0.817	0.847	1.140
As	ND	ND	ND

Table 4. WHO recommended permissible limits of the metals (Pb, Zn, Cr, Cu and As in mg/L in fruits.

Metal	Permissible limit
Pb	0.050
Zn	<5.000
Cr	<0.050
Cu	<1.000
As	0.010

Table 5. Translocation factors (TF) in mango and cashew fruit samples.

Metals	TF in mango fruit	TF in cashew fruit
Pb	6.698	1.038
Zn	1.784	3.086
Cr	0.704	0.926
Cu	1.037	1.395

Zn, Cu and Cr more than the mango fruit. Cu has the highest concentration in the soil (0.817 mg/L) and fruits (0.847 mg/L in mango and 1.140 mg/L in cashew). This could be due to the presence of different forms of copper in the soil e.g. water soluble, exchangeable, organically bound, associated with carbonates and hydrous oxides of Fe, Mn, Al, and residual (Ioannou et al., 2003). Moreover the levels of Zn and Cu were found to be higher in the cashew fruit (younger trees) than in the mango fruit (older trees). This suggests that younger plants bio accumulate heavy metals more than older ones. In both soil and fruits the average concentrations of heavy metals follow the order: Cu > Zn >Pb> Cr. Arsenic was not detected in any of the fruit samples.

Table 3 also shows concentration of metals to be higher in fruits than in the soil. This could be due to loss in soil by leaching and uptake by the plant growing there including the plants under investigation. The sampled plants showed high translocation factor values (TF > 1in almost all cases) implying that the plants could be labeled as accumulators of pollution (Table 5). Trends in translocation factor are: for Mango (*M. indica* L.), Pb>Zn>Cu>Cr, and for Cashew (*A. occidentale*), Zn > Cu > Pb > Cr. The high TF values could be attributed to low retention rate of the metals in the soil. Pearson's product moment correlation reveal a very strong correlation between soil and mango (R_m^2 = 0.925) and between soil and cashew (R_c^2 = 0.933) fruits. This implies that the source of the extra metal concentrated by the fruits is most likely the soil.

Conclusion

This study revealed that the soil in the rehabilitated Mokaba site carry significant levels of heavy metals. The buildup of heavy metals in soil profile may prove detrimental not only to plants and animals which bio accumulate them, but also to consumers of the harvested fruits from the farms. It is therefore suggested that other economic trees such as rubber (for rubber), acacia and eucalypti (for charcoal) be cultivated in the rehabilitated lands to reduce human health risk to metal pollution.

Conflict of Interest

The authors have not declared any conflict of interest.

REFERENCES

Anderson M C, Ferguson JF, Gavis J (1976). Arsenate adsorption on amorphous aluminum hydroxide. J. Colloid Interface Sci. 54:391-399.

Aziz HA, Yusoff MS, Adlan MN, Adnan NH, Alias S (2004). Physico-chemical removal of iron from semi-aerobic landfill leachate by limestone filter. Waste Manag. 24:353-358.

Baker AJM, Brooks RR (1989). Terrestrial higher plants which hyperaccumulate metallic elements – a review of their distribution, ecology and phytochemistry. 1:81–126.

Bartlett RJ, Kimble JM (1976). Behavior of chromium in soils: II. hexavalent forms. J. Environ. Qual. 5:383-386.

Cavallaro N, McBride MB (1978). Copper and cadmium adsorption characteristics of selected acid and calcareous soils. Soil Sci. Soc. Am. J. 42:550-556.

Cooper EM, Sims JT, Cunningham SD, Huang JW, Berti WR (1999). Chelate-assisted phytoextraction of lead from contaminated soils. J Environ Qual. 28:1709-1719.

Djingova R, Kuleff I, Andreev N (1993). Comparison of the ability of several vascular plants to reflect environmental pollution. Chemosphere. 27:1385–1396.

El-Ghawi UM, Bejey MM, Al-Fakhri SM, Al-Sadeq AA, Doubali KK (2005). Analysis of Libyan arable soils by means of thermal and epithermal NAA. Arabian J. Sci. 30:147-154.

Elvingson P, Agren C (2004). Air and the environment by Elvingson and Christer. The Swedish. NGO Secretariat.

http://www.airclim.org/sites/default/files/documents/Air_and_the_Environment_0.pdf
FAO/WHO (2011). Joint FAO/WHO food standards programme. Codex Committee on 10 contaminants in foods. 5th session 64-89.
Gildon PB, Tinker A (1981). Heavy metal tolerant strain of mycor-rhizal fungus. Trans. Br. Mycol. Soc. 77:648–649.
Gimmler H, Carandang J, Boots A, Reisberg E, Woitke M (2002). Heavy metal content and distribution within a woody plant during and after seven years continuous growth on municipal solid waste (MSW) bottom slag rich in heavy metals. Appl. Bot. 76:203-217.
Griffin RA, Shimp NF (1978). Attenuation of pollutants in municipal landfill leachate by clay minerals. EPA-600/ 2-78-157.
Hickey MG, Kittrick JA (1984). Chemical partitioning of cadmium, copper, nickel, and zinc in soils and sediments containing high levels of heavy metals. J. Environ. Qual. 13:372-376.
Huffmann EW Jr, Allaway HW (1973). "Chromium in plants: distribution in tissues, organelles and extracts and availability of bean leaf Cr to animals". J. Agric. food Chem. 21(6):982–986.
Ioannou A, Tolner L, Dimirkou A, Füleky GY (2003). Copper adsorption on bentonite and soil as affected by pH. Bulletin of the SzentIstván University, Gödöllo Hungary. pp. 74-84.
Keane B, Collier MH, Shann JR, Rogstad SH (2001). Metal contents of dandelion (*Taraxacum officinale*) leaves in relation to soil contamination and airborne particulate matter. Sci. Total Envion. 281:63−78.
Martinez CE, Motto HL (2000). Solubility of lead, zinc and copper added to mineral soils. Environ. Pollut. 107:153-158.
Murray KS, Cauvet D, Lybeer M, Thomas JC (1999). Particle size and chemical control of heavy metals in bed sediment from the Rouge River, southeast Michigan. Environ. Sci. Technol. 33:987-992.
Nabulo G (2006) Assessment of heavy metal contamination of food crops and vegetables grown in and around Kampala city, Uganda. Ph.D. Dissertation, Makerere University.
Naidu R, Kookana RS, Sumner ME, Harter RD, Tiller KE (1997). Cadmium sorption and transport in charge soils: A review. Environ. Qual. 26:602–617.
O'Brien J (2008). What are heavy metals. Plant nutrition newsletter. April. 9(3).
Oscarson DW, Huang PM, Liaw WK, Hammer UT (1983). Kinetics of oxidation of arsenite by various manganese dioxides. Soil Sci. Soc. Am. J. 47:644-648.
Rai D, Sass BM, Moore DA (1987). Chromium (III) hydrolysis constants and solubility of chromium (III) hydroxide. Inorg. Chem. 26:345-349.
Rensing C, Maier RM (2003). Issues underlying use of biosensors to measure bioavailability. Ecotoxicol. Environ. Sat. 56:140–147.
Roselli W, Koker C, Boschi K (2003). Phytoextraction capacity of trees growing on metal contaminated soil. Plant soil. 256:265−272.
Sal jasir M, Shaker A, Khalid MA (2005). "Deposition of heavy metals on green leafy vegetables sold on road sides of Riyadh city, Saudi Arabia", bulletin of environmental contamination and toxicology, 75(5):1020-1027.
Sandrin TR, Hoffman DR (2007). Bioremediation of organic and metal contaminated environments: effects of metal toxicity, speciation and bioavailability or biodegradation. Environ. Bioremed. Technol. pp. 1–34.
Slagle A, Skousen J, Bhumbla D, Sencindiver J, McDonald L (2004). Trace Element Concentrations of Three Soils in Central Appalachia. Soil Sur. Hortic. 45(3).
Sridhara N, Chary CT, Kamala D, Samuel SR (2008). Assessing risk of heavy metals from consuming food grown on sewage irrigated soils and food chain transfer. Ecotoxicol. Environ. Safe. 695:13-524.
Tiwar KK, Dwivedi S, Singh NK, Rai UN, Tripathi RD (2009). "Chromium (VI) phytotoxicity and oxidative stress in pea (*Pisum sativum* L): biochemical changes and translocation of essential nutrients". J. Env. Biol. 30(3):389–394.
Twiss MR, Errecalde O, Fortin C, Campbell PGC, Jumarie C, Denizeau F, Berkelaar E, Hale B, Van Rees K (2001). Coupling the use of computer Chemical speciation models and culture techniques in laboratory investigations of trace metal toxicity. Chem. Spec. Bioavailab. 13:9–24.
Woolson EA (1977). Fate of arsenicals in different environmental substrate. Environ. Health Perspect. 19:73- 81.
Zarchara JM, Cowan CE, Schmidt RL, Ainsworth CC (1988). Chromate adsorption on kaolinite. Clays Clay Miner. 36:317-326.

Dichloro-diphenyl-trichloroethane (DDT) residue levels in marketed Silver Fish (*Rastreneobola argentea*) caught from major water bodies in Uganda

Proscovia Nnamuyomba[1], Jolocam Mbabazi[2] and Muhammad Ntale[2]

[1]Department of Chemistry, Gulu University, P. O. Box 166, Gulu, Uganda.
[2]Department of Chemistry, Makerere University, P. O. Box 7062, Kampala, Uganda.

Silver fish (*Rastreneobola argentea*) samples from Lake Victoria and Lake Kyoga were analyzed for dichloro-diphenyl-trichloroethane (DDT) and metabolites using a Gas Chromatograph equipped with an Electron Capture Detector. Confirmation of results was by Gas Chromatography equipped with Mass Spectrophotometer (GC-MS). DDT and metabolites were detected in 90% of the total samples (63 out of 70). *pp'* DDT was the most frequently detected residues in the samples whereas *op'* DDE was detected in none of the samples. The mean ΣDDT concentrations in fish from Lake Victoria varied from 0.147 to 0.396 µg/kg while that in Lake Kyoga varied from 0.207 to 0.506 µg/kg dry wt of fish sample. The concentration of total DDT in the fish samples ranged from 0.147 to 0.506 µg/kg dry wt. This was far below the recommended FAO limit of 5000 µg/kg for fish (edible portion), implying that the fish are safe for human consumption.

Key words: Dichloro-diphenyl-trichloroethane (DDT) and metabolites, *Rastreneobola argentea*, Uganda water bodies.

INTRODUCTION

1,1,1-Trichloro-2,2-bis(p-chlorophenyl)ethane dichloro-diphenyl-trichloroethane (DDT) is one of the "dirty dozen" persistent organic pollutants (UNEP, 2008) whose manufacture and use was banned in many countries worldwide in the 1970's (Tsuda, 2012) except in cases of public health emergency (ASTDR, 2002). In Uganda, the use of DDT was prohibited in the mid 1980's (Kasozi et al., 2006), however; it was used in the recent past for controlling malaria in the districts of Oyam and Apac in Northern Region. DDT residues are still circulating in Ugandan ecosystems that are simultaneously being exposed to increasing pollution loads resulting from rapid national population growth (Rusongoza, 2003) and urbanization. This is the case in the region surrounding Lake Victoria, a lake that is regarded as the most important natural resources in East Africa, with fishing as the main economic activity. Fish is an important object for research of water ecosystem contamination by substances due to its high position in the food chain.

Lake Kyoga on the other hand may receive DDT and metabolites through surface run-off and atmospheric drifts, being the nearest water body to the districts of

Apac and Oyam, where indoor residual spray (IRS) was carried out. Uganda re-introduced the use of DDT for malaria control in the districts of Oyam and Apac in 2008 in accordance with the recommendations and guidelines of WHO and in line with the Stockholm convection (UNEP, 2008). The Northern Region is the most vulnerable to malaria in the whole country and possibly in the whole world: it lies within the tropical Lake Kyoga basin which provides an ideal breeding ground for mosquitoes (Bimenya et al., 2009). In spite of the previous use of the pesticide in the districts of Oyam and Apac, no monitoring of its levels has been carried out. However, a recent study identified op' and pp' DDT isomers in vegetables grown in homes which participated in IRS (Mukasa, 2012), a clear indication of recent exposure.

Lake Victoria (UBOS/UDHS, 2006) and Lake Kyoga (LAKIMO, 2004) are the most important fisheries in Uganda. The fisheries sector contributes greatly to the Ugandan economy in terms of income, employment and export revenue. For example in 2001 fish products exports were valued at US $ 80.4 million, in 2002 at US $ 87.57 million (Nyombi and Bolwig, 2004) and in 2005 at US $ 142 million (UBOS/UDHS, 2006). Fish is a protein source for 24.7 million people in Uganda (UBOS, 2002) and other consumers in the East African Region. The demand for Ugandan fish has increased drastically over the last two decades due to rapid domestic population growth and the emergence of an export market for especially Nile perch and Nile tilapia (Odada et al., 2004). The decline in Nile perch and Nile tilapia has increased the demand for non-export small fish species on the domestic market. Silver fish, *Rastreneobola argentea* is the most abundant non-export small fish species found in Lakes. It is predominantly found in Lake Kyoga and to a lesser extent in Lake Victoria (Mbabazi and Wasswa, 2010). The silver fish is caught by light attraction, where Kerosene pressure lamps are floated on water attracting fishes to light and are pulled out of the water in their thousands with scoop-nets (FAO). They are then sun-dried directly on the beaches.

Previous studies indicate that DDT residues are still detected in various compartments of the environment such as Lake Victoria sediments (Wasswa, 2009) and fish (Kasozi et al., 2006; Kyarimpa, 2007; Ogwok et al., 2009; Sebugere et al., 2008; Simon et al., 2004) from different water bodies in the country. DDT residues have previously been found in *Lates niloticus* and *Oreochromis nioticus* in Lake Victoria (Kasozi et al., 2006). Similarly, Bimenya et al. (2007) and Sebugere et al. (2008) found DDT residues in numerous fish species namely: *Bagnus docmac, Protopterus aethiopinus, Haplochromi nigripinnis, oreochromis niloticus* and *Clarius gariepinus* in Lake Edward. The presence of *op'* and *pp'* DDT isomers in Lake Victoria sediments (Wasswa, 2009) and fish from various water bodies in the country suggests that, Ugandan water bodies still receive considerable inputs of DDT residues from various anthropogenic sources such as direct discharges or indirectly from river flows and run-off, as well as from long distance transport through the atmosphere. Many studies have been conducted on the DDT residues contamination of the aquatic environment in Lake Victoria and Lake Edward. However, no studies have investigated DDT residues in non-export small fish species which are considered the most important protein source for the local population. The present work was undertaken to find out the status of DDT and its metabolites as well as to determine their residue levels in silver fish from Lake Victoria and Lake Kyoga. In this study, we incorporated modifications in the sample extraction procedure which necessitated method validation before application in the determination of DDT residues in fish samples. The validated method was then used to determine DDT residues in fish from Lake Victoria and Lake Kyoga so as to establish the extent of environmental contamination by this pesticide and also to evaluate the toxicological significance to the health of the consumers.

MATERIALS AND METHODS

Study area

The study area comprised of, seven landing sites, four on Lake Victoria and three on Lake Kyoga. Sampling was conducted for 1 year starting March 2012. Four stations (Gaba, Port Bell, Kaseny and Kigungu) were selected on Lake Victoria and three (Namasale Biko and Kayago) on Lake Kyoga. The various sampling stations on Lake Victoria and Lake Kyoga are shown in Figure 1. Lake Victoria is located along the equator at 0° 30' N, 3° 00' E and 31° 39' N, 34° 53' E, while Lake Kyoga is situated between longitudes 32° 10' and 34° 20' East, and between 1° 00' and 2° 00' North.

Samples

Silver fish, *R. argentea* were bought from the selected stations on Lake Victoria (V1-V4) and Lake Kyoga (K1-K3). Equal amounts of silver fish, were bought from four different people at each landing site and pooled to obtain a representative sample. Samples were then wrapped in aluminium foil, placed in air tight bags and appropriately labeled. They were then kept in cold boxes containing ice during transportation to the laboratory where they were kept at -18°C until they were extracted for DDT residues.

Chemicals and standards

Pesticide residue grade solvents: ethyl acetate, hexane and acetonitrile were supplied by the British Drug Houses (BDH, UK). The chemicals were used for extraction, analysis and confirmation. The purity of the solvents was checked by using a GC-ECD to ensure that no detectable traces of pesticides were contributed by solvents. The organochlorine pesticides standards (DDT and metabolites) as well as internal standard parathion ethyl were obtained from Dr. Ehrenstorfer GmbH (Augsburg, Germany).

Sample extraction and delipidation

The extraction of DDT residues from fish samples was performed

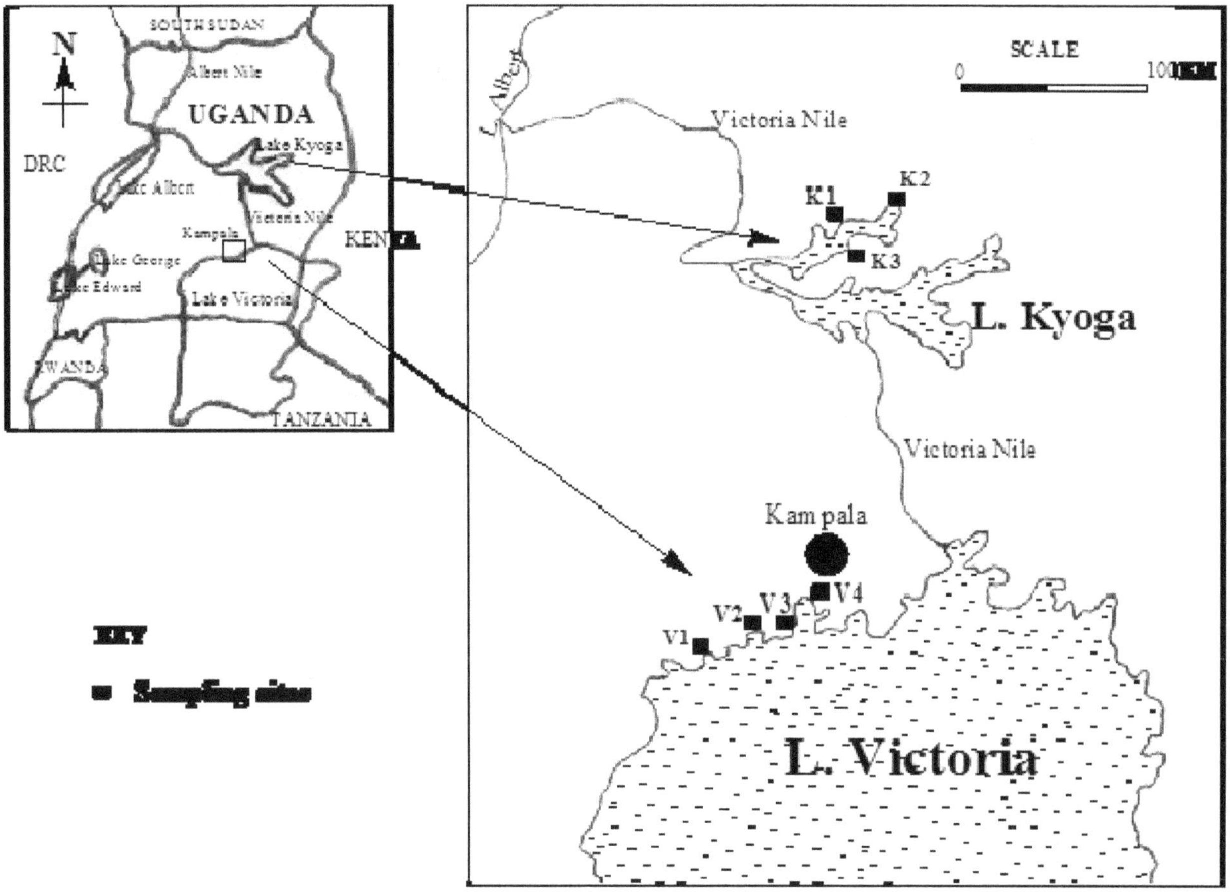

Figure 1. Map showing sampling sites on Lakes Victoria and Kyoga. Inset is the map of Uganda showing the location of Lake Victoria and Lake Kyoga.

according to the method described by Åkerblom (1995). Portions of 10 g of dry fish was weighed and ground with a known quantity of sodium sulphate using a mortar and a pestle. The powder was then extracted with ethyl acetate (50, 20, 20 and 20 ml). The combined extract was concentrated using a rotary evaporator at a temperature of 40°C. The concentrated extract was then dissolved in 30 ml of acetonitrile and stored in a freezer at -18°C and allowed to stand overnight for lipids precipitation and separation. The cold extract at -18°C was immediately filtered with glass wool to remove frozen lipids. The precipitated lipids on the flask surface were re-dissolved in 10 ml acetonitrile to perform filtration again by the same procedure. The filtered extract was concentrated to 1 ml by rotary evaporation and dried under nitrogen atmosphere to follow Florisil clean up procedure.

Sample clean up by florisil

A 15 cm long glass column of 4 mm id was plugged with glass wool and washed with hexane (10 ml). It was then packed with florisil (10.0 g) followed by anhydrous sodium sulfate (4 g). The column was lightly tapped to compact the florisil bed and then rinsed with hexane (5 ml) to remove any impurities. The column was conditioned with 5 ml of hexane/acetone 9:1 and then with 5 ml hexane. The concentrated sample was applied to the column and eluted with 10 ml hexane/acetone 9:1 mixture. The eluate was then concentrated at 40°C with a rotary evaporator to about 1 ml and then to complete dryness by nitrogen stream. The residue was dissolved in hexane (1 ml) for GC-ECD analysis.

Analysis of samples

For GC analysis, a Varian (CP-3800, Palo Alto, CA, USA) gas chromatograph equipped with a 63Ni ECD (Electron Capture Detector) fitted with both semi-polar (CP-Sil 19 CB, J & W Scientific, Folsom, CA, USA), and non-polar (CP-Sil 8 CB, J &W Scientific, Folsom, CA, USA) fused-silica capillary columns (30 m length, 0.25 mm i.d. and 0.25-µm film thickness) was used. Hydrogen (99.9% purity) with a flow rate electronically set at 1.2 ml min^{-1} was used as the carrier gas, while Nitrogen was used as an auxiliary gas for the ECD at a flow rate of 30 ml min^{-1}.

An oven temperature was programmed as follows: 90°C for 1 min, 30°C min^{-1} to 180°C, 4°C min^{-1} to 260°C, and maintained at this temperature for 16 min for both categories of columns. The temperatures of the injector and ECD detector were 250 and 300°C, respectively. A Turbochrom (Perkin–Elmer Corporation, 1989-1995, Norwak, CT, USA) 4.0 Chromatography station was used for chromatographic data processing. The GC was operated in a splitless mode and the injection volume was 1 µl for each injection. Identification and quantification were accomplished by comparison with reference standards obtained from Dr.

Table 1. Mean values of slopes, intercepts and regression coefficients from calibration curves.

Pesticide	Slope	Intercept	Regression coefficient (r^2)
*pp'*DDE	1.001	-0.006	1.000
*op'*DDE	1.002	+0.102	0.998
*op'*DDD	0.990	+0.047	0.999
*pp'*DDD	1.014	-0.003	0.999
*pp'*DDT	0.971	+0.035	0.998
*op'*DDT	1.100	-0.005	0.997

EhrenstorferGmbH, Augsburg, Germany.

The concentration (µg/kg) of each component in the sample was calculated from its chromatogram using the peak area and concentration of the reference standards using the formula below.

$$\frac{1000 \times Concentration\ of\ standard \times Peak\ area\ of\ sample \times Volume\ of\ concentrated\ extract}{Peak\ area\ of\ standard \times Weight\ extracted}$$

Confirmation of results

An Agilent 6890N GC–MS, USA version with a fused silica capillary column (HP-5MS) of 30 m length, 0.25 mm i.d. and 0.25 µm film thickness was used for confirmation of the results. The GC-MS used was equipped with a selective mass detector (Agilent 5975 inert XL Quadrupole, Palo Alto, CA, USA). The operating conditions were: An oven temperature was programmed as follows: 90°C for 1 min, 30°C min^{-1} to 180°C, 4°C min^{-1} to 260°C, and maintained at this temperature for 16 min. The temperatures of the injector and MS detector were 250 and 300°C, respectively. Helium was used as the carrier gas at 1 ml min^{-1} flow rate. The GC-MS was operated in a splitless mode with a purge-off of 1 min and the injection volume was 1 µl for each injection. The MS solvent delay time was 3.57 min and the scanned mass range was 50 to 550 m/z. The full scan ion monitoring mode was used for the determination of DDT and metabolites. Identification of the analytes was done using the internal standards method. Data acquisition and processing was achieved using GC-MSD Chemstation Software (G1701dad.02.0sp1, JAS CWA, USA).

Method validation

Linearity

The linearity of the method was evaluated by analyzing in triplicate, a set of 6 DDT and metabolites standard solutions in hexane ranging in concentration from 0.03 to2.7 µg/ml. Peak area ratios of the analytes to the internal standard were plotted against concentrations. The regression equations with the slopes, intercepts and correlation coefficient (r^2) were generated with Microsoft Excel software and are presented in Table 1.

Recovery and precision

Recovery and precision were determined by standard addition method (ICH, 1995) at 3 spiking levels of 0.05, 0.1 and 0.15 µg/ml of mixtures of *pp'* DDE, *op'* DDE, *op'* DDD, *pp'* DDD, *op'* DDT and *pp'* DDT and analyzing the samples in triplicates under repeatable and internal reproducible conditions. The mean recovery ranged between 81 to 124% and the coefficients of variation (CV) were ≤ 10%.

Limit of detection (LOD) and limit of quantification (LOQ)

In order to determine the sensitivity of the GC-ECD, nine independent fish samples (from control site) were fortified at three different low concentrations of each pesticide (ICH, 1995) and analyzed in triplicates. The values of LOD and LOQ were obtained by calibration curve method (they were based on the residual standard deviation and slope of the regression line) (ICH, 1994, 1995). The LOD and LOQ were calculated as 3.3 and 10SD/b respectively; where SD is the Standard Deviation of the GC chromatographic area (response) and b is the slope of the regression line of each pesticide. The values of LOD obtained were 0.013, 0.021, 0.030, 0.021, 0.024 and 0.058 µg/ml for *pp'* DDE, *op'* DDE, *op'* DDD, *pp'* DDD, *op'* DDT and *pp'* DDT respectively; meanwhile the values of LOQ were 0.044 µg/ml (*pp'* DDE), 0.071 µg/ml (*op'* DDE), 0.099 µg/ml (*op'* DDD), 0.069 µg/ml (*pp'* DDD), 0.074 µg/ml (op' DDT) and 0.065 µg/ml (*pp'* DDT).

Statistical analysis

The data was subjected to Analysis of variance (one way, ANOVA) using Graph pad Instat package program to test site differences.

RESULTS AND DISCUSSION

Levels of DDT residues in fish samples

70 samples of *R. argentea* from Lake Victoria and Lake Kyoga were analyzed for DDT and metabolites. Residues of DDT and its metabolites were detected in 90% (63 out of 70) of all the samples analyzed. Figures 2 and 3 show some of the selected GC chromatograms of DDT and metabolites detected in *R. argentea* from Lake Victoria and Lake Kyoga respectively. The mean concentrations on dry weight basis of *R. argentea* of *op'* DDE, *pp'* DDE, *op'* DDD, *pp'* DDD, *op'* DDT, *pp'* DDT and ΣDDT in the fish samples in different parts of Lake Victoria and Lake Kyoga are shown in Table 2. The proportion of DDT and its metabolites in the fish samples showed predominance of *pp'* DDT (39.4%), followed by *pp'* DDD (24%), *pp'* DDE (21%), *op'* DDT (8.3%), *op'* DDD (7.3%) and *op'* DDE was detected in none of the samples. Predominance of *p,p'*-DDT in the fish samples suggests recent exposure of the aquatic ecosystems to DDT, may be due some illegal usage or dumping. This pattern is similar to that reported in fish from Lake Songkhla, Thailand (Kumblad et al., 2001). The moderate percentage composition of *pp'*DDD

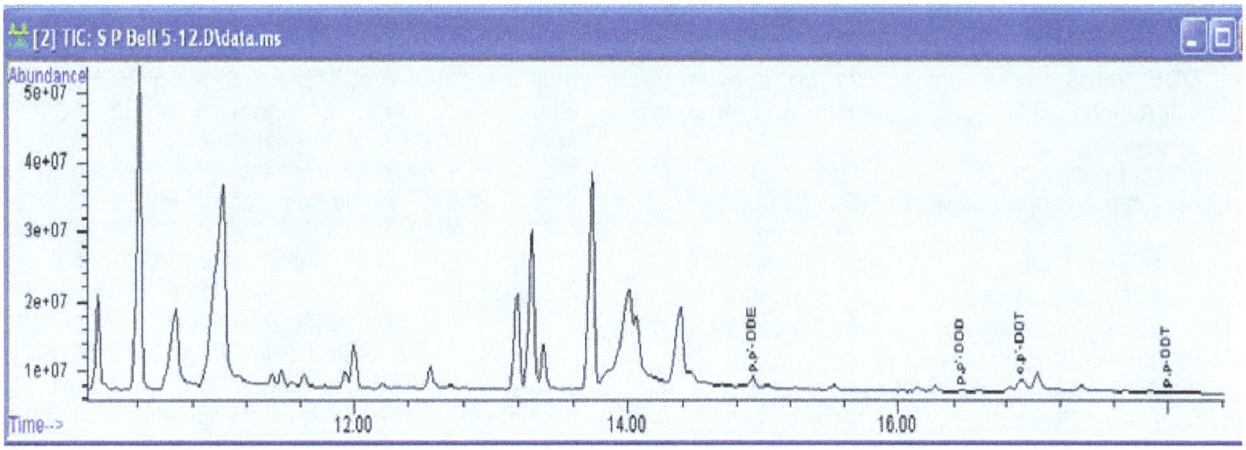

Figure 2. GC Chromatogram of DDT and metabolites in *R. argentea* from Lake Victoria.

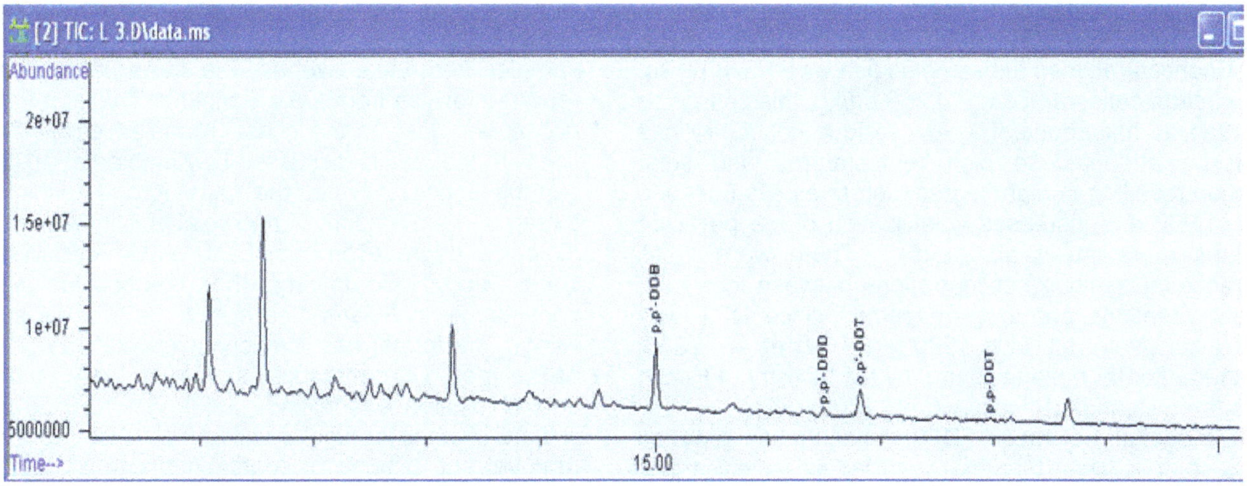

Figure 3. GC Chromatogram of DDT and metabolites in *R. argentea* from Lake Kyoga.

Table 2. Concentrations of DDT residues (µg/kg) in fish collected from different water bodies in Uganda.

Sampling location	*op'*DDE	*pp'*DDE	*op'*DDD	*pp'*DDD	*op'*DDT	*pp'*DDT	ΣDDT
Lake Victoria landing sites							
Kasenyi	ND	ND	ND	0.091±0.008	ND	0.077±0.007	0.168
Port Bell	ND	0.057±0.018	0.082±0.004	0.088±0.022	0.076±0.026	0.094±0.031	0.397
Gaba	ND	ND	0.062±0.013	ND	ND	0.085±0.009	0.147
Kigungu	ND	0.072±0.089	ND	ND	ND	0.097±0.024	0.169
Lake Kyoga landing sites							
Namasale	ND	ND	ND	ND	0.087±0.088	0.120±0.031	0.207
Biko	ND	0.197±0.018	ND	0.202±0.108	ND	0.107±0.048	0.506
Kayago	ND	0.083±0.008	ND	0.093±0.058	ND	0.193±0.056	0.369

Concentrations of DDT residues (µg/kg) expressed as Mean ± SD, ΣDDT (µg/kg) = *op'*DDE + *pp'*DDE + *op'*DDD + *pp'*DDD + *op'*DDT + *pp'*DDT, ND- Not Detected.

and *pp'* DDE metabolites may indicate slow biotransformation of DDT (ASTDR, 2002). The absence of *op'* DDE in all the samples may be due to the quantification level (0.099 µg/ml) which was relatively high.

The concentrations of DDT metabolites in *R. argentea* from Lake Victoria ranged between ND-0.090 µg/kg, ND-0.095 µg/kg, ND-0.099 µg/kg, ND-0.156 µg/kg, and ND-0.136 µg/kg for *pp'*DDE, *op'*DDD, *pp'*DDD, *op'*DDT and *pp'* DDT respectively, meanwhile their concentrations in the same fish species from Lake Kyoga ranged between ND-0.234 µg/kg (*pp'*DDE), ND-0.422 µg/kg (*pp'*DDD), ND-0.288 µg/kg (*op'*DDT), ND-0.244 µg/kg (*pp'*DDT). The mean ΣDDT concentrations in Lake Victoria varied from 0.147 to 0.396 µg/kg while that in Lake Kyoga varied from 0.207 to 0.506 µg/kg dry weight of fish sample. There was no significant difference in mean ΣDDT concentration between Lake Victoria and Lake Kyoga. The mean ΣDDT concentrations in the fish samples were far below the FAO action levels for fish (edible portion) of 5000 µg/kg (FAO-WHO, 1997). The maximum mean ΣDDT concentration in the sampled fish was 1.328 µg/kg, with a mean concentration of 0.266 µg/kg; this cannot be regarded as hazardous. The low residue concentrations might be attributed to high temperatures and solar radiation resulting in high degradation rates (Samuel and Pillai, 1989) and increased volatilization of the pesticide residues (Larsson et al., 1995). There were small differences in residue concentrations between locations. Spatial variations are to be expected since DDT was used extensively until mid 1980's (Kasozi et al., 2006) and thereafter for malaria control by the Ministry of Health in Northern Uganda (Bimenya et al., 2009).

The magnitude of mean ΣDDT concentration reported in this study is lower than that reported in earlier studies in Uganda. Sebugere et al. (2008) reported ΣDDT concentrations of 33 µg/kg in *Bagrus docman*, 20 µg/kg in *Protopterus gethiopinus* and 33 µg/kg in *Oreochromis niloticus* from Lake Edward. Kasozi et al. (2006) on the other hand found ΣDDT of 11.67 µg/kg wet weight in Nile perch and 20 µg/kg wet weight in Nile tilapia from Lake Victoria. The species in these studies are at higher trophic levels in the food chain and have large fat content compared to *R. argentea*. Low concentrations reported in this study could be attributed to the bottom positions of the studied species in the trophic level, small fish species lay at the bottom of the food chain, implying that they feed mainly on plant material and small insects from which they accumulate the pollutants. Many factors determine bioaccumulation of organic pollutants in aquatic organisms, position in trophic level being one of them (Feng et al., 2003; Kumblad et al., 2001)

Although the levels of total DDT in fish in the present study appear to be within the acceptable FAO-WHO limits, over consumption of silver fish from those water bodies may lead to increased levels of the pesticide residues over time. Increased levels of DDE in the blood of pregnant women increase the chances of having a pre term baby. Long-term exposure to small amounts of DDT may affect the liver. Children being smaller than adults would accumulate a higher dose of DDT residues (amount of DDT ingested per kilogram of body weight) than adults once exposed over the same period, implying that those effects would be more severe in the former (ASTDR, 2002).

In general, the concentrations of ∑DDT observed in this study of Ugandan fish were lower than, and in other cases comparable with those reported in edible and marketed fish collected from other parts of the world as shown in Table 3. Fish sampled from different locations presented varying relative compositions of the different DDT residues as shown in the Figure 4.

The most striking feature is the high relative amount of *p,p'* DDT found in fish samples from all sampling locations, actually it was the major constituent of ΣDDT in *R. argentea.* This implies that DDT products are still circulating in the Ugandan environment. There was notable relative amount of *pp'*DDD concentration in fish samples from Lake Kyoga. The same pattern has been reported for fish from Lake Songkha, Thailand in nutrient rich areas close to the coast (Kumblad et al., 2001). In line with the results discussed by Kumblad et al. (2001), it can be suggested that the Lake Kyoga, ecosystem is extreme in the amount of microbial activity leading to fast biological degradation of pp'DDT to its metabolites in this water body. Comparing the composition of DDT constituents in *R. argentea* from Lake Victoria and Lake Kyoga, the former has a much lower relative contribution of the most DDT metabolites. This is likely to be either due to dilution effect since Lake Victoria has a larger volume of water or high organic matter that might result in removal and binding of contaminants to the sediments. High organic matter in sediments has by several authors been reported to drastically increase the adsorption of contaminants on to sediments (Kumblad et al., 2001; Wasswa, 2009).

Conclusion

Only five DDT metabolites: *pp'*DDE, *op'*DDD, *pp'*DDD, *op'*DDT, *pp'* DDT were detected in the fish samples caught from the different parts of Lake Victoria and Lake Kyoga. The present study indicates that mean ΣDDT concentrations in the fish were far below the FAO standards. This implies that the fish species studied is safe for human consumption. However, over consumption or long time exposure may lead to health effects especially in pregnant women and children. The detection of notable relative amounts of *op'* DDT and *pp'* DDT suggests that Ugandan ecosystems still receive considerable inputs of DDT residues from various anthropogenic sources. This could be due to weak laws leading to inadequate management, use, distribution and disposal of the pesticides.

Table 3. ∑DDT concentration (μg/kg) in fish collected from different parts of the world.

Country	Fish species	∑DDT (μg/kg)	References
Burundi	*Limnothrissa miodon*	95.7	Manirakiza et al. (2002)
(L. Tanganyika)	*Stolothrissa tanganyikae*	794.7	
Malawi	*Bagrus meridional*	13.1	Kidd et al. (2001)
(L. Malawi)	*Buccochromis nototaenia*	3.4	
	Clarius sp.	1.4	
Egypt	*Oreochromis niloticus*	5.13	Said et al. (2008)
(L. Burullus)	*Clarius sp.*	12.54	
Italy			
(L.Como como)	Shad	1010	Bettinetti et al. (2008)
Norway			
(L. Oyangeu)	Archit char	60	Evenset et al. (2004)
(L. Stuorajavri)	White fish	0.35	Christensen et al. (2007)
USA			
(L.Washington)	Cutthroat trout	168	McIntyre and Beauchamp (2007)
Canada			
(L. Kusawa)	Lake trout	40.85	Ryan et al. (2005)
Uganda			
(L.Victoria)	*Rasteneobola argentea*	0.88	Present study
(L.Kyoga)		1.08	

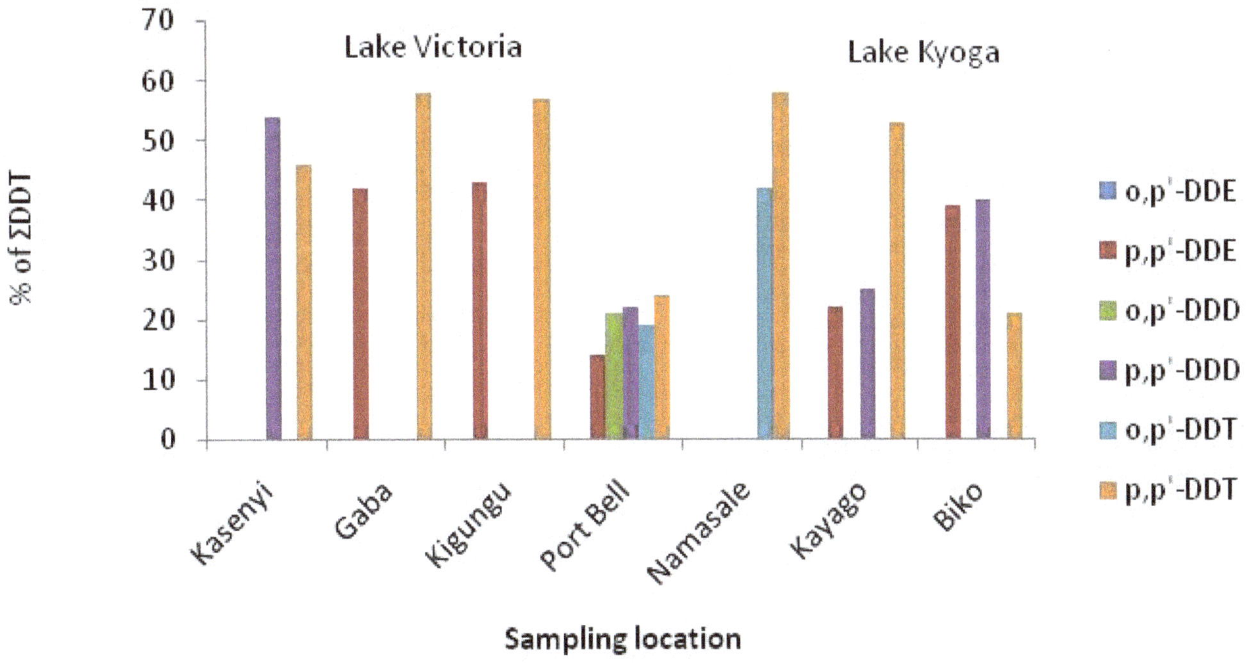

Figure 4. Relative composition of DDT metabolites in the fish samples from different locations in Lake Victoria and Lake Kyoga.

Conflict of Interests

There were no conflicts of interest which influenced the study.

ACKNOWLEDGEMENTS

The authors would like to thank Dr. Wasswa, J., Mr. Mulinda, S and Mr. Bitainensha, C. all from the Department of chemistry, Makerere University for their support during sample preparation and analysis. This work was financially supported by Gulu University.

REFERENCES

Åkerblom M (1995). Environmental monitoring of pesticide residues: Guideline for SADAC Region, SADC/ELMS, monitoring techniques series, Uppsala, Sweden.

ASTDR (2002). Public Health Statement for DDT/DDD/DDE. US Department of Health and Human Services.

Bettinetti R, Quadroni S, Galassi S, Bacchetta R, Bonardi L, Vailati G (2008). Is water from Alpine glaciers a secondary DDT source for lakes? Chemosphere. 73:1027-1031. http://dx.doi.org/10.1016/j.chemosphere.2008.08.017

Bimenya M, Byarugaba W, Baterana BB, Lugemwa M, Okwi AL, Biagini BG (2007). The case for spraying with DDT as a strategy against malaria. In: Bimenya, S.G. (Ed.), Malaria Control and Prevention Strategies and Policy Issues. Forum on Health and Nutrition, Uganda Natl. Acad. Sci. pp. 83-96.

Bimenya M, Habarulema H, Okot AL, Okwi KD, Lugemwa M, Lugemwa M (2009). Kidney function and DDT value comparison in pre- and post-spray plasma of the spray personnel in Northern Uganda in 2008. Int. J. Biol. Chem. Sci. 3(5):948-956.

Christensen GN, Savinova T (2007). Screening studies of POP levels in fish from selected lakes in the Paz watercourse, Akvaplan-niva report APN 514-3365.02.

Evenset A, Christensen GN, Skotvold T, Fjeld E, Schlabach M, Wartena E, Gregor D (2004). A comparison of organic contaminants in two high Arctic Lake ecosystems, Bjørnøya (Bear Island), Norway. Sci. Total Environ. 318:125-141. http://dx.doi.org/10.1016/S0048-9697(03)00365-6

FAO-WHO (1997). Codex Maximum Residue Limits for Pesticides. FAO. Rome.

Feng K, Yu BY, Ge DM, Wong MH, Wang XX, Wang W (2003). Organo-chlorine pesticide (DDT and HCH) residues in the Taihu Lake Region and its movement in soil-water system. Field survey of DDT and HCH residues in ecosystem of the region. Chemosphere 50:683-687. http://dx.doi.org/10.1016/S0045-6535(02)00204-7

ICH (1994). Harmonised Tripartite Guidelines (1994) prepared with the Third International Conference on Harmonisation of technical Requirements for the Registration of Pharmaceuticals for Human Use (ICH), Text on Validation of Analytical Procedures.

ICH (1995). Harmonised Tripartite Guidelines (1995) on Validation of Analytical Procedures: Text and Methodology.

Kasozi GW, Kiremire BT, Bugenyi WB, Kirsch NH, Nkedi-Kizza P, Gallicano K (2006). Organochlorine residues in fish and water samples from Lake Victoria, Uganda. J. Environ. Qual. 35:584-589. http://dx.doi.org/10.2134/jeq2005.0222

Kidd KA, Bootsman HA, Hesslein RH (2001). Biomagnification of DDT through the benthic and pelagic food webs of Lake Malawi, East Africa: Importance of trophic level and carbon source. Environ. Sci. Technol. 35:14-20. http://dx.doi.org/10.1021/es001119a

Kumblad L, Olsson A, Koutny V, Berg H, Causon B, Audran M (2001). Distribution of DDT residues in fish from the Songkhla Lake. J. Environ. Poll. 112:193-200. http://dx.doi.org/10.1016/S0269-7491(00)00118-4

Kyarimpa C (2007). Distribution of pesticide residues in parts of fish species (*Oreochromis niloticus* and *Lates niloticus*) from Murchison Bay - Lake Victoria. M.Sc Thesis, Department of Chemistry, Makerere University, Kampala, Uganda.

LAKIMO (2004). Lake Kyoga Management Plan, 2004 – 2007, LAKIMO/LKMP/2004. LAKIMO, Uganda.

Larsson P, Berglund O, Backe C, Bremle G, EkloÈv A, JaÈrn-mark C, Persson A (1995). DDT fate in tropical and temperate regions. Naturwissenschaft. 82:559-561. http://dx.doi.org/10.1007/BF01140245

Manirakiza P, Covaci P, Nizigiyman A, Ntakimaz L, Schepens G, Adran M (2002). Persistent chlorinated pesticides and polychlorinated biphenyls in selected fish species from Lake Tanganyika, Burundi. Africa. Environ. Pollut. 117:447-455. http://dx.doi.org/10.1016/S0269-7491(01)00188-9

Mbabazi J, Wasswa J (2010). Contamination by heavy metals in silver fish (Rastreneobola argentea) caught from Lakes Kyoga and Victoria,Uganda. Int. J. Environ. Stud. 67(4):543-556. http://dx.doi.org/10.1080/00207233.2010.499000

McIntyre JK, Beauchamp DA (2007). Age and trophic position dominate bioaccumulation of mercury and organochlorines in the food web of Lake Washington. Sci. Total Environ. 372:571-584. http://dx.doi.org/10.1016/j.scitotenv.2006.10.035

Mukasa P (2012). Assessment of levels of dichlorodiphenyltrichloroethane and its metabolites in soil and selected food crops from Apac and Oyam districts in Northern, Uganda M.Sc Thesis. Makerere University, Department of Chemistry, Kampala, Uganda.

Nyombi K, Bolwig S (2004). A qualitative evaluation of alternative development strategies for Ugandan fisheries.

Odada EO, Olago DO, Kulindwa K, Ntiba M, Wandiga S, Wandiga S (2004). Mitigation of Environmental Problems in Lake Victoria, East Africa: Causal Chain and Policy Options Analyses. J. Human. Environ. 33(1):13-23.

Ogwok P, Muyonga JH, Serunjogi ML (2009). Pesticide residues and heavy metals in lake Victoria nile perch. Lates niloticus belly falp oil. Bull. Environ. Contam. Toxicol. 82:529-533. http://dx.doi.org/10.1007/s00128-009-9668-x

Rusongoza PK (2003). Urban Disaster Mitigation and Environmental Impact Assessment of urbanisation in Uganda, In: New Technologies for Urban Safety of Mega Cities in Asia (Japan, Tokyo). pp. 371-378.

Ryan MJ, Stern, GA, Diamond M, Groft MV, Roach P, Kidd K (2005). Temporal trends of organochlorine contaminants in burbot and lake trout from three selected Yukon lakes. Sci. Total Environ. 351:501-522. http://dx.doi.org/10.1016/j.scitotenv.2004.08.022

Said TO, Moselhy KM, El-Rashad AA, Shreadah MA, Saida AT, Blessborn D (2008). Organochlorine contaminants in water, sediment and fish of Lake Burullus, Egyptian Mediterranean Sea. Bull. Environ. Contam. Toxicol. 81:136-146. http://dx.doi.org/10.1007/s00128-008-9422-9

Samuel T, Pillai M (1989). The effect of temperature and solar radiations on volatilisation, mineralisation and degradation of [14C]-DDT in Soil. Environ. Pollut. 5:63-77. http://dx.doi.org/10.1016/0269-7491(89)90130-9

Sebugere P, Kiremire BT, Kishimba M, Wandiga SO, Nyanzi SA,Wasswa J (2008). DDT and metabolites in fish from Lake Edward, Uganda. Chromosphere 76:212-217. http://dx.doi.org/10.1016/j.chemosphere.2009.03.049

Simon H, Afema AJ, Sikarskie J, Kineene WW, Audran M, Berg H (2004). Mercury and persistent organic pollutant concentration in African fish Eagles, marabou stocks and nile tilapia in Uganda. J. Wildlife Dis. 40(3):501-514. http://dx.doi.org/10.7589/0090-3558-40.3.501

Tsuda T (2012). Residues of DDT and HCH in fish from Lakes and Rivers in the World. http//dx.doi.org/105772/46842.

UBOS (2002). Uganda Population and Housing Census Provisional Results. Printed by Intersoft Business Services Ltd.

UBOS/UDHS (2006). Uganda Demographic and Household Statistics, Uganda Bureau of Statistics, Kampala, Uganda.

UNEP (2008). National Implementation Plan of the Stockholm Convention on Persistent Organic Pollutants for Uganda.

Wasswa J (2009). Characterisation of selected pesticide residues in sediments of the Ugandan side of Lake Victoria. PhD Thesis. Makerere University.

Preconcentration of benzene and phenolic compounds in water sample by adsorption on Carbon nanotubes coated fiber

Ali Moghimi[1] and Mahdi Alijanianzadeh[2]

[1]Department of Chemistry, Islamic Azad University Varamin (Pishva) Branch, Varamin,, Iran.
[2]Department of Biochemistry and Biophysics, School of Biological Sciences, Varamin-Pishva Branch, Islamic Azad University, Varamin, Iran.

A simple method has been developed for the preconcentration of benzene and phenolic compounds (BPC) by the adsorption of its carbon nanotubes (CNTs). The preconcentration of benzene and phenolic compounds (BPC) are ubiquitous pollutants, and many of them are carcinogenic or mutagenic. Carbon nanotubes (CNTs) are a kind of novel and interesting carbon material which can be used for separation and purification. In this investigation, commercial solid-phase microextraction (SPME) fibers (PDMS) were coated with single-wall nano tubes (SWNTs) and multi-wall nano tubes (MWNTs) to study the adsorption and extraction ability of hydrophobic BPC. While MWNTs adsorbed more hydrophobic BPC than SWNTs, the fibers coated with CNTs had advantages over traditional SPME fibers in selectivity and sensitivity. They could be used to separate hydrophobic BPC. The results show that the selectivity, sensitivity and reproducibility of this method are good for real sample analysis.

Key words: Carbon nanotubes (CNTs), Solid-phase microextraction (SPME), preconcentration of benzene and phenolic compounds (BPC).

INTRODUCTION

Many organic pollutants, such as phenol, o-chlorophenol, and benzene, are present in the environment in trace amounts. Due to their high toxicity (Zhang et al., 2001) and capability of accumulating in the environment, these pollutants are hazardous to living organisms (Bianchi et al., 2002). Due to their toxicity, persistence, and unpleasant organoleptic properties, both the US Environmental Protection Agency (EPA) and the European Union (EU) have classified several phenols and chlorophenols as priority pollutants (Kovacs et al., 2008; Puig and Barcelo, 1996). Phenol is a pollutant widely occurring in surface waters, in which it originates mainly from sewage effluents of chemical, pharmaceutical, and dye industries (Cernakova and Zemanovicova, 1998). Phenol derivatives, (Cernakova and Zemanovicova, 1998; Li et al., 2008) including chlorophenols, show genotoxic, mutagenic, and carcinogenic properties, and are characterized by high stability (Saitoh et al., 2008).

Benzene can naturally occur in crude oil; therefore, it can be found in refinery products (Schnatter, 2007; Roma-Torres et al., 2006). Since it has been classified as a first class carcinogenic factor (Eining and Dehnen, 1995) targeting bone marrow, more and more limits are being placed on its use. The above-mentioned pollutants are commonly present in surface waters. Since surface waters, are used by humans and in animal breeding after their treatment, (Manini et al., 2006) it is necessary to control the levels of organic pollutants (Michaowicz and Duda, 2007), particularly those that are present in trace amounts and show toxic and carcinogenic activities. Therefore, research on efficient methods for purification and preconcentration of analytes before their determination by means of an appropriate analytical technique is of great importance (Fattahi et al., 2007).

Among the many techniques developed for the isolation and preconcentration of different organic compounds and metal ions, solid phase extraction (SPE) is the most frequently used (Rendle, 2000). The extraction process to the solid phase is based on the retention of an analyte on a sorbent (Pic´o et al., 2007). Originally, the SPE application area was usually restricted to preconcentration of analytes showing strongly hydrophobic properties, whereas the recovery rates for polar compounds were not satisfactory (Saitoh et al., 2004). In our previous papers, a successful application of silica modified with ketoimine to the determination of trace amounts of bisphenol-A (BPA) in mineral water and powdered milk samples was reported (Rykowska et al., 2008; Rykowska 2007). The packing of ketoimine groups bonded with complexes of transition metals was also studied (Rykowska et al., 2004).

Carbon nanotubes (CNTs) are a kind of novel and interesting carbon material first found in 1991 by Iijima. CNTs are divided into single-wall nanotubes (SWNTs) and multi-wall nanotubes (MWNTs) according to their numbers of graphite sheets. The unique electrical, mechanical, and chemical properties of CNTs have aroused great interest among research workers (Pic´o et al., 2007; Saitoh et al., 2004; Rykowska et al., 2004). CNTs have a curved surface (composed of two fullerene halves and a cylinder made of a rolled up graphite sheet), and are thus expected to show a stronger binding affinity for hydrophobic molecules compared with a planar carbon surface. Furthermore, the internal pores of the CNTs are large enough to allow molecules to penetrate. The adsorption can occur on the inner hollow cavity of CNTs, on the outside surface, and on the interstitial spaces between the nanotube bundles. All these indicate that CNTs have strong physical adsorption ability to hydrophobic compounds. With the great process in the methods of preparing CNTs, large efforts have been devoted to the fields of application, such as gas storage (Rykowska, 2007), preconcentration of volatile organic compounds (Rykowska et al., 2004), removal of chemical and toxic wastes from water (Moghimi, 2013) and GC (Moghimi et al., 2009) to name but a few. It is therefore conceivable that CNTs may have great analytical potential as an effective solid-phase microextraction adsorbent for some suitable compounds. It is known that the interaction between BPC and surfaces are complex, involving many types of noncovalent forces that are electrostatic, hydrogen bonding, hydrophobic, or entropic in nature related to surfaces as well as surrounding water molecules (Mrksich et al., 1997; Brash, 1991). Nevertheless, this is a topic crucial to many areas of biological and medical science and technology. It is reported that the internal surface of CNTs interacts strongly with the enzyme BPC and the CNTs appear to act as a benign host to encapsulate protein molecules in their internal tube cavity.

There are many ways to separate the target protein from the sample matrix (Johansson et al., 2008), and few research papers focus on the subject of solid phase extraction of acidic and basic BPC using CNTs, CNTs immobilization, or CNTs functionalization (Hu et al., 2009). However, to the best of our knowledge, there is hardly any research on the SPME process of BPC using CNTs. The purpose of our present work was to develop a novel SPME technique using CNTs (SWNTs and MWNTs) as fiber coating for the extraction of two important hydrophobic also, the concentration of hydrophobic BPC was roughly determined using our method. It presents many advantages, that is, it is relatively inexpensive and is a parallel fabrication technique, allowing the production of a large number of samples during the same process.

MATERIALS AND METHODS

Reagents and instrumentation

Phenol, o-chlorophenol, and benzene were obtained from Fluka (Buchs, Switzerland). Individual stock solutions at concentrations of 100 mg/ml each were prepared in dichloromethane–acetone (50:50). The working standard solution was prepared by mixing each of the PAH stock solutions and diluting with dichloromethane–acetone (50:50) for a final concentration of 2 mg/ml. *p*-Terphenyl (analytical-reagent grade) from Tokyo Kasei was used as internal standard (I.S.) in gas chromatography–mass spectrometry (GC–MS) analysis. The standard solution for this compound was prepared in benzene at a concentration of 8 µg/ml. All the standard solutions were stored in the dark at 4°C to prevent photolysis of the compounds.

Raw SWNTs and MWNTs were purchased from Shenzhen Nanotech Port Co. Ltd and used without further purification. The specific surface area of SWNTs and MWNTs are 450 m^2/g and 40-300 m^2/g, were obtained from E.Merck, Darmstat, Germany respectively. A commercial 30 µm PDMS fiber and SPME holder were purchased from Supelco (Bellefonte, PA, USA). Other materials were of analytical grade and used as purchased without further treat. The SEM images were taken by Field Emission Scanning Electron Microanalyser JSM-6700F (JEOL, Japan). The UV-Vis spectra were determined by a UV-Vis spectrophotometer TU-1901. The mixtures of BPC were separated by SDS-polyacrylamide gel electrophoresis (SDSPAGE) and detected by a fluorescence spectrometer.

The pH measurements were carried out by an ATC pH meter (EDT instruments, GP 353).

Preparation of CNTs fiber for SPME

First, 0.5 g of SWNTs was dispersed in 125 ml of 1,2- dichloroethane

Table 1. Retention times and quantitative ions for GC–MS (SIM) analysis of BPC.

Compound	Retention time (min)	Quantitative ion (*m/z*)
Phenol	7.24	128
o-chlorophenol	9.92	152
Benzene	10.24	153

[a] Internal standard BPC from a sample of 500 ml of pond water.

by ultrasonic agitation for 30min. Then, 7.5 g of organic binder, made up of 90% terpineol, 5% ethylcellulose, and 5% dibutyl phthalate (by mass), was added to the solution and sonicated for another 30 min. This mixture was filtered through a 300-mesh net to remove big clusters. Finally, it was heated to 90°C to totally evaporate the 1,2-dichloroethane and obtain a sticky paste.

The used SPME fiber was a PDMS fiber (30 µm diameter) which was stuck in the SPME holder. It was coated with SWNTs paste by carefully spinning the fiber on a sheet of sticky SWNTs paste. After it was dried by hot air (about 150°C), the thickness of the coated layer was a few micrometers. The above procedure was repeated three times, so the final thickness of the coated SWNTs layer was about 15 µm (determined by SEM). The diameter of the coated SPME fiber was about 45 µm. Finally, the coated fiber was sintered at 400°C for 30 min in a nitrogen atmosphere to remove the organic binder. The coated SPME fiber was used directly in the extraction process. The MWNTs coated SPME fiber was fabricated in exactly the same way.

Preparation of hydrophobic BPC solutions

Different volumes of K_2HPO_4 solution (0.1 mol L^{-1}) and KH_2PO_4 solution (0.1 mol L^{-1}) were mixed to get 0.1 mol L^{-1} phosphate buffer of various pH (standard method). Then, the pH meter was used to adjust the pH to exactly the required value (NaOH or HCl was used). 5 mg of hydrophobic BPC were dissolved in 10 ml of various phosphate buffers with different pH, and the prepared solutions were all stored in a refrigerator until use.

Solid-phase microextraction process

20 ml glass vial was used as a sample container, and 5 ml of sample solution was placed into it. The SPME holder was fixed at a suitable height above the sample vial so that the coated SPME fiber was totally immerged in the solution and then the bottle was sealed. It was shaken for a required time at ambient temperature. After extraction, the fiber was removed from the sample vial and immediately inserted into the desorption buffer. The desorption process was conducted in various buffers in the same way as the SPME process. A commercial SPME fiber (PDMS) was conducted as a SPME process in the same way for comparison.

Recovery calculations

Extraction efficiency (E_{ex}) was calculated from the hydrophobic BPC concentrations before (C_0) and after extraction (C_1).

$$E_{ex} = (C_0 - C_1)/C_0 \times 100\%$$

Recovery yield (E_r) was calculated from the hydrophobic BPC concentration in the desorption buffer (C_2) and the adsorbed BPC.

$$(C_0 - C_1).\ E_r = C_2/(C_0 - C_1) \times 100\%$$

Concentrations were determined from the maximum absorbance around 200 nm by UV-Vis spectra.

Gas chromatography mass spectrometry (GC–MS) analysis

A Shimadzu QP-5000 GC–MS system equipped with a Model AOC-17 auto sampler was used for analysis. Chromatographic separation of the 16 BPC and the internal standard was accomplished with a DB-5 (J & W, Folsom, CA, USA) fused-silica capillary column (30 m × 0.32 mm I.D., 0.25 mm film thickness). Helium was the carrier gas at a flow-rate of 2.0 ml/min, and an inlet split ratio of 1:20 was used. Sample injection was in the splitless mode with an injection volume of 2 µl and an injection time of 2 min. The GC oven temperature was programmed as follows. An initial temperature of 50°C was held for 2 min and then ramped at 15°C/min to 220°C followed by another ramp of 58°C/min to 300°C, held for 1 min. The temperatures of the injection port and the interface with the MS system were set at 300°C. For selected ion monitoring (SIM), the voltage of the detector was 1.5 kV, and for each compound, one ion was chosen for quantification while two other ions were for identification. The retention times and quantitative ions for the analytes and I.S. are listed in Table 1. Linear calibration curves could be obtained for all the PAH components across a 50-fold concentration range; typically, R^2 0.999. The instrumental detection limit was measured to be from 2 (for the two-ring and three-ring BPC) to 10 pg/ injection (for five-ring and six-ring BPC).

Sample analysis

Our method was firstly applied to pond water. To prove the feasibility of our method in real samples, 1 ml pond water was diluted to 10 ml by phosphate buffer (0.1 mol L^{-1}) and various amounts of BPC were added to it to get various BPC concentrations (such as 0.1, 0.2 and 1 mg mL^{-1} and so on). Then the SPME experiments were conducted in 5 ml of these prepared samples under optimized conditions. 1 ml BPC was diluted to 10 ml by phosphate buffer (0.1 mol L^{-1}), and then 5 ml was taken to conduct the SPME procedure under optimized conditions. The concentration of BPC was determined by the obtained concentration curve.

RESULTS AND DISCUSSION

Characterization of CNTs and CNTs coated SPME fiber

The diameters of the MWNTs were within 10 to 30 nm and the lengths were in the range of 5 to 15 µm. The Raman spectra of both were also studied (Figure 1). There were three major lines of MWNTs, which were around 1360 cm^{-1} (D line), 1570 cm^{-1} (G line) and 2930

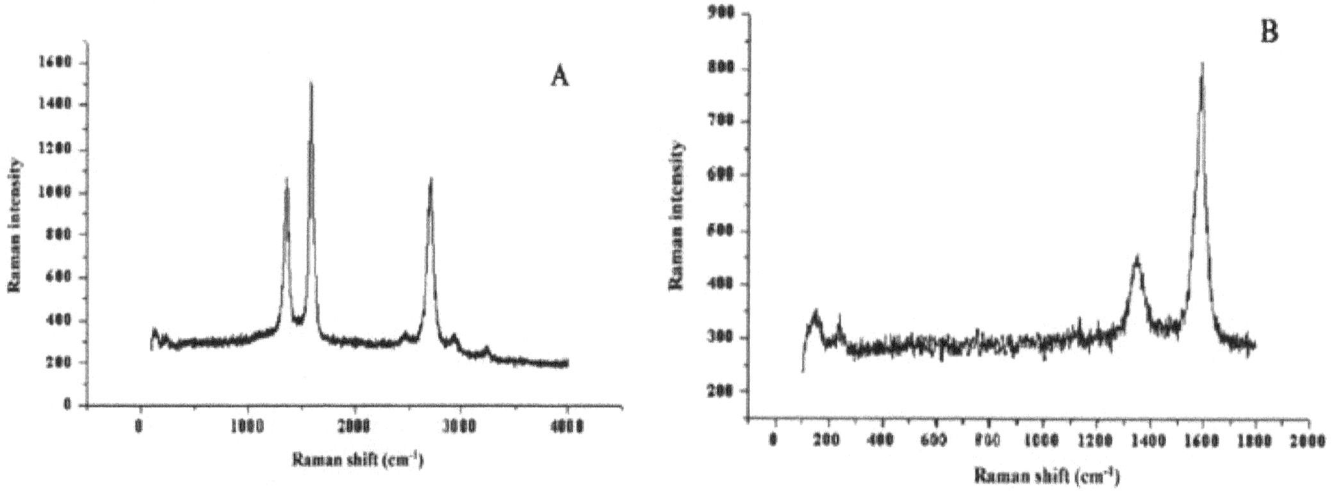

Figure 1. Raman spectra of MWNTs (A) and SWNTs (B).

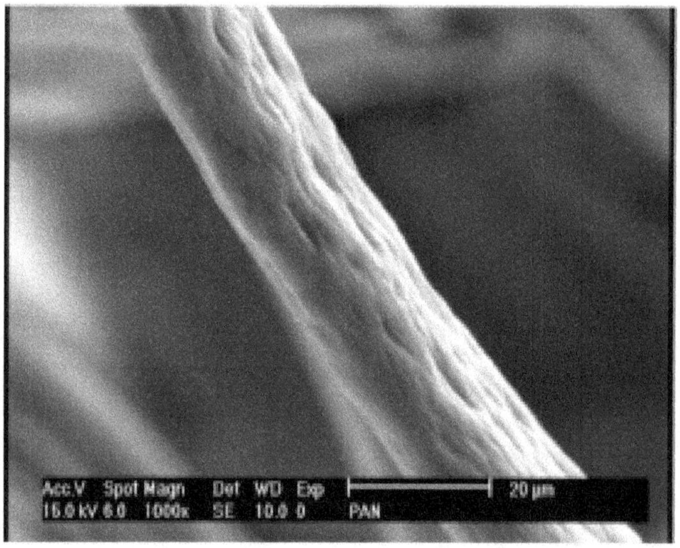

Figure 2. SEM images of coated MWNTs (lower magnification); Acceleration voltages and magnifications were indicated on the images.

cm^{-1} (D+G line). There were two major lines of SWNTs, which were around 1350 and 1600 cm^{-1}. In order to examine the coating state of CNTs on the SPME fiber using our method, the SEM images of MWNTs-coated fiber were studied (Figure 2). It demonstrated the presence of CNTs as a homogenous coating on the surface of the fiber and the coating possessed a rough surface, which resulted in larger surface areas and higher extractive capacity than conventional polymeric phases. The film thickness of the carbon nanotubes coating was proven to be about 15 µm by SEM. Also, it showed that the MWNTs were cut short by sonication, which might affect the extraction efficiency. However, the sonication step was necessary for dispersion.

Optimization of SPME procedure

It is known that SPME is an equilibrium process. The principle behind SPME is the partitioning of analytes between the sample matrix and the extraction medium. In order to obtain good sensitivity, several experimental parameters related to both extraction and desorption steps needed to be optimized.

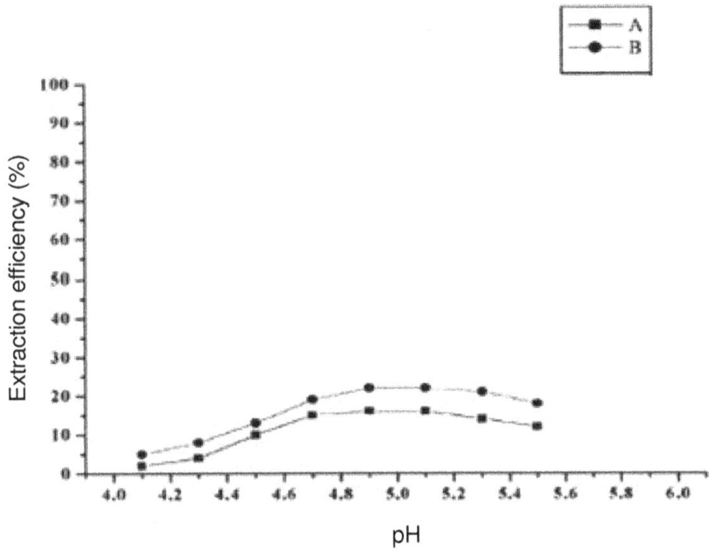

Figure 3. Effect of pH value of the extraction buffer on extraction efficiency (A: MWNTs extraction of BPC; B: SWNTs extraction of BPC). Other extraction conditions: phosphate concentration of extraction buffer: 0.1 mol L^{-1}; extraction time: 120 min for BPC 5 ml; BPC concentration: 0.5 mg mL^{-1}.

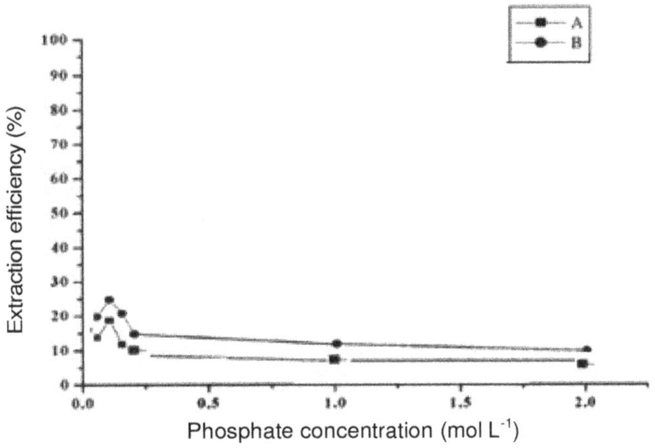

Figure 4. Effect of extraction buffer ionic strength on extraction efficiency (A: MWNTs extraction of BPC; B: SWNTs extraction of BPC). Other extraction conditions: pH of extraction buffer: 5.0 for BPC; extraction time: 120 min for BPC; extraction buffer volume: 5 ml; BPC concentration: 0.5 mg mL^{-1}.

The electrical property of BPC is dependent on the pH value of the solution. First of all, the pH value of the extraction buffer was optimized (Figure 3). As can be seen from Figure 3, MWNTs and SWNTs adsorbed BPC most strongly between a pH of 4.9 and 5.1. It is known that the pI value of BPC is around 4.7. When the pH value of the buffer is around the pI value of the BPC, the BPC is nearly neutral and more hydrophobic than at other pHs. Also, CNTs are hydrophobic and they are inclined to adsorb hydrophobic compounds. That is to say CNTs adsorb BPC most strongly around the pI value of the BPC. A pH of 5.0 for BPC was chosen in the next experiments. Other literatures report that the ionic strength of the extraction buffer also has some impact on the extraction efficiency (Mrksich et al., 1997; Brash, 1991). Thus, the effect of ionic strength of the extraction buffer is investigated (Figure 4). When the concentration of phosphate increased, the extraction efficiency

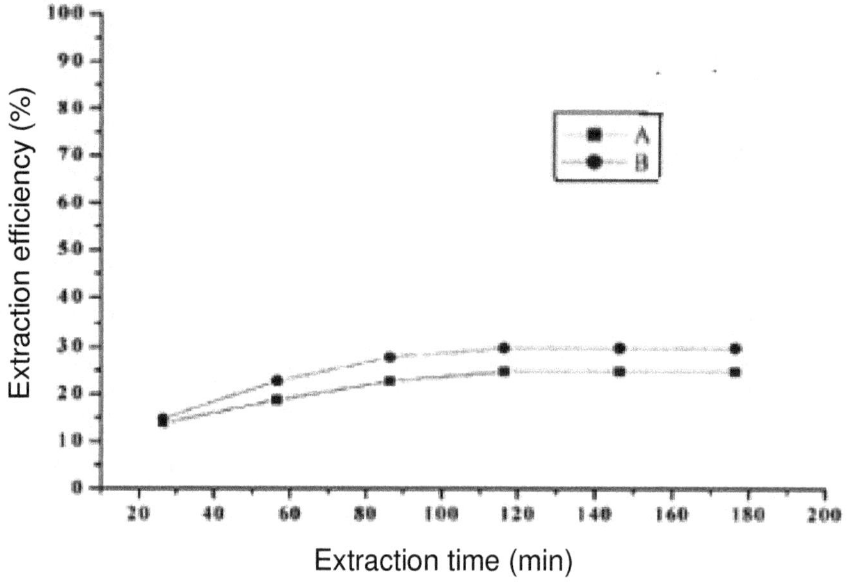

Figure 5. Effect of extraction time on extraction efficiency (A: MWNTs extraction of BPC; B: SWNTs extraction of BPC). Other extraction conditions: pH of extraction buffer: 5.0 for BPC; phosphate concentration of extraction buffer: 0.1 mol L^{-1}; extraction buffer volume: 5 ml; BPC concentration: 0.5 mg mL^{-1}.

Table 2. Effect of coating times on extraction efficiency BPC, concentration: 0.5 mg mL-1; pH of extraction buffer: 5.5; phosphate concentration of extraction buffer: 0.1 mol L^{-1}; extraction time: 150 min; extraction buffer volume: 5 ml).

Coating times	1	2	3	4	5	6
MWNTs' E_{ex} for BPC (%)	78.2	84.1	88.8	92.7	97.6	98.3
SWNTs' E_{ex} for BPC (%)	60.2	62.9	66.9	68.0	73.5	76.1

increased accordingly at first and reached a maximum value. But then, the extraction efficiency dropped. This may be explained as follows. On the one hand, an increase of phosphate can reduce the amount of water available to dissolve analyte molecules due to the formation of hydration spheres around the ionic salt molecules (Boyd-Boland and Pawliszyn 1995), which improved the extraction efficiency for the investigated compounds. Thus, phosphate anions can interact with a positive charge on BPC molecules and form "ionic atmosphere", which increases with phosphate concentration.

This interaction could cause the reduction of BPC adsorption on CNTs. 42 0.1 mol L^{-1} was chosen as the extraction buffer concentration. The equilibration time in SPME for given analytes was determined by constructing an adsorption-time profile by exposing the fiber to the same analytes concentration solution for different times (Figure 5). As can been seen from Figure 5, MWNTs and SWNTs extraction could reach equilibrium at the same time (after 120 min) when BPC was the analyte. For further experiments, 120 min for BPC was chosen as extraction time. The coating thickness of CNTs could also affect the extraction efficiency (Table 2). As can be seen from Table 2, when coating times increased, the extraction efficiency also increased a little. When the coating times increased, the coating thickness also increased which made the coating surface area increase accordingly. Therefore, more BPC could be adsorbed on the coating. Coating carried out three times was sufficient in our study.

Some reports show that increasing the ionic strength and decreasing the pH value of the extraction buffer can wash off the adsorbed BPC (Mrksich et al., 1997; Brash, 1991). The pH value of the desorption buffer would not affect the recovery yield if it was low enough. That might be because the ionic strength effect was dominant and much larger than the pH effect. Desorption time is also an important parameter (Figure 6). As can be seen from Figure 6, MWNTs and SWNTs could reach their highest recovery yield at the same time (after 60 min) when BPC was the analyte. Similarly, MWNTs and SWNTs could

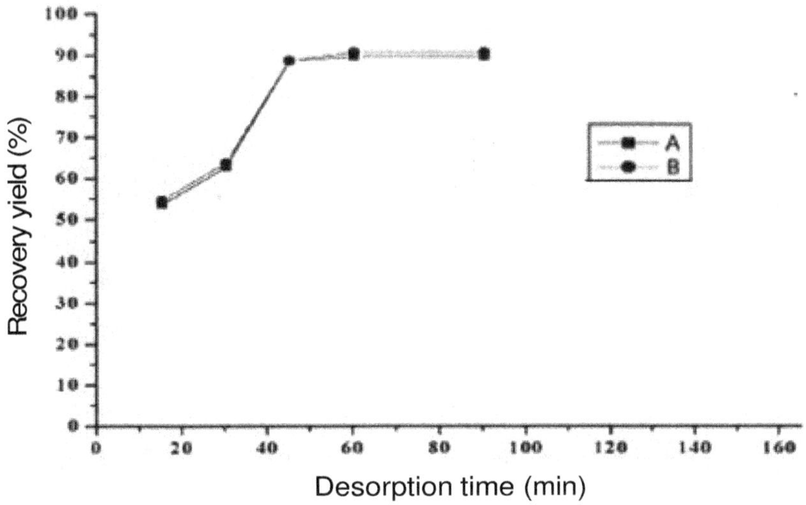

Figure 6. Effect of desorption time on recovery yield (A: MWNTs extraction of BPC; B: SWNTs extraction of BPC). Extraction conditions: pH of extraction buffer: 5.0 for BPC; phosphate concentration of extraction buffer: 0.1 mol L^{-1}; extraction time: 120 min for BPC; extraction buffer volume: 5 ml; BPC concentration: 0.5 mg mL^{-1}. Other desorption conditions: pH of desorption buffer: 2.0; NaCl concentration of desorption buffer: 3 mol L^{-1}; desorption buffer volume: 5 ml.

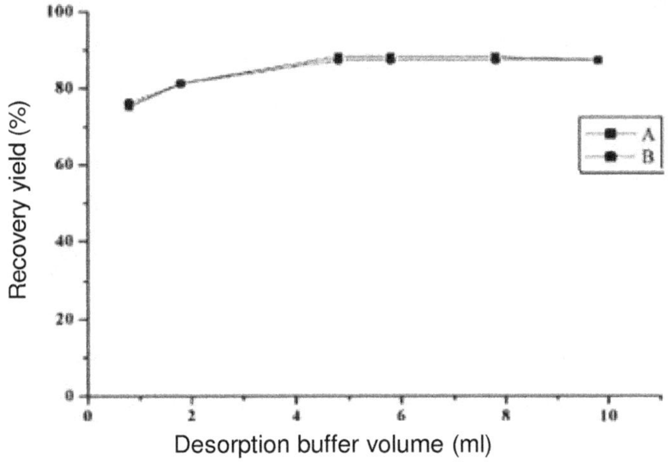

Figure 7. Effect of desorption buffer volume on recovery yield (A: MWNTs extraction of BPC; B: SWNTs extraction of BPC). Extraction conditions: pH of extraction buffer: 5.0 for BPC; phosphate concentration of extraction buffer: 0.1 mol L^{-1}; extraction time: 120 min for BPC; extraction buffer volume: 5 mL; BPC concentration: 0.5 mg mL^{-1}. Other desorption conditions: pH of desorption buffer: 2.0; NaCl concentration of desorption buffer: 3 mol L^{-1}; desorption time: 120 min for BPC.

reach their highest efficiency at the same time (after 120 min) when BPC was adsorbed much stronger on CNTs. Either using MWNTs or SWNTs, BPC reached extraction equilibrium and desorption equilibrium at the same time respectively, which may demonstrate the same mechanism of MWNTs and SWNTs adsorption and desorption of BPC. Only the adsorption ability was different. 60 min for BPC was chosen as the desorption time in further experiments.

The volume of desorption buffer is also important for high recovery yield (Figure 7). As can be seen from Figure 7, all experiments could reach their highest

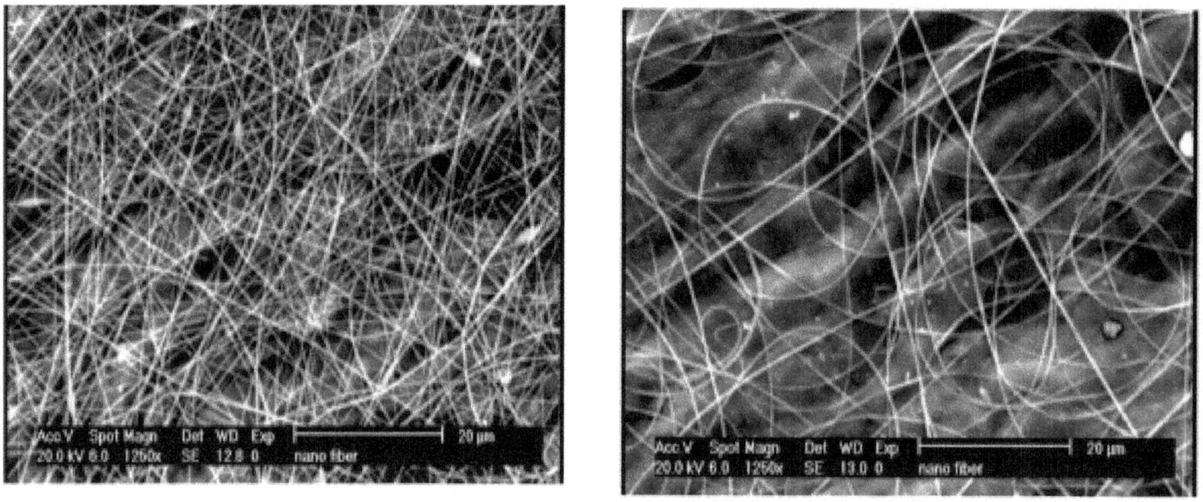

(a) (b)

Figure 8. SEM images of BPC -adsorbed MWNTs coated fiber (A: lower magnification; B: higher magnification). Acceleration voltages and magnifications were indicated on the images.

recovery yield when using the same desorption buffer volume as the extraction buffer (5 ml). The desorption buffer volume barely had an effect, which might be because the adsorption and desorption of BPC in our experiment are determined by the amount of BPC, not the concentration. 5 ml was chosen as the desorption buffer volume in all the following experiments. The results showed that the adsorbed BPC could not be totally desorbed. The desorption process was repeated, and a little more BPC was desorbed. It was difficult to totally desorb the adsorbed BPC. This might be because the adsorbed BPC was denatured and adsorbed very strongly on the CNTs (Sandeep et al., 2004). Thus, the carry over effect was a problem in our study. However, using the coated fibers of the same fabricating process, the reproducibility was good enough. So in every experiment, a newly fabricated fiber was used to avoid the carry over effect. From these results, it is concluded that CNTs adsorbed BPC. SWNTs adsorbed more BPC than MWNTs, which might be because SWNTs have larger specific surface area than MWNTs.

Characterization of BPC adsorbed MWNTs

The surface characterization of MWNTs coated fiber that had adsorbed BPC was also investigated by the SEM technique (Figure 8). As can be seen from Figure 8, it was confirmed that BPC had been adsorbed on the MWNTs coated fiber. The MWNTs were buried in the BPC. Some reports show that the secondary structure and activity of BPC adsorbed onto nanotubes are changed as a result of the protein interaction with the nanotube surface.

Analytical figures of merit

As can be seen from the former work, CNTs adsorbed BPC. When the concentration of BPC reached some degree, CNTs could only adsorb BPC, which was useful for real sample analysis. BPC is a much larger hydrophobic molecule. Since CNTs are hydrophobic, they adsorb hydrophobic compounds more strongly than hydrophilic compounds. As a result of more BPC being adsorbed on CNTs, it was used as the experimental protein in a further study. The fiber to-fiber reproducibility was investigated on five replicate experiments using CNTs coated fibers prepared in the same way under the optimized conditions. MWNTs' E_{ex} for BPC was 87.8 ± 1.20% (n = 5, RSD = 2.15%) and SWNTs' E_{ex} for BPC was 69.8 ± 1.94% (n = 5, RSD = 2.16%). All these results indicated that our method was feasible and reliable.

Comparison of the SPME efficiency of the coated fiber with PDMS fiber

To prove the advantage of our method to extract BPC, our method was compared with commercial SPME fibers (PDMS). PDMS fiber is relatively a blank reference to the proposed CNTs coating fiber because CNTs in this study were coated on it. The PDMS fiber's E_{ex} was 35.5 ± 0.90% (n = 3) for BPC under optimized SPME conditions. The results showed that PDMS adsorbed less BPC but more BPC than CNTs. The selectivity between BPC was

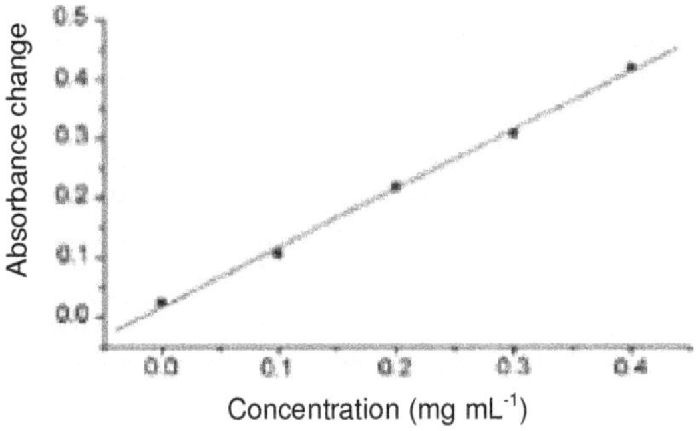

Figure 9. Absorbance change of different spiked BPC concentrations in bovine serum. The inserted figure is the liner relation of BPC concentration and absorbance change. Extraction conditions: pH of extraction buffer: 5.0; phosphate concentration of extraction buffer: 0.1 mol L^{-1}; extraction time: 120 min; extraction buffer volume: 5 ml.

not as obvious as that of CNTs. This might be because PDMS are less hydrophobic than CNTs. Thus, CNTs have some advantages in BPC analysis as compared to PDMS.

Sample analysis

Our method was applied to the analysis of pond water to study the substrate effect. The extraction efficiency of the BPC water sample was a little lower than in the phosphate buffer, around 80% when using MWNTs. And the recovery yield of BPC in the desorption buffer was 65 to 75%. The results showed that when the concentration of BPC was under 0.49 mg mL^{-1} (5 ml sample), the absorbance change was linear to the concentration (Figure 9). The regression equation is $y = 0.025 + 0.97x$ ($R^2 = 0.99884$, SD = 0.00956, N = 5, $p < 0.0001$). Thus, the amount of BPC can also be determined in these samples, and our method is applicable to water samples of various concentrations. The detection limit of this method in water was 0.078 mg mL^{-1} determined according to 3 S_b/m (S_b is the blank standard deviation and m is slope of the linear fit line). Finally, our method was applied to determine the BPC concentration in a pond water sample. The results showed that the concentration of BPC was 2.64 ± 0.18 mg mL^{-1} (n = 3, RSD = 2.99%). This data was about 3.0 mg mL^{-1} determined by the method of GC-MS. This result demonstrated the feasibility of our method to roughly determine the concentration of BPC in the water sample.

Conclusion

In this paper, SWNTs and MWNTs were coated on commercial SPME fibers to study the adsorption effect of BPC on CNTs. First, raw CNTs and CNTs coated with SPME fiber were characterized. The results showed that the surface of CNTs coating was propitious to BPC adsorption. The SEM image of BPC adsorbed MWNTs coated fiber confirmed this assumption. Then, the SPME procedure was optimized. From the results, it was observed that both MWNTs and SWNTs had much stronger adsorption ability than BPC. SWNTs have larger specific surface area, so their adsorption ability is stronger than MWNTs. However, in the situation of BPC, the much larger size relative to the diameter of SWNTs is dominant, which creates small interaction area with the nanotubes. Thus, MWNTs have stronger affinity for BPC. Also, the reproducibility is good for real sample analysis. From our sample analysis, the concentration of BPC was linear to the absorbance change when the concentration of BPC was under 0.49 mg mL^{-1} (5 ml sample). When compared with commercial SPME fiber (PDMS), our method was more selective and sensitive.

In conclusion, our method is advantageous for BPC extraction from the water sample. It is simple, inexpensive and time-saving, which may make it a good way to determine BPC concentration. The method developed was simple, reliable, and precise for determining BPC in water. Also, the proposed method was free of interference compared to conventional procedures to determine copper (Zhang et al., 2001; Manini et al., 2006; Bianchi et al., 2002). The method can be successfully applied to the separation and determination of BPC in binary mixtures.

ACKNOWLEDGMENT

This study is supported by the Department of Chemistry, Varamin (Pishva) Branch Islamic Azad University, Iran.

REFERENCES

Bianchi F, Careri M, Mucchino C, Musci M (2002). Improved determination of chlorophenols in water by solid-phase microextraction followed by benzoylation and gas chromatography with electron capture detection. Chromatogra. *55*:595-600.

Boyd-Boland AA, Pawliszyn J (1995). Solid-phase microextraction of nitrogen-containing herbicides J. Chromatogra. A.704:163.

Brash JL (1991). Mechanism of Adsorption of BPC to Solid Surfaces. *In* Biocompatible Polymers; Szycher, M., Eds.;Technomic Publication: Lancaster, pp.35-52.

Cernakova M, Zemanovicova A (1998). Healthy Soils in Macadamia orchards. Folia Microbiol. 43:411-416.

Eining T, Dehnen W (1995). Development of Liquid Chromatography Electrospray Ionization-Tandem Mass Spectrometry Methods for Determination of Urinary Metabolites of Benzene in Humans J. Chromatogr. A 697:371-375.

Fattahi N, Samadi S, Assai Y, Hosseini MRM (2007). An Investigation of Genetic Diversity among Some Almond Genotypes and Species by Morphological Traits J. Chromatogra. A, 1169:63-69.

Hu C, Xu YJ, Duo SW (2009). Non-Covalent Functionalization of Carbon Nanotubes with Surfactants and Polymers. J. Chin. Chem. Soc. 56:234-239.

Johansson HO, Ishii M, Minaguti M, Feitosa E, Penna TCV, Pessoa A

(2008). Phase Diagrams of the Aqueous Two-Phase Systems of Poly(ethylene glycol)/Sodium olyacrylate/Salts. Sep. Purif. Technol. 62:166-170.

Kovacs A, Kende A, Mortl M, Volk G, Rikker T, Torkos KJ (2008). Determination of phenolic compounds in water and urine samples using solid-phase microextraction based on sol–gel technique prior to GC-FID). Chromatography *A*, 1194:139-142.

Li J, Zhao X, Shi Y, Cai Y, Mou S, Jiang G (2008). Healthy Soils in Macadamia orchards. J. Chromatogr. A, 1180:24-31.

Manini P, De Palma G, Andreoli R, Poli D, Mozzoni P, Folesani G, Mutti A, Apostoli P (2006). Occupational exposure to low levels of benzene: Biomarkers of exposure and nucleic acid oxidation and their modulation by polymorphic xenobiotic metabolizing enzymes. Toxicol. Letts. 167:142-151.

Michaowicz J, Duda W (2007). Evaluation of chemical and physico-chemical indicators of water and bottom macrofauna the Starzyc Lake on the basis of the European Union Water Framework Directive. J. Chemosphere. 66:657-663.

Moghimi A (2013). Separation of lead(II) paraffin-embedded tissues from liver loggerhead turtles specimens by organicsolutionprocessable functionalized-nano graphene prior to determination by flame atomic absorption spectrometry (FAAS). Afr.J. Pure Appl. Chem. 7(2):79-90.

Moghimi A, Ghiasi R, Abedin AR, Ghammamy S (2009). Solid phase extraction of Cd(II) using mesoporous organosilicas and determination by FAAS. Afr. J. Pure Appl. Chem. 3(3):051-059.

Mrksich M, Whitesides GM, Harris JM, Zalipsky S (1997). Surveying for Surfaces that Resist the Adsorption of Proteins *PEG Chemistry and Biological Applications*. Am. Chem. Soc. Washington, D.C. 361-373.

Pic´o Y, Fern´andez M, Ruiz MJ, Font G (2007). Application of molecularly imprinted polymers to solid-phase extraction of analytes from real samples. J. Biochem. Biophys. Methods. 70:117-131.

Puig D, Barcelo D (1996). Determining mycotoxins and mycotoxigenic fungi in food and feed Trends. Anal. Chem. 15:362-365.

Rendle DF (2000). SPE and GC methods of preconcentration and determination of phenol, o-chlorophenol, and benzene by means of chemically modified silica. Talanta 51:1235-1235.

Roma-Torres J, Teixeira JP, Silva S, Laffon B, Cunha LM, M´endez J, Mayan O (2006). Cytogenetic and molecular biomonitoring of a Portuguese population exposed to pesticides. Mutat. Res. 604:19-27.

Rykowska I (2007). PREPARING SAMPLES FOR ANALYSIS- THE KEY TO ANALYTICAL SUCCESS. Trends in Chromatography. 3:11-20.

Rykowska I, Byra J, Wasiak W (2008). Chemically Bondem Phases for the Analysis of Trace Amounts of Organic Pollutants. Toxicology Mechanisms and Methods. Chem. Papers 62:255-259

Rykowska I, Szymaski A, Wasiak W (2004). PROPERTIES, THREATS, AND METHODS OF ANALYSIS OF BISPHENOL A AND ITS DERIVATIVES. Chem. Papers. 58:382-385.

Saitoh T, Kondo T, Hiraide M (2007). Admicelle-Based Solid Phase Extraction of Phenols Using Dialkylammonium Surfactant in the Hydroxide Form. J. Chromatogr. A, 1164:40-47.

Saitoh T, Matsuhima S, Hiraide M (2004). Application of Self-Assembled System Based on Ionic Surfactants Adsorbed onto Oxides Surface in. Chem. J. Chromatogr. A, 1040:185-191.

Sandeep S, Karejanagi AA, Vertegel RS, Kane J (2004). Carbon Nanotubes Coated Fiber for Solid-phase Microextraction of Bovine Fibrinogen and Bovine Serum Albumin. J. Langmuir. 20:11594-11598.

Schnatter RJ (2007). Assessment of DNA Damage by Comet Assay in Lymphocytes of Workers Occupationally Exposed to Petroleum fumes. Environ. Health. Part A, 61:433-437.

Zhang A, Li Q, Chen J, Fei Z, Long C, Li W (2001). Evaluation of solid-phase extraction sorbent with ketoimine groups for the preconcentration of benzene and phenolic compounds in water. React. Funct. Polym. 49:225-233.

Permissions

All chapters in this book were first published in AJPAC, by Academic Journals; hereby published with permission under the Creative Commons Attribution License or equivalent. Every chapter published in this book has been scrutinized by our experts. Their significance has been extensively debated. The topics covered herein carry significant findings which will fuel the growth of the discipline. They may even be implemented as practical applications or may be referred to as a beginning point for another development.

The contributors of this book come from diverse backgrounds, making this book a truly international effort. This book will bring forth new frontiers with its revolutionizing research information and detailed analysis of the nascent developments around the world.

We would like to thank all the contributing authors for lending their expertise to make the book truly unique. They have played a crucial role in the development of this book. Without their invaluable contributions this book wouldn't have been possible. They have made vital efforts to compile up to date information on the varied aspects of this subject to make this book a valuable addition to the collection of many professionals and students.

This book was conceptualized with the vision of imparting up-to-date information and advanced data in this field. To ensure the same, a matchless editorial board was set up. Every individual on the board went through rigorous rounds of assessment to prove their worth. After which they invested a large part of their time researching and compiling the most relevant data for our readers.

The editorial board has been involved in producing this book since its inception. They have spent rigorous hours researching and exploring the diverse topics which have resulted in the successful publishing of this book. They have passed on their knowledge of decades through this book. To expedite this challenging task, the publisher supported the team at every step. A small team of assistant editors was also appointed to further simplify the editing procedure and attain best results for the readers.

Apart from the editorial board, the designing team has also invested a significant amount of their time in understanding the subject and creating the most relevant covers. They scrutinized every image to scout for the most suitable representation of the subject and create an appropriate cover for the book.

The publishing team has been an ardent support to the editorial, designing and production team. Their endless efforts to recruit the best for this project, has resulted in the accomplishment of this book. They are a veteran in the field of academics and their pool of knowledge is as vast as their experience in printing. Their expertise and guidance has proved useful at every step. Their uncompromising quality standards have made this book an exceptional effort. Their encouragement from time to time has been an inspiration for everyone.

The publisher and the editorial board hope that this book will prove to be a valuable piece of knowledge for researchers, students, practitioners and scholars across the globe.

List of Contributors

Magut Hillary
Department of Chemistry, University of Eastern Africa, Baraton, P. O. Box 2500 Eldoret, Kenya

Terer Erick Kipngetich
Department of Chemistry, University of Eastern Africa, Baraton, P. O. Box 2500 Eldoret, Kenya

Mandina Shadreck
Department of Chemical Technology, Midlands State University, P. Bag 9055, Gweru, Zimbabwe

Tawanda Mugadza
Department of Chemical Technology, Midlands State University, P. Bag 9055, Gweru, Zimbabwe

Jyoti Nema
Department of Applied Chemistry, Government Engineering College, Jabalpur-482011, (MP), India

S. K. Shrivastava
Department of Applied Chemistry, Government Engineering College, Jabalpur-482011, (MP), India

N. G. Mitra
Department of Soil Science and Agriculture Chemistry, Jawahar Lal Nehru Krishi Vishwa Vidyalaya, Jabalpur (MP), India

S. A. Osemeahon
Department of Chemistry, Federal University of Technology, Yola, Nigeria

D. I. Esenowo
Department of Chemistry, Federal University of Technology, Yola, Nigeria

M. H. Shagal
Department of Chemistry, Modibbo Adama University of Technology, P.M.B. 2076, Yola, Nigeria

D. Kubmarawa
Department of Chemistry, Modibbo Adama University of Technology, P.M.B. 2076, Yola, Nigeria

K. Tadzabia
Department of Chemistry, Umar Suleiman College of Education, P.M.B. 2 Gashua, Yobe State, Nigeria

K. I. Dennis
Department of Chemistry, Modibbo Adama University of Technology, P.M.B. 2076, Yola, Nigeria

I. K. Shuaibu
Department of Pharmacy, School of Health Technology, Kankia, Katsina State College of Health Sciences, Katsina, Nigeria

M. Yahaya
Department of Pharmacy, School of Health Technology, Kankia, Katsina State College of Health Sciences, Katsina, Nigeria

U. K. Abdullahi
Department of Pure and Applied Chemistry, Umaru Musa Yar' Adua University, Katsina, Nigeria

Bernard Thole
Ngurdoto Defluoridation Research Station, P. O. Box 3020, Usa River, Arusha, Tanzania

Felix Mtalo
College of Engineering and Technology, University of Dar Es Salaam, P. O. Box 35131, DSM, Tanzania

Wellington Masamba
Harry Oppenheimer Okavango Research institute, University of Botswana, P/Bag 285, Maun, Botswana

Olatunji Olatunde Stephen
Department of Chemistry, Faculty of Applied Sciences, Cape Peninsula University of Technology, Bellville, Western Cape, South Africa

Osibanjo Oladele
Department of Chemistry, Faculty of Science, University of Ibadan, Ibadan, Oyo State, Nigeria

Ghulam Murtaza Mastoi
Centre for Environmental Sciences, University of Sindh, Jamshoro, Sindh, Pakistan

Khalid Hussain Lashari
Department of Fresh Water Biology and Fisheries, University of Sindh, Jamshoro, Sindh, Pakistan

Zameer Ali Palh
Department of Fresh Water Biology and Fisheries, University of Sindh, Jamshoro, Sindh, Pakistan

Anila Naz Soomro
Department of Fresh Water Biology and Fisheries, University of Sindh, Jamshoro, Sindh, Pakistan

Zulfiqar Ali Laghari
Department of Physiology, University of Sindh, Jamshoro, Sindh, Pakistan

Khalida Faryal Almani
Centre for Environmental Sciences, University of Sindh, Jamshoro, Sindh, Pakistan

Gul- e-Rana Abdul Waheed Mastoi
Centre for Environmental Sciences, University of Sindh, Jamshoro, Sindh, Pakistan

Abdul Rasool Abbasi
Department of Fresh Water Biology and Fisheries, University of Sindh, Jamshoro, Sindh, Pakistan

Z. Ladan
National Research Institute for Chemical Technology, Private Mail Bag 1052, Zaria, Nigeria

J. O. Amupitan
Department of Chemistry, Ahmadu Bello University, Zaria, Nigeria

O. A. Oyewale
Department of Chemistry, Ahmadu Bello University, Zaria, Nigeria

R. G. Ayo
Division of Agricultural Colleges, Ahmadu Bello University, Zaria, Nigeria

E. Temple
Department of Chemistry, Ahmadu Bello University, Zaria, Nigeria

E. O. Ladan
National Agricultural Extension and Research Liaison Services, Ahmadu Bello University, Zaria, Nigeria

P. Alexander
Department of Chemistry, Adamawa State University, Mubi, Nigeria

W. H. Ubandoma
Department of Chemistry, Adamawa State University, Mubi, Nigeria

S. Gupta
Department of Applied Chemistry, Jabalpur Engineering College, Jabalpur-482011(M.P.) India

S. K. Shrivastava
Department of Applied Chemistry, Jabalpur Engineering College, Jabalpur-482011(M.P.) India

M. shrivastava
Department of Chemistry Govt. M. H. College of Home Science and Science for Women Jabalpur (M. P.) India

Moriam Dasola Adeoye
Industrial and Environmental Chemistry Unit, Chemical Sciences Department, Fountain University, Osogbo Osun State, Nigeria

Azeez, Lukman Adeyemi
Industrial and Environmental Chemistry Unit, Chemical Sciences Department, Fountain University, Osogbo Osun State, Nigeria

Lawal, AbdulAzeez Tunbosun
Industrial and Environmental Chemistry Unit, Chemical Sciences Department, Al-Hikmah University, Ilorin Kwara state, Nigeria

Olayiwola, Olajumoke Abidemi
Chemical Sciences Department. Oduduwa University, Ipetumodu, Ile-Ife, Osun State, Nigeria

Shitta, Olanrewaju Razaq
Industrial and Environmental Chemistry Unit, Chemical Sciences Department, Fountain University, Osogbo Osun State, Nigeria

I. S. Singh
Department of Physical Sciences, School of Mathematics and Natural Sciences, The Copperbelt University, P. O. Box 21692, Kitwe, Zambia

Aderonke A. Okoya
Institute of Ecology and Environmental Studies, Obafemi Awolowo University, Ile - Ife, Nigeria

Abimbola B. Akinyele
Institute of Ecology and Environmental Studies, Obafemi Awolowo University, Ile - Ife, Nigeria

Ifeanyi E. Ofoezie
Institute of Ecology and Environmental Studies, Obafemi Awolowo University, Ile - Ife, Nigeria

Omotayo S. Amuda
Ladoke Akintola University of Technology, Ogbomoso, Nigeria

Oluwagbemiga S. Alayande
Center for Energy and Research Development, Obafemi Awolowo University, Ile - Ife, Nigeria

Oladotun W. Makinde
Center for Energy and Research Development, Obafemi Awolowo University, Ile - Ife, Nigeria

Faizan Danish Khaleel
School of Study in Chemistry and Biochemistry, Vikram University Ujjain (M.P)-456010, India

S. M. Zuber
Division of Aquatic Sciences, The Himalayan Ecological and Conservation Research Foundation, J&K, 180006, India

B. K. Mehta
School of Study in Chemistry and Biochemistry, Vikram University Ujjain (M.P)-456010, India

Darshina Mehta
School of Study in Chemistry and Biochemistry, Vikram University Ujjain (M.P)-456010, India

S. R. Kolisetty
School of Study in Chemistry and Biochemistry, Vikram University Ujjain (M.P)-456010, India

Krishna Vaidya
Department of Biology, K.B.S. College, Vapi, Gujarat, India

Mohini Gadhia
Veer Narmad South Gujarat University, Surat, Gujarat, India

Aimé Serge Ello
Université Félix Houphouët-Boigny de Cocody, Laboratoire de Chimie physique 22 bp 582 Abidjan 22, Cote d'ivoire

Jacques Aboua Yapo
Université Félix Houphouët-Boigny de Cocody, Laboratoire de Chimie physique 22 bp 582 Abidjan 22, Cote d'ivoire

Albert Trokourey
Université Félix Houphouët-Boigny de Cocody, Laboratoire de Chimie physique 22 bp 582 Abidjan 22, Cote d'ivoire

H. Ayedun
Department of Science Laboratory Technology, Federal Polytechnic Ilaro, P. M. B. 50 Ilaro, Ogun State, Nigeria

R. T.Oyede
Department of Science Laboratory Technology, Federal Polytechnic Ilaro, P. M. B. 50 Ilaro, Ogun State, Nigeria

B. G. Osinfade
Department of Science Laboratory Technology, Federal Polytechnic Ilaro, P. M. B. 50 Ilaro, Ogun State, Nigeria

B. K. Oguntade
Department of Science Laboratory Technology, Federal Polytechnic Ilaro, P. M. B. 50 Ilaro, Ogun State, Nigeria

B. F. Umar
Department of Science Laboratory Technology, Federal Polytechnic Ilaro, P. M. B. 50 Ilaro, Ogun State, Nigeria

C. V Abiaziem
Department of Science Laboratory Technology, Federal Polytechnic Ilaro, P. M. B. 50 Ilaro, Ogun State, Nigeria

A. H. Santuraki
Department of Chemical Sciences, Federal University Kashere, Gombe State - Nigeria

A. A. Muazu
Department of Chemical Sciences, Federal University Kashere, Gombe State - Nigeria

Sheraz Ahmed
Department of Chemistry, Government College University, Katchery Road, Lahore-54000, Pakistan

Fahim A. Qureshi
Department of Chemistry, Government College University, Katchery Road, Lahore-54000, Pakistan

Amjad Islam
Department of Chemistry, Government College University, Katchery Road, Lahore-54000, Pakistan
Industrial Chemicals Division, Guides Corporation, Davis Road, Lahore-54000, Pakistan

Ahmed Adnan
Industrial Chemicals Division, Guides Corporation, Davis Road, Lahore-54000, Pakistan

John A. M. Mahugija
Chemistry Department, University of Dar es Salaam, P. O. Box 35061 Dar es Salaam, Tanzania

S. T. Ubwa
Department of Chemistry, Benue State University, Makurdi, Nigeria

G. H. Atoo
Department of Chemistry, Benue State University, Makurdi, Nigeria

J. O. Offem
Department of Pure and Applied Chemistry, University of Calabar, Cross River State, Nigeria

J. Abah
Department of Chemistry, Benue State University, Makurdi, Nigeria

K. Asemave
Department of Chemistry, Benue State University, Makurdi, Nigeria

A. K. M. Lutfor Rahman
Department of Chemistry, Jagannath University, Dhaka 1100, Bangladesh

M. Islam
Department of Chemistry, Jagannath University, Dhaka 1100, Bangladesh

M. Z. Hossain
Department of Chemistry, Jagannath University, Dhaka 1100, Bangladesh

M. A. Ahsan
Bangladesh Council of Scientific and Industrial Research (BCSIR), Dhaka, Bangladesh

Ali Moghimi
Department of Chemistry, Varamin (Pishva) Branch Islamic Azad University, Varamin, Iran

Hesham R. Lotfy
Chemistry and Biochemistry Department, Science Faculty, University of Namibia, P. Bag 13301, Windhoek, Namibia

Jane Misihairabgwi
Chemistry and Biochemistry Department, Science Faculty, University of Namibia, P. Bag 13301, Windhoek, Namibia

Mary Mulela Mutwa
Chemistry and Biochemistry Department, Science Faculty, University of Namibia, P. Bag 13301, Windhoek, Namibia

M. I. Yahaya
Department of Chemistry, Sokoto State University, Sokoto State, Nigeria

Z. M. Agbendeh
Department of Applied Chemistry, Federal University Dutsin-Ma, Katsina State, Nigeria

P. O. Egbenda
Department of Chemistry, Fourah Bay College, University of Sierra Leone, Freetown, Sierra Leone

F. Thullah
Department of Chemistry, Fourah Bay College, University of Sierra Leone, Freetown, Sierra Leone

I. Kamara
Department of Chemistry, Fourah Bay College, University of Sierra Leone, Freetown, Sierra Leone

Proscovia Nnamuyomba
Department of Chemistry, Gulu University, P. O. Box 166, Gulu, Uganda

Jolocam Mbabazi
Department of Chemistry, Makerere University, P. O. Box 7062, Kampala, Uganda

Muhammad Ntale
Department of Chemistry, Makerere University, P. O. Box 7062, Kampala, Uganda

Ali Moghimi
Department of Chemistry, Islamic Azad University Varamin (Pishva) Branch, Varamin,, Iran

Mahdi Alijanianzadeh
Department of Biochemistry and Biophysics, School of Biological Sciences, Varamin-Pishva Branch, Islamic Azad University, Varamin, Iran

www.ingramcontent.com/pod-product-compliance
Lightning Source LLC
Chambersburg PA
CBHW080633200326
41458CB00013B/4613